AUGUSTUS DE MORGAN, POLYMATH

Augustus De Morgan, Polymath

New Perspectives on his Life and Legacy

Edited by
Karen Attar, Adrian Rice and Christopher Stray

https://www.openbookpublishers.com
©2024 Karen Attar, Adrian Rice and Christopher Stray
Copyright of individual chapters are maintained by the chapter author(s)

This work is licensed under an Attribution-NonCommercial 4.0 International (CC BY-NC 4.0). This license allows you to share, copy, distribute and transmit the text; to adapt the text for non-commercial purposes of the text providing attribution is made to the authors (but not in any way that suggests that they endorse you or your use of the work). Attribution should include the following information:

Karen Attar, Adrian Rice and Christopher Stray (eds), *Augustus De Morgan, Polymath: New Perspectives on his Life and Legacy*. Cambridge, UK: Open Book Publishers, 2024, https://doi.org/10.11647/OBP.0408

Copyright and permissions for the reuse of many of the images included in this publication differ from the above. This information is provided in the captions and in the list of illustrations. Where no licensing information is provided in the caption, the figure is reproduced under the fair dealing principle. Every effort has been made to identify and contact copyright holders and any omission or error will be corrected if notification is made to the publisher.

Further details about CC BY-NC licenses are available at
http://creativecommons.org/licenses/by-nc/4.0/

All external links were active at the time of publication unless otherwise stated and have been archived via the Internet Archive Wayback Machine at https://archive.org/web

Any digital material and resources associated with this volume will be available at https://doi.org/10.11647/OBP.0408#resources

ISBN Paperback: 978-1-80511-326-3
ISBN Hardback: 978-1-80511-327-0
ISBN Digital (PDF): 978-1-80511-328-7
ISBN Digital eBook (EPUB): 978-1-80511-329-4
ISBN HTML: 978-1-80511-330-0
DOI: 10.11647/OBP.0408

Cover image: Portrait of Augustus De Morgan, in Sophia Elizabeth De Morgan, *Memoir of Augustus De Morgan* (1882), https://commons.wikimedia.org/wiki/File:Augustus_De_Morgan_1850s.jpg. Background: Nico Baum, White round light on gray textile (2020), https://unsplash.com/photos/white-round-light-on-gray-textile-xZroI5V_dxc.
Cover design: Jeevanjot Kaur Nagpal.

Contents

Introduction   xi

De Morgan: Polymath   xi
Karen Attar, Adrian Rice and Christopher Stray

PART I: SCIENTIFIC WORK   1

1. De Morgan and Mathematics   3
Adrian Rice

2. De Morgan and Logic   31
Anna-Sophie Heinemann

3. Augustus De Morgan, Astronomy and Almanacs   57
Daniel Belteki

4. De Morgan, Periodicals and Encyclopaedias   83
Olivier Bruneau

5. Augustus De Morgan: Meta-Scientific Rebel   107
Lukas M. Verburgt

PART II: BEYOND SCIENCE   151

6. De Morgan and Mathematics Education   153
Christopher Stray

7. De Morgan's *A Budget of Paradoxes*   175
Adrian Rice

8. Augustus De Morgan and the Bloomsbury Milieu   197
Rosemary Ashton

9. De Morgan's Family: Sophia and the Children   221
Joan L. Richards

PART III: THE BIBLIOGRAPHIC RECORD   247

10. Augustus De Morgan's Library Revisited: Its Context and Its Afterlife   249
Karen Attar

11. Augustus De Morgan: The Archival Record   279
Karen Attar, Alexander Lock, Katy Makin, Jane Maxwell, Virginia Mills and Diana Smith

12. Bibliography of the Works of Augustus De Morgan   305
William Hale

List of Illustrations   331
Notes on Contributors   333
Index   337

Fig. 1 Augustus De Morgan pictured in the 1850s. (Public domain, via Wikimedia Commons, https://commons.wikimedia.org/wiki/File:Augustus_De_Morgan_1850s.jpg)

Introduction

De Morgan: Polymath

Karen Attar, Adrian Rice and Christopher Stray

> Mathematicians must accept that their talent does not confer on them any particular competence outside their own domain.
>
> — Jean Dieudonné [1]

The field of intellectual history abounds with names of individuals whose accomplishments in a particular domain were significant, and whose learning and expertise moreover spanned an astonishing array of disciplines.[2] Several epithets exist for such scholars, from the Latin 'homo universalis' to the English 'Renaissance man'. Perhaps the word that best describes this phenomenon is that with the oldest linguistic roots: *polymath*. From the Greek, literally meaning 'much learning', it first appeared in the philosopher Johann von Wowern's *De polymathia tractatio* of 1603, which defined 'polymathy' as 'a knowledge of various things, collected from every kind of study, … roaming freely and at an unbridled pace through all the fields of learning'.[3] Since then, the term 'polymath' has come to mean an accomplished scholar of deep and wide-ranging expertise in a variety of disciplines. The names of Aristotle, Thomas Aquinas, Leonardo da Vinci, René Descartes,

---

1 Jean Dieudonné, *Mathematics—The Music of Reason* (Berlin: Springer, 1992), p. 11.
2 See, for example, Peter Burke, *The Polymath: A Cultural History from Leonardo da Vinci to Susan Sontag* (New Haven: Yale University Press, 2021).
3 'Perfectam Polymathian intelligo, notitiam variarum rerum, ex omni genere studiorum collectam … Vagatur enim libero & effreni cursu per omnes disciplinarum campos.' Johann von Wowern, *De polymathia tractatio* (Basel: Officina Frobeniana, 1603), p. 16.

Blaise Pascal, Gottfried Leibniz, Benjamin Franklin, Thomas Jefferson, Mary Somerville, Henri Poincaré, Bertrand Russell and Alan Turing may spring to mind as examples of polymathic thinkers. To these we propose to add the name of another, less renowned figure, yet one who, as the content of this book will demonstrate, has every right to the title: Augustus De Morgan.

The genesis of this book, edited by scholars from three distinct academic disciplines, testifies to De Morgan's polymathic status. Christopher Stray, a classicist by training and a historian of education, encountered De Morgan via his many writings on educational matters, on which he was an acknowledged authority for many years. The volume began with an approach from Stray to Karen Attar about the feasibility of editing one of De Morgan's bibliographical essays; Attar came to De Morgan as a book and library historian and the rare books librarian at the University of London Library, cataloguing and writing about De Morgan's mathematical library at the University. It quickly became apparent that De Morgan merited broader treatment than the study of a single essay. Stray and Attar consequently approached an expert on De Morgan's mathematics, Adrian Rice, who had first encountered De Morgan through his mathematical studies at University College London (UCL). There he had learned about 'De Morgan's Laws' during lectures on algebra and logic, and also that De Morgan had been the college's first professor of mathematics when it opened for classes in 1828. The three of us started work on *Augustus De Morgan, Polymath* to re-evaluate De Morgan's multiple achievements, galvanised particularly by the approaching 150th anniversary of his death and with it the gift of his mathematical library to the University of London.

De Morgan was a mathematician, educationalist and bibliophile who furthermore published ground-breaking research in logic, the history of mathematics and scientific biography, and who exhibited substantial expertise in matters related to astronomy, almanacs, calendar computation and actuarial science. A skilled expositor, he wrote countless popular articles and surveys for the general reader. He was an influential and admired teacher, an office holder in several learned societies, an indefatigable letter writer, and a prominent and respected member of the early Victorian intelligentsia. Indeed, examination of the period in Great Britain between the passage of the first (1832) and second (1867) Reform Acts, reveals De Morgan to have been involved to some extent in almost every area of British intellectual life during the middle third of the nineteenth century. Yet the very multiplicity of De Morgan's talents militated against his renown, since studies of different aspects of his work have appeared in widely scattered publications, including books and journals devoted to mathematics, education, and book history. Only in biographical accounts has his life and work been considered as a whole, and such accounts are relatively brief.[4] By uniting different aspects of De Morgan's activity and environment for the first time in a single volume, we invite scholars to reconsider a remarkable and inspiring individual.

De Morgan's Life

Augustus De Morgan was born in Madurai, southern India, on 27 June 1806, the fifth child and eldest surviving son of Elizabeth (née Dodson) and John De Morgan, a colonel in the British army. His mother was the granddaughter of James Dodson, an eighteenth-century English mathematician of note at the time, due to his publication of the then-unique *Anti-Logarithmic Canon* (1742) and other mathematical works. As his great-grandson would later do, Dodson earned his living as a mathematics teacher, rising to the position of master at the prestigious Royal Mathematical School at Christ's Hospital in London; the two men also shared an interest in the mathematics of insurance, with De Morgan's great-grandfather being credited for foundational work in the embryonic discipline of actuarial mathematics. Indeed, when the Equitable Life Assurance Society was launched in 1762, it based its

---

[4] The principal general source for Augustus De Morgan's life remains an uncritical monograph published by his widow about a decade after his death as a memorial: Sophia Elizabeth De Morgan, *Memoir of Augustus De Morgan* (London: Longmans, Green, 1882). More impartial but briefer accounts appear in standard biographical dictionaries: for example, Leslie Stephen, 'Morgan, Augustus De (1806–1871)', rev. by I. Grattan-Guinness, *Oxford Dictionary of National Biography* (Oxford: Oxford University Press, 2004); John M. Dubbey, 'De Morgan, Augustus', in *Dictionary of Scientific Biography*, vol. 3 (New York: Charles Scribner's Sons, 1970), pp. 35–37. Most recently, De Morgan is one of the key figures in Joan L. Richards's study of his extended family: Joan L. Richards, *Generations of Reason: A Family's Search for Meaning in Post-Newtonian England* (New Haven: Yale University Press, 2021).

insurance premiums on actuarial methods and calculations pioneered by Dodson prior to his death in 1757.[5]

On his father's side, the De Morgan family were British descendants of Huguenot refugees who, unlike their French forebears, insisted on spelling their surname with a capital D. As De Morgan later wrote to a friend:

> De Morgan—not de Morgan—when I was at Cambridge, I used to *get* out of my misery in *viva voce* examinations sooner by M—D than I should otherwise have done, by insisting on this capital arrangement.[6]

For three generations since 1710, De Morgan's male forebears had been officers in the employment of the East India Company, stationed at various posts in southern India, including Madras (now called Chennai), Masulipatam (Machilipatnam) and Pondicherry (Puducherry). By the time of Elizabeth De Morgan's fifth pregnancy in 1806, her husband was in command of a battalion in the city of Madura (now Madurai in modern-day Tamil Nadu).[7]

When the young Augustus was born that summer, he was found to have the use of only one eye: his left. Many years later he recounted:

> When I was in preparation, my mother attended much to a favourite native servant (in India) who had the ophthalmia, which they call the country sore eyes. When I was born it was found I had had it too, and one eye was not destroyed, but never completely formed: it is only a rudiment, with a discoloration in the centre, which shows that nature intended a pupil. ... Accordingly I have always been strictly unocular. I have seen as much with my right eye as with any one finger - no more, and no less.[8]

This distinctive physical peculiarity would soon result in his concentration on mental rather than physical activities.

---

5   G. J. Gray, 'Dodson, James (c.1705–1757)', rev. by Anita McConnell, *Oxford Dictionary of National Biography* (Oxford: Oxford University Press, 2004).
6   Robert Perceval Graves, *Life of Sir William Rowan Hamilton*, vol. 3 (Dublin: Hodges Figgis, 1889), p. 364. Unless otherwise stated, all italics in quotations are original. 'M—D' presumably stood for 'Mr De Morgan'.
7   UCL Special Collections, MS. ADD. 7, Augustus De Morgan, 'Memorandums on the Descendants of Captain John De Morgan ...', ff. 115–16.
8   Graves, *Life*, pp. 612–13.

At the time of De Morgan's birth, tensions between the British officers and their native troops—which were frequently strained—had reached critical levels, with mutiny a constant threat. It was for this reason that Colonel De Morgan broke with family tradition and took his young family back to the relative safety of England. On 22 October 1806, they set sail on the *Jane, Duchess of Gordon* in a convoy of nearly forty ships. After a voyage of nearly six months, their ship landed at Deal in Kent on 12 April 1807. 'At this period,' the younger De Morgan commented, 'I had passed three-fifths of my life on the water.'[9] He was later to use this voyage as an excuse for his subsequent aversion to travelling: 'I consider I had my share of it in my nurse's arms, in which I began life with a journey of 11,000 miles, crossed the line twice, and knew nothing about it all—Heaven be praised.'[10]

After some time in London, Colonel De Morgan settled his family at Worcester so that his wife might be close to her sister. He returned to India alone in 1808, for a period of two years. On his return, the family moved to north Devon, first to Appledore, and then to Bideford. It was here that, at the age of just over four years old, the education of Augustus De Morgan began with lessons from his father in 'reading and numeration'.[11] In 1812, the family moved again, this time to Barnstaple. The Colonel's imminent departure to India for another tour of duty occasioned a final move to Taunton in Somerset, from where he departed on 29 January 1813. He never saw his family again, dying of a liver complaint somewhere near St. Helena on his way home in 1816.[12]

Meanwhile, the young Augustus was receiving a solid but unremarkable education in a number of private schools in the southwest of England. In common with most school teaching at the time, in addition to arithmetic and a little algebra, his learning was dominated by classical studies of Latin and Greek, augmented with a little Hebrew.[13] For two and a half years from the age of fourteen, De Morgan attended a boarding school in Redland, near Bristol, run by the Reverend John Parsons who, by all accounts, was a good teacher, although 'not a

---

9   A. De Morgan, 'Memorandums', f. 116.
10   Graves, *Life*, p. 525.
11   S. E. De Morgan, *Memoir*, p. 3.
12   A. De Morgan, 'Memorandums', f. 128.
13   A. De Morgan, 'Memorandums', f. 155.

high mathematician'.[14] It was around this time that the boy's hitherto uncultivated mathematical skill was first recognised, though not by Parsons. We are told, in a verbose manner typical of the period, that 'the first suspicion of Augustus having inherited the ostensibly reprehensible proclivity of his maternal forbear was due to a mere chance',[15] the propensity being 'accidentally developed, and indeed made known to its possessor'[16] by a family friend who, on finding him making an elaborate drawing of a figure from Euclid with ruler and compass, initiated him into the concept of a mathematical proof.

From this point, De Morgan's mathematical progress was rapid, as a school-friend, Robert Reece, later testified:

> It seems an odd thing to record, but I well remember that I was advanced in 'Bland's Quadratic Equations'[17] when De Morgan took up that well-known elementary book, 'Bridge's Algebra,'[18] for the first time. But it was so. He read Bridge's book like a novel. In less than a month he had gone through that treatise and dashed into Bland, and so got out of sight, as far as I was concerned.[19]

The final stage of De Morgan's intellectual development began on 1 February 1823, when he entered Trinity College, Cambridge, at the age of just over sixteen and a half.[20] This early start to his university career is probably explained by his rapid progress at Parsons's school where, in mathematics at least, he had 'soon left his teacher behind'.[21] However, neither Parsons nor De Morgan's mother intended mathematics to be his principal subject of study at Cambridge, the former advising concentration on Classics to comply with the latter's wish that her son should ultimately enter the church. This aspiration would soon be frustrated by two major factors: firstly, De Morgan's insatiable appetite

---

14 S. E. De Morgan, *Memoir*, p. 3.
15 Anna M. W. Stirling, *William De Morgan and his Wife* (London: Thornton Butterworth, 1922), p. 25.
16 S. E. De Morgan, *Memoir*, p. 4.
17 Miles Bland, *Algebraical Problems, Producing Simple and Quadratic Equations, With Their Solutions* (Cambridge: J. Smith, 1812).
18 Bewick Bridge, *An Elementary Treatise on Algebra* (London: T. Cadell & W. Davies, 1815).
19 S. E. De Morgan, *Memoir*, p. 7.
20 Walter William Rouse Ball & John A. Venn, eds, *Admissions to Trinity College, Cambridge*, vol. 4 (London: Macmillan, 1911), p. 216.
21 S. E. De Morgan, *Memoir*, p. 4.

for mathematics; and secondly, the intellectual environment he quickly encountered at Cambridge.

De Morgan's principal tutor for the entirety of his undergraduate career was John Philips Higman, but he found himself influenced by all of his college teachers to some extent. In particular, it is highly probable that he acquired his interest in algebra from the algebraist George Peacock and his love of astronomy from the future Astronomer Royal George Airy. It is also entirely conceivable that his passion for the history of science was inspired (and certainly encouraged) by Peacock and the scientific philosopher William Whewell, both of whom had strong interests in that area.[22] There is also a suggestion that it was from Whewell that De Morgan inherited his great fascination for logic,[23] although the link here is less obvious. Nevertheless, the fundamental contribution of all of these teachers was to confirm De Morgan's intention to concentrate on the study of mathematics while at college, and ultimately to determine the course of his professional career.

He was by nature a compulsive reader on almost any topic and, when not consuming mathematical books, would devote his leisure hours to the study of works on philosophy, theology, literature and history. Towards the end of his life, he wrote to a friend: 'I did with Trinity College library what I afterwards did with my own—I foraged for relaxation.'[24] A result of this discursive reading was the development of an almost encyclopaedic knowledge of an impressive range of scientific subjects. His wife Sophia recalled, for example, that as early as their meeting in 1827, he was already an expert in the history of science, being 'well informed in Eastern astronomy and mythology' and critical of writers on the subject, pointing out 'the insufficiency of their theories to account for all that they have tried to explain'.[25]

In January 1827, De Morgan sat the prestigious and highly demanding 'Tripos' examination, on the basis of which candidates were awarded

---

22 Peacock's article 'Arithmetic' in the *Encyclopaedia Metropolitana* (vol. 1., 369–523), written in 1825, was the best historical account of the subject to date. Whewell was later famous for, amongst many other things, his three-volume *History of the Inductive Sciences*, first published in 1837.
23 Alexander MacFarlane, *Lectures on Ten British Mathematicians of the Nineteenth Century* (London: Chapman & Hall, 1916), p. 20.
24 S. E. De Morgan, *Memoir*, p. 393.
25 S. E. De Morgan, *Memoir*, p. 21.

their degrees. Graduates were divided into several classes: the lowest were known as *poll men* who, while awarded a degree, did not receive honours; above them were the *junior* and *senior optimes*, while those who achieved first-class status were called *wranglers*, from the word meaning to dispute. Of these, the student in first place was known as the Senior Wrangler, and competition for this distinction was intense. In De Morgan's year, there was a widespread expectation that this coveted position would be his. However, when the results were announced, he was disappointed to learn that he had only achieved the rank of Fourth Wrangler, a place which, as it was later said, 'failed to declare his real power or the exceptional aptitude of his mind for mathematical study'.[26] Ironically, it was his exhaustive programme of reading which was principally to blame for this disappointing result, since it often distracted him from the course required for examination. The realisation that wide and discursive mathematical study had actually been *detrimental* to his performance imbued a thorough distrust of competitive examinations that was to last for the rest of his life.

It was at this point that De Morgan's firmly held nonconformist religious beliefs came to the fore, a reaction to the strict evangelical education he had received in childhood. This had started at an early age with his father: 'A rigid Evangelical in tenets and practice—a heritage, doubtless, from his Huguenot ancestry—Colonel De Morgan was known to his fellow officers by the nickname of "Bible John".'[27] His wife shared his beliefs and, after his death, had continued to administer the same discipline. As a child, De Morgan had been taken to church twice in the week, three times on Sunday, and required to give an abstract of every sermon he heard. Not surprisingly, this left him with a lifelong inability to listen to any speaking or lecturing for a prolonged period. The 'dreary sermons',[28] combined with the logical inconsistencies which formed part of the arguments used to convince him, made it inevitable that he would rebel at the first opportunity, though he never became an atheist.

---

26 Arthur Cowper Ranyard, Obituary Notice of Augustus De Morgan, *Monthly Notices of the Royal Astronomical Society*, 32 (1871–72), 112–18 (pp. 113–14).
27 Stirling, *William De Morgan*, p. 24.
28 S. E. De Morgan, *Memoir*, p. 11.

While admitting a personal faith in Jesus Christ, he subjected all religious arguments to the same unbending rigour of rational thought that he devoted to his other intellectual pursuits. 'My opinion of mankind,' he wrote, 'is founded upon the mournful fact that, so far as I can see, they find within themselves the means of believing in a thousand times as much as there is to believe in.'[29] Rejecting anything that smacked of hypocrisy or sectarianism, he refused to join any church, regarding himself throughout life as a 'Christian Unattached'.[30] For him, religious belief was a strictly personal experience and nobody else's concern. Moreover, he believed that one should be able to achieve one's goals in life regardless of religious persuasion. As he later wrote in his will, he refrained from any open profession of faith 'because in my time such confession has always been the way up in the world'.[31] Such conviction and commitment to principle was to be a constant feature of De Morgan's life.

An immediate consequence of his religious nonconformity was his departure from Cambridge, for, although his degree result was more than sufficient to win him a college fellowship, it was first necessary to swear adherence to the tenets of the Church of England (a requirement not fully abolished at Oxford and Cambridge until 1871) which, due to his religious convictions, he refused to do.[32] De Morgan now had to decide on a profession, since 'few, if any, occupations in England in the early nineteenth century required much training in mathematics or involved mathematics at all'.[33] An academic career thus closed to him, he toyed briefly with the idea of a medical or legal career, before his attention was drawn to the newly established London University (now called University College London, or UCL), which was then in the process of recruiting professors. Inspired by the progressive aims and explicit secular character of 'the godless institution on Gower Street', De

---

29   Augustus De Morgan, *A Budget of Paradoxes* (London: Longmans, Green, 1872), p. 70.
30   Ranyard, Obituary Notice, p. 114.
31   S. E. De Morgan, *Memoir*, p. 368.
32   It is worth mentioning that De Morgan's doctrinal scruples, strong though they undoubtedly were, did not prevent him actually taking his B.A. degree, which required acceptance of the thirty-nine Articles of Faith. It can only be assumed that he took the oath under (silent) protest.
33   Philip C. Enros, 'The Analytical Society (1812-1813): Precursor of the Renewal of Cambridge Mathematics', *Historia Mathematica*, 10 (1983), 24–47 (p. 41).

Morgan applied for the mathematics chair. Despite his relative youth and lack of experience, he was unanimously elected as the founding professor of mathematics on 23 February 1828.[34]

However, his academic career nearly ended as prematurely as it had begun. Being a new institution, UCL experienced considerable instability during its early years, due to the poor state of its finances, student discipline and general morale. The relationship between the professors and the college's ruling council was particularly uneasy. Matters finally came to a head in 1831 with the dismissal of the professor of anatomy, Granville Sharp Pattison, whose alleged incompetence had resulted in student unrest. De Morgan, being a man of principle, immediately resigned in support of his colleague.[35] But five years later, shortly before the beginning of the 1836–37 academic year, his successor was accidentally drowned while on a family holiday in the Channel Islands. With the beginning of term only days away, De Morgan offered himself as a temporary replacement and, after he had received assurances that the circumstances that had led to his resignation could not recur, this arrangement became permanent. He was to remain at the college for another thirty years.[36]

He was now secure enough financially to propose marriage, after ten years of courtship, to Sophia Elizabeth Frend, the daughter of William Frend, a social reformer and fellow liberal nonconformist, with whom he had become acquainted on moving to London in 1827, due to their common interests in mathematics, their actuarial work, and their mutual membership and involvement in learned bodies such as the Royal Astronomical Society and the Society for the Diffusion of Useful Knowledge. De Morgan's wedding to Sophia, on 3 August 1837, was one of the first in England to take place in a registry office, after the practice was legalised earlier that year.[37] As well as being progressively-minded intellectuals, the Frend family had good connections to a wide range of liberally-inclined social reformers, into whose orbits De Morgan was now introduced, including Lady Byron, Elizabeth Fry, and John

---

34   Adrian Rice, 'Inspiration or Desperation? Augustus De Morgan's Appointment to the Chair of Mathematics at London University in 1828', *British Journal for the History of Science*, 30 (1997), 257–74 (p. 268).
35   S. E. De Morgan, *Memoir*, pp. 34–39.
36   S. E. De Morgan, *Memoir*, pp. 69–74.
37   A. De Morgan, 'Memorandums', ff. 29, 30.

Stuart Mill. No doubt encouraged by his wife, he used his mathematical abilities in the service of the wider community, for example serving for twelve years as the manager of a savings bank, as 'he thought this the best way in which he could be useful to his poorer neighbours'.[38]

He was also supportive of the first steps towards providing higher education for women, giving 'lectures or lessons on arithmetic and algebra'[39] for the first two terms when the Ladies' College, Bedford Square (later to become Bedford College) opened for classes in the autumn of 1849.[40] But by the end of his life, his social liberality, so progressive in the 1830s and 1840s, began to appear less broadminded, drawing the line, for example, at votes for women. As he wrote in 1868 to John Stuart Mill, who famously proposed such a measure in Parliament:

> To be a voter is sometimes dangerous. A man ought to face the danger, but you have no right to enforce it on women; in principle you might as well enforce the militia on them. Many women think exemption from politics is one of their rights.[41]

In general, however, De Morgan tended to steer clear of political matters, largely adopting an attitude of total indifference. As he wrote in 1852: 'I never gave a vote in my life.'[42] He went on to say:

> I hate the system. Given two persons of whom I know nothing; required which is the best qualified to manage matters of which I know next to nothing. The presumption is that 5000 incompetent persons, by a contest of opposite incompetencies, will produce a competent decision. This absurdity fills the House of Commons.[43]

His lack of interest in parliamentary democracy also extended to sightseeing and tourism:

> I never was in the House of Commons, or in the Tower, or in Westminster Abbey. I spent only one and three-quarter hours in the Great Exhibition. ... I never got further north than Cambridge, and never while at Cambridge penetrated to the

---

38 S. E. De Morgan, *Memoir*, p. 248.
39 S. E. De Morgan, *Memoir*, p. 174.
40 Margaret J. Tuke, *A History of Bedford College for Women 1849–1937* (London: Oxford University Press, 1939), p. 65.
41 S. E. De Morgan, *Memoir*, p. 384.
42 Graves, *Life*, p. 377.
43 Graves, *Life*, p. 385.

northern extremity of the town. So much for me as a sight-seer and traveller.[44]

In fact, De Morgan loved city life so much that, apart from the occasional trip to France and the odd reluctantly taken family holiday in the countryside, he rarely left London. He once said of himself,

> Ne'er out of town; 'tis such a horrid life:
> But duly sends his family and wife.[45]

De Morgan was a man of many eccentricities. In 1859, when offered an honorary law doctorate by Edinburgh University, he declined it, saying that he 'did not feel like an LL.D.'[46] In fact, he once styled himself:

> Augustus De Morgan,
> H.O.M.O. P.A.U.C.A.R.U.M. L.I.T.E.R.A.R.U.M.[47]

De Morgan also refused to allow himself to be proposed as a Fellow of the Royal Society, as he considered the body to be more concerned with social standing than scientific attainment.[48] 'Whether I could have been a Fellow,' he later said, 'I cannot know; as the gentleman said who was asked if he could play the violin, I never tried.'[49] But nowhere is his unconventionality better illustrated than by his endearingly whimsical sense of humour, which is curiously reminiscent of a blend of Lewis Carroll, W. S. Gilbert and Monty Python. His writings abound with witticisms, anecdotes, jokes, puns, parodies and conundrums, either of his own invention or, just as frequently, acquired from other people. It is even possible that he was the first to express a precursor of 'Murphy's

---

44 Graves, *Life*, p. 462.
45 A. De Morgan, *Budget*, p. 82.
46 S. E. De Morgan, *Memoir*, p. 269.
47 'Augustus De Morgan, Man of Few Letters.' MacFarlane, *Lectures*, p. 24.
48 Although membership of the Royal Society certainly included Fellows of the highest scientific calibre, under the leadership of Joseph Banks (President from 1778–1820) the Society had obtained a not unjustified reputation for admitting wealthy patrons and valuing privilege as much as high scientific attainment. This conflicted with the ideals of more progressive scientific 'professionalisers' such as De Morgan. Thus, although he was certainly an eminently suitable candidate for a Fellowship, he repeatedly refused to be put forward for the honour, despite the urging of friends and colleagues. See Rebekah Higgitt, 'Why I don't FRS my tail: Augustus De Morgan and the Royal Society', *Notes and Records of the Royal Society*, 60 (2006), 253–59.
49 A. De Morgan, *Budget*, p. 18.

Law', namely, that 'anything that can go wrong will go wrong', although De Morgan's version is considerably broader: 'whatever can happen will happen'.[50]

Above all, he appears to have been a warm and generous individual, with firmly held principles and a fierce intellect, who inspired great affection and loyalty among his friends. The lawyer and diarist Henry Crabb Robinson said of him that 'He is the only man whose calls, even when interruptions, are always acceptable. He has such luminous qualities, even in his small-talk.'[51] These qualities were clearly in evidence in the professors' common room at UCL, as a junior colleague wrote in 1865:

> I never met a man who enjoyed telling a funny story more than de Morgan [sic] and he tells them well. It would be worth while to keep a record of some of them. ... [For example], Mr. Stirling Coyne, a barrister, and Albert Smith (of Mont Blanc celebrity) [who had died five years previously] married two sisters who were as like each other as two peas. Coyne was in court one very hot day with a friend. The latter afterwards repaired to the Crystal Palace; there he met a lady whom he took to be Mrs. Coyne. After shaking hands she remarked, 'How hot it is here.' 'Yes,' replied the gentleman, 'but your husband is in a far hotter place I can assure you.' The horror with which this remark was received was inexplicable to the gentleman. It was only afterwards that he discovered he had been addressing the widow of the late mountaineer.[52]

By 1866, De Morgan had been associated with University College for nearly four decades, making him one of its longest-serving professors, a distinction which brought him considerable pride. But these feelings changed dramatically when the college's governing council refused to appoint a candidate to the vacant chair of philosophy because he was a controversial Unitarian minister. To De Morgan, the college's decision was not only an affront to his view that religious beliefs should have no bearing on professional advancement, but more importantly it was

---

50 A. De Morgan, *Budget*, p. 171.
51 Henry Crabb Robinson, *Diary, Reminiscences, and Correspondence of Henry Crabb Robinson*, ed. by Thomas Sadler, vol. 2 (Boston: Fields, Osgood, 1869), p. 489.
52 William H. Brock and Roy M. MacLeod, eds, *Natural Knowledge in Social Context: The Journal of Thomas Archer Hirst, F.R.S.* (London: Mansell, 1980), pp. 1759–60.

a fundamental betrayal of its founding principle of religious neutrality. He resigned his professorship on 10 November 1866 and, after his last lecture in the summer of 1867, never returned. He even refused a request to sit for a portrait or bust to be placed in the college library. As far as he was concerned, 'our old College no longer exists'.[53]

The years following his retirement were beset by illness and bereavement. The circumstances surrounding his final resignation had put De Morgan under tremendous emotional stress, which now took a toll on his health. His previously robust constitution began to deteriorate, with the untimely death of his son George from tuberculosis in October 1867 further weakening his spirits. After suffering a stroke in 1868, De Morgan never fully recovered, and a final decline in his health followed the premature death of another child, Helen Christiana, in August 1870. He died of kidney failure at his home in northwest London on 18 March 1871 and was buried at Kensal Green Cemetery five days later.[54]

De Morgan's death prompted the publication of numerous memorials and obituaries, each paying tribute to his many and varied achievements. One of the comments most frequently made regarded the sheer breadth and volume of his published work. The notice in *The Athenæum* asserted that if all his articles for periodicals and encyclopaedias were collected together, there would be found 'such a mass of literary achievement as seldom comes from the pen of a man whose sole business it is to write for journals'.[55] *The Spectator* no doubt spoke for many of his former students when it declared that 'no testimonial which can be raised to Professor De Morgan will adequately express his many pupils' deep sense of intellectual and moral obligation'.[56]

But perhaps the most perceptive and candid judgement came nearly half a century later from the historian of mathematics Walter William Rouse Ball, who, although he had never known De Morgan, was able to encapsulate his personality and character in a paragraph which serves as a fitting epitaph to a remarkable man:

---

53   S. E. De Morgan, *Memoir*, p. 360.
54   *The Times*, 20 March 1871, 1a; 21 March 1871, 5c; Brock and MacLeod, *Natural Knowledge*, p. 1896.
55   *The Athenæum*, 25 March 1871, p. 370.
56   *The Spectator*, 13 May 1871, p. 563.

> That De Morgan was obstinate and somewhat eccentric I readily admit, and I do not consider he was a genius, but he leaves on my mind the impression of a lovable man, with intense convictions, of marked originality, having many interests, and possessing exceptional powers of exposition. In those cases where his actions were criticized it would seem that the explanation is to be found in his determination always to take the highest standard of conduct without regard to consequences; he hated suggestions of compromise, expediency, or opportunism. Such men are rare, and we do well to honour them.[57]

De Morgan's Work and Legacy

For all his many interests and areas of expertise, Augustus De Morgan remained first and foremost a mathematician—for which reason the opening chapter of this volume surveys his mathematical work. As a mathematician, his most significant contribution lay arguably as a catalyst in the birth of modern abstract algebra; but algebra was by no means his sole mathematical interest. In covering his work in multiple branches of mathematics, Adrian Rice grapples with the demise of De Morgan's reputation. How could somebody be lauded at the time of his death as one of the country's major mathematicians and largely forgotten half a century later? Were De Morgan's contemporaries overly generous or his successors inaccurately harsh? In a new evaluation, Rice demonstrates that neither is the case and that the nature of De Morgan's achievements as a supporter more than a trailblazer, and as a polymath within mathematics instead of a one-track researcher, both made his name and allowed it to fade.

It is significant in connection with De Morgan's diminished reputation that, in his lifetime, and for some time afterwards, he was acknowledged principally as a great mathematics teacher. His students praised him highly, their recollections revealing an idiosyncratic but talented professor whose lectures were at once thought-provoking, intriguing and challenging. He was particularly critical of student examinations, preferring independent thought to

---

57 Walter William Rouse Ball, 'Augustus De Morgan', *The Mathematical Gazette*, 8 (1915–16), 42–45 (p. 45).

the mere regurgitation of proofs in an exam,[58] while his rigorous and uncompromising attitude towards academic standards would establish UCL as the centre for advanced mathematical instruction in London. Christopher Stray's chapter discusses De Morgan's strong opinions on mathematical education and his numerous articles on the subject. Stray further enters new territory in his discussion of De Morgan's own undergraduate education.

Later described by the American philosopher Charles Sanders Peirce as 'the greatest formal logician that ever lived',[59] De Morgan is best remembered as a logician for the famous De Morgan's Laws and for his logic of relations, which appeared later in his career. He was one of the few mathematicians of his time to realise the importance of logic to mathematics, and vice versa:[60]

> We know that mathematicians care no more for logic than logicians for mathematics. The two eyes of exact science are mathematics and logic: the mathematical sect puts out the logical eye, the logical sect puts out the mathematical eye; each believing that it sees better with one eye than with two. [61]

De Morgan attempted to bring mathematical ideas into his logic by introducing a numerically precise method of 'quantifying the predicate'.[62] His consequent controversy with the Scottish philosopher

---

58 One ex-student later wrote: 'All cram he held in the most sovereign contempt. I remember, during the last week of his course which preceded an annual College examination, his abruptly addressing his class as follows: "I notice that many of you have left off working my examples this week. I know perfectly well what you are doing; you are cramming for the examination. But I will set you such a paper as shall make all your cram of no use."' S. E. De Morgan, *Memoir*, pp. 100–01.
59 Peter Heath, 'Editor's Introduction', in Augustus De Morgan, *On the Syllogism, and Other Logical Writings* (London: Routledge & Kegan Paul, 1966), vii–xxxi (p. xxx).
60 *The Athenæum*, 18 July 1868, p. 71.
61 De Morgan always had an eye for a *bon mot*; but, recalling his forementioned ocular disability, perhaps no passage in all of his writings better illustrates his sublime sense of humour than this.
62 This rather technical term can be explained as follows. In logical statements such as 'All men are mortal', the word 'men' is the subject and 'mortal' is its predicate—a characteristic or attribute of the subject. In traditional Aristotelian logic, problems arise with statements like 'Some men are dead', because we are told neither how many men are dead nor the total quantity of dead things. To rectify this defect, De Morgan introduced more precise notions of number and quantity into his logic. This was known as 'quantifying the predicate'.

Sir William Hamilton,[63] who mistakenly accused him of plagiarism,[64] served to stimulate his contemporary George Boole to publish his ideas on logic in 1847.[65] Anna-Sophie Heinemann, in her chapter on De Morgan's logic, focuses on his early research on the subject, particularly on logical 'quantification'. She argues that, despite its relative lack of influence on later developments, it still represented a notable departure from traditional syllogistic methods and anticipated the modern understanding of quantification in logic.

De Morgan's logic was also innovative in its attempt to develop a coherent system of symbolic notation to facilitate logical deductions. Indeed, one of Hamilton's objections to De Morgan's work on the subject was the latter's introduction of mathematical ideas and concepts into a discipline then regarded purely as an area of philosophy. In both his research and in his teaching, De Morgan's mathematics was often very philosophical in nature, although he always retained a healthy sense of humour about philosophical modes of inquiry:

> I would not dissuade a student from metaphysical inquiry; on the contrary, I would rather endeavour to promote the desire of entering upon such subjects: but I would warn him, when he tries to look down his own throat with a candle in his hand, to take care that he does not set his head on fire.[66]

He was, however, keenly interested in matters of 'meta-science', an area of the philosophy of science relating to methodology. Lukas Verburgt explains in his chapter how the dominant underlying scientific methodology of Victorian Britain was grounded on an appreciation of the work of the seventeenth-century philosopher Francis Bacon, and how,

---

63 Not to be confused with the Irish mathematician Sir William Rowan Hamilton, who was one of De Morgan's great friends and a regular correspondent.

64 Anna-Sophie Heinemann, *Quantifikation des Prädikats und numerisch definiter Syllogismus. Die Kontroverse zwischen Augustus De Morgan und Sir William Hamilton: Formale Logik zwischen Algebra und Syllogistik* (Münster: Mentis, 2015); Luis María Laíta, 'Influences on Boole's Logic: The Controversy between William Hamilton and Augustus De Morgan', *Annals of Science*, 36 (1979), 45–65 (pp. 51–60).

65 De Morgan strongly encouraged Boole's own research in this area; see Gordon C. Smith, *The Boole-De Morgan Correspondence 1842–1864* (Oxford: Clarendon Press, 1982).

66 Augustus De Morgan, *Formal Logic* (London: Taylor & Walton, 1847), p. 27.

via his correspondence with Whewell and in various publications, De Morgan revealed himself to be one of a relatively small group of British scientists who were anti-Baconian in outlook. Thus, the contemporary debate about the merits of Baconianism in British science provides a further example of De Morgan going against the grain—this time in opposition to what was then mainstream meta-scientific thinking.

De Morgan's knowledge of the history of science in general, and mathematics in particular, was encyclopaedic. His historical publications are characterised by their extensive use of primary sources, particularly archival documents, and an obvious desire to set the historical record straight. Significant contributions included his recognition of the earliest known printed work to contain the + and − signs, as well as extensive research into the infamous calculus priority dispute between Isaac Newton and Gottfried Leibniz. He drew attention to previously hidden flaws in Newton's character and initiated the rehabilitation of Leibniz's reputation in Britain, thereby leading scientific biography away from hagiographical studies and towards the more measured style of modern historiography.

Another area of prolonged interest was astronomy and its history. Except perhaps for his writings on the calendar, De Morgan's astronomical work has received little attention. Daniel Belteki redresses this to foreground his contributions to that subject, through the publication of a host of learned papers, biographical studies, book reviews, popular articles and encyclopedia entries throughout his career, and through his organisational role in the Royal Astronomical Society. Belteki's chapter shows De Morgan as a prominent member of the British astronomical community, despite his inability to participate in observational astronomy due to his visual impairment. In particular, we see De Morgan's merging of his astronomical knowledge with his historical interest in almanacs and calendar reckoning, particularly with regard to the calculation of the date of Easter, which is in itself a noteworthy achievement.

De Morgan's historical scholarship and his eccentric sense of humour came together in *A Budget of Paradoxes*, a collection of humorous writings and witty reviews originally featured in the weekly periodical *The Athenæum*. Its wealth of witticisms, anecdotes and sayings included his famous remark that 'I was X years old in A.D. X^2,' a peculiarity unique to those born in years such as 1640, 1722, 1806, 1892, 1980, and so on.[67] Adrian Rice's chapter delves into the pages of this book, which spanned a 375-year period from the invention of printing from moveable type to the mid-nineteenth century. Cheerfully lampooning scientific ignorance in all its many forms, the *Budget* gives perhaps the best insight into De Morgan's intellect, revealing alongside his comedic ability and love of the absurd his vast erudition and extensive knowledge of a broad range of topics from mathematics to theology.

De Morgan applied the same combination of historical scholarship and anecdotal wit in his bibliography of nearly 400 published works on arithmetic, *Arithmetical Books from the Invention of Printing to the Present Time* (1847). Whereas the *Budget* was based entirely on works De Morgan owned, in *Arithmetical Books*, De Morgan used his own books alongside others. From a modern bibliographer's point of view, *Arithmetical Books* is not a good work. It provides too little in terms of bibliographical description, for example, failing to record pagination or foliation. Despite the published presence of British Museum cataloguing rules which De Morgan could have used, it does not note when books are in black letter (Gothic type), and it applies the terms folio, quarto and octavo anachronistically to size rather than bibliographical format, as De Morgan himself discusses.[68]

Nonetheless, *Arithmetical Books* gained De Morgan a reputation as a bibliographer, with his most detailed obituary noting his interest in such matters of physical bibliography as watermarks, colophons and catchwords.[69] Bibliographically, it stood out for De Morgan's insistence on seeing the books he described in order to ensure accuracy, a concern he also expressed elsewhere,[70] and it also drew attention to the relationships between editions. Idiosyncratically, he spelt out dates of publication in words: a decision also made to promote accuracy by avoiding errors

---

67 We leave it as an exercise for the reader to discover the value of X.
68 Augustus De Morgan, *Arithmetical Books from the Invention of Printing to the Present Time* (London: Taylor & Walton, 1847), pp. xi–xiii. For a modern assessment of *Arithmetical Books*, see David McKitterick, *Readers in a Revolution: Bibliographical Change in the Nineteenth Century* (Cambridge: Cambridge University Press, 2022), pp. 88–91.
69 Ranyard, Obituary Notice, p. 117; see also A. De Morgan, *Arithmetical Books*, pp. xii–xiii.
70 Augustus De Morgan, 'Mathematical Bibliography', *Dublin Review*, 41 (Sept. 1846), 1–37.

that can arise when copying or printing figures. As a list, *Arithmetical Books* is incomplete because, as De Morgan notes frankly, inclusion depended on his personal examination of works.[71] Yet it quickly became a standard reference tool, as references to it in Victorian sale catalogues of mathematical books in and beyond Britain demonstrate.[72] In 1908, David Eugene Smith was able to write of it, in terms of its overview of its subject matter, as 'still one of our best single sources, although sixty years have elapsed since it first appeared', while, in a 1967 reprint, A. Rupert Hall called it 'a minor classic', still of use, on the same basis.[73]

De Morgan's personal library comprised nearly 4,000 items and was known as one of the most impressive collections of mathematical books in Britain, although it was not in fact the largest mathematical library of its time. Karen Attar has written elsewhere about De Morgan's library and his annotations on a significant minority of the books therein. In her chapter here, she tests the various nineteenth- and twentieth-century statements about its excellence by comparing and contrasting it with contemporary mathematical collections such as those of Francis Baily, Charles Babbage and John Thomas Graves. She demonstrates its unique importance through the connection between the books and their owner, a feature absent from the other collections. The second part of her chapter treads further new ground by chronicling the library's fate after Lord Overstone purchased and gifted it to the University of London Library (now Senate House Library, University of London), which opened in 1877.

The words De Morgan left behind him are not only those he published and the printed words he collected but are also contained in an enormous amount of archival material: mathematical manuscripts, and in particular personal letters, scattered among several repositories. De Morgan's entry in the *Oxford Dictionary of National Biography* lists these institutions, all of which have archival catalogues. The penultimate chapter of this volume, written by curators, brings the material in these

---

71 A. De Morgan, *Arithmetical Books*, pp. ii, ix–x.
72 See for example *Catalogue de Livres Astronomiques, Mathématiques et Physiques Provenant des Bibliothèques de Feu M. A.C. Petersen ... dont la Vente Publique se Fera à Berlin le Lundi 17 Decembre 1855 Et Jours Suivants* (Berlin, 1855), lot 1789.
73 David Eugene Smith, *Rara Arithmetica: A Catalogue of the Arithmetics Written before the Year MDCI, with a Description of Those in the Library of George Arthur Plimpton of New York* (Boston: Ginn, 1908), p. xii; A. Rupert Hall, 'Introduction', in A. De Morgan, *Arithmetical Books* (London: Hugh K. Elliott, 1967), p. vii.

archives to life. The descriptions demonstrate, as the catalogues cannot, how the papers held illumine De Morgan's work, character and life; his personal and professional relationships—and also why edited extracts do not substitute for the originals.

De Morgan spent his entire working life in London's Bloomsbury, where University College was situated on Gower Street. Now a fashionable district in central London, Bloomsbury in the early nineteenth century was a relatively uninspiring neighbourhood on the city's northernmost edge. Yet notwithstanding the area's aesthetic shortcomings, UCL students could benefit from its flourishing intellectual atmosphere, as Richard Holt Hutton, one of De Morgan's erstwhile pupils, recalled:

> It is sometimes said that it needs the quiet of a country town remote from the capital, to foster the love of genuine study in young men. But of this at least I am sure: that Gower Street, and Oxford Street, and the New Road, and the dreary chain of squares from Euston to Bloomsbury, were the scenes of discussions as eager and as abstract as ever were the sedate cloisters or the flowery river-meadows of Cambridge or Oxford.[74]

Rosemary Ashton puts her expertise as leader of the UCL Bloomsbury Project[75] to good use in her chapter to contextualise De Morgan in his physical and intellectual surroundings. She paints an evocative picture of Bloomsbury in the early years of the nineteenth century, when social reformers like Henry Brougham and George Birkbeck founded 'The London University' on one of its main thoroughfares, thereby creating the possibility of an academic career for De Morgan in the capital and initiating Bloomsbury's strong and enduring association with higher education, culture and the arts.

While Ashton discusses De Morgan's geographical environment, Joan Richards places De Morgan in his familial context. His wife Sophia and two of the De Morgans' children, the writer Mary and especially the novelist and ceramic artist William (creator of 'De Morgan tiles') have

---

74 Walter Bagehot, *Literary Studies*, ed. by Richard Holt Hutton, vol. 1 (London: Longmans, Green, 1879), p. xiii.
75 University College London, *UCL Bloomsbury Project*, https://www.ucl.ac.uk/bloomsbury-project/index.htm

received their own studies,[76] while Richards herself has done much to shed light on the family in her recent monograph *Generations of Reason*. Her chapter here balances De Morgan's intellectual legacy with his familial legacy through his children's achievements. Richards's exploration of Sophia's fascination with the development of her offspring's powers of reasoning underlines that interest in education was the preserve of both husband and wife, not of Augustus De Morgan alone.

De Morgan published no magnum opus, proved no major theorem and made no major scientific discovery. Yet it has been said: 'Were the writings of De Morgan published in the form of collected works, they would form a small library.'[77] His published output comes to more than 2,200 individual items, including 1,400 papers, articles and surveys, and over 700 encyclopedia entries. These varied in length from just a few lines to scores of pages, appearing in newspapers, literary magazines, proceedings of learned societies, and some of the premier research journals of the day. The wide range of De Morgan's publication venues and the anonymity under which many of his articles appeared renders the task of listing every item herculean, and previously unknown De Morgan-authored texts will probably be unearthed by subsequent literature searches. But in our final chapter, William Hale has produced the fullest and most detailed De Morgan bibliography to date. Olivier Bruneau's chapter discusses some of this raw material, considering the length as well as the quantity of these publications to balance the areas of his work. He analyses De Morgan's scholarly and journalistic output, showing the extent of his writings for a general audience, and revealing him to have been one of early Victorian Britain's most prolific and gifted mathematical expositors.

---

76   Sophia Elizabeth De Morgan, *Threescore Years and Ten: Reminiscences of the Late Sophia Elizabeth De Morgan to which are Added Letters to and from her Husband the Late Augustus De Morgan, and Others*, ed. by Mary A. De Morgan (London: Bentley, 1895); Marilyn Pemberton, *Out of the Shadows: The Life and Work of Mary De Morgan* (Newcastle upon Tyne: Cambridge Scholars Publishing, 2012); Anna M. W. Stirling, *William De Morgan and His Wife* (London: Thornton Butterworth, 1922); William Gaunt and M.D.E. Clayton-Stamm, William De Morgan (London: Studio Vista, 1971); Jon Catleugh, *William De Morgan Tiles* (New York: Van Nostrand Reinhold, 1983); Mark Hamilton, *Rare Spirit: A Life of William De Morgan 1839–1911* (London: Constable, 1997); Rob Higgins and Christopher Stolbert Robinson, *William De Morgan: Arts and Crafts Potter* (New York: Bloomsbury, 2010).

77   MacFarlane, *Lectures*, p. 24.

As a polymath, De Morgan was no dilettante: his erudition was deep and his knowledge wide, and the chapters that follow reflect the richness and diversity of his intellectual output. To do justice to his extraordinarily multifaceted career, our team of fifteen authors have covered a wide area, from the relatively well known to the more obscure. Naturally, this volume does not claim to be the last word—page constraints and the nature of scholarly interest mean that there are inevitable omissions. For example, although this book includes information on De Morgan as a professor and as a historian, no specific chapters are devoted exclusively to these subjects.[78] Detailed studies of his work as a bibliographer and as an actuary and of his religious views, yet to be undertaken, will further nuance our picture of De Morgan.[79] Yet this volume brings together examinations of various facets of his life and legacy as no previous book has done. As the 150th anniversary of the opening of the University of London Library and the 200th anniversary of the commencement of De Morgan's career at University College London approach, we hope that this volume will spark interest in, and provide the impetus for, further research into a significant but largely overlooked figure in British intellectual and cultural history.

Bibliography

Bagehot, Walter, *Literary Studies*, ed. by Richard Holt Hutton, 2 vols. (London: Longmans, Green, 1879).

Bland, Miles, *Algebraical Problems, Producing Simple and Quadratic Equations, with their Solutions* (Cambridge: J. Smith, 1812).

Bridge, Bewick, *An Elementary Treatise on Algebra* (London: T. Cadell & W. Davies, 1815).

---

78   However, studies do exist. For example, for De Morgan as a professor, see Adrian Rice, 'What Makes a Great Mathematics Teacher? The Case of Augustus De Morgan', *American Mathematical Monthly*, 106 (1999), 534–52; and on De Morgan as historian, see Joan L. Richards, 'Augustus De Morgan, the History of Mathematics, and the Foundations of Algebra', *Isis*, 78 (1987), 7–30; Adrian Rice, 'Augustus De Morgan: Historian of Science', *History of Science*, 34 (1996), 201–40; and Rebekah Higgitt, *Recreating Newton: Newtonian Biography and the Making of Nineteenth-Century History of Science* (London: Pickering & Chatto, 2007), Chapters 4–6.

79   For early gestures towards a bibliographical study, see McKitterick, *Readers in a Revolution*, pp. 87–91.

Brock, William H., and Roy M. MacLeod, ed., *Natural Knowledge in Social Context: The Journal of Thomas Archer Hirst, F.R.S.* (London: Mansell, 1980).

Burke, Peter, *The Polymath: A Cultural History from Leonardo da Vinci to Susan Sontag* (New Haven: Yale University Press, 2021). https://doi.org/10.12987/9780300252088

Catleugh, Jon, *William De Morgan Tiles* (New York: Van Nostrand Reinhold, 1983).

De Morgan, Augustus, *A Budget of Paradoxes* (London: Longmans, Green, 1872).

—, *Arithmetical Books from the Invention of Printing to the Present Time* (London: Taylor & Walton, 1847; repr. London: Hugh K. Elliot, 1967).

—, *Formal Logic* (London: Taylor & Walton, 1847).

—, 'Mathematical Bibliography', *Dublin Review*, 41 (Sept. 1846), 1–37.

—, 'Memorandums on the Descendants of Captain John De Morgan ...', MS. ADD. 7, UCL Special Collections.

De Morgan, Sophia E., *Memoir of Augustus De Morgan* (London: Longmans, Green, 1882).

—, *Threescore Years and Ten: Reminiscences of the Late Sophia Elizabeth De Morgan to which are Added Letters to and from her Husband the Late Augustus De Morgan, and Others*, ed. by Mary De Morgan (London: Bentley, 1895).

Dieudonné, Jean, *Mathematics—The Music of Reason* (Berlin: Springer, 1992).

Enros, Philip C., 'The Analytical Society (1812–1813): Precursor of the Renewal of Cambridge Mathematics', *Historia Mathematica*, 10 (1983), 24–47.

Gaunt, William, and M.D.E. Clayton-Stamm, *William De Morgan* (London: Studio Vista, 1971).

Graves, Robert Perceval, *Life of Sir William Rowan Hamilton*, vol. 3 (Dublin: Hodges, Figgis, 1889).

Gray, G. J., 'Dodson, James (c.1705–1757)', rev. Anita McConnell, *Oxford Dictionary of National Biography* (Oxford: Oxford University Press, 2004). https://doi.org/10.1093/odnb/9780192683120.013.7756

Hall, A. Rupert, 'Introduction', in Augustus De Morgan, *Arithmetical Books*, repr. (London: Hugh K. Elliot, 1967), pp. [i–vii].

Hamilton, Mark, *Rare Spirit: A Life of William De Morgan 1839–1911* (London: Constable, 1997).

Heath, Peter, 'Editor's Introduction', in Augustus De Morgan, *On the Syllogism, and Other Logical Writings* (London: Routledge & Kegan Paul, 1966), pp. vii–xxxi.

Heinemann, Anna-Sophie, *Quantifikation des Prädikats und numerisch definiter Syllogismus. Die Kontroverse zwischen Augustus De Morgan und Sir William Hamilton: Formale Logik zwischen Algebra und Syllogistik* (Münster: Mentis, 2015).

Higgins, Rob, and Christopher Stolbert Robinson, *William De Morgan: Arts and Crafts Potter* (New York: Bloomsbury, 2010).

Higgitt, Rebekah, *Recreating Newton: Newtonian Biography and the Making of Nineteenth-Century History of Science* (London: Pickering & Chatto, 2007). https://doi.org/10.4324/ 9781315653068

—, 'Why I don't FRS my tail: Augustus De Morgan and the Royal Society', *Notes and Records of the Royal Society*, 60 (2006), 253–59. https://doi.org/10.1098/rsnr.2006.0150

Laíta, Luis María, 'Influences on Boole's Logic: The Controversy between William Hamilton and Augustus De Morgan', *Annals of Science*, 36 (1979), 45–65.

MacFarlane, Alexander, *Lectures on Ten British Mathematicians of the Nineteenth Century* (London: Chapman & Hall, 1916).

McKitterick, David, *Readers in a Revolution: Bibliographical Change in the Nineteenth Century* (Cambridge: Cambridge University Press, 2022). https://doi.org/10.1017/ 9781009200882

Pemberton, Marilyn, *Out of the Shadows: The Life and Work of Mary De Morgan* (Newcastle upon Tyne: Cambridge Scholars Publishing, 2012).

Ranyard, Arthur Cowper, Obituary Notice of Augustus De Morgan, *Monthly Notices of the Royal Astronomical Society*, 32 (1871–72), 112–18.

Rice, Adrian, 'Augustus De Morgan: Historian of Science', *History of Science*, 34 (1996), 201–40. https://doi.org/10.1177/007327539603400203

—, 'Inspiration or Desperation? Augustus De Morgan's Appointment to the Chair of Mathematics at London University in 1828', *British Journal for the History of Science*, 30 (1997), 257–74. https://doi.org/10.1017/s0007087497003075

—, 'What Makes a Great Mathematics Teacher? The Case of Augustus De Morgan', *American Mathematical Monthly*, 106 (1999), 534–52. https://doi.org/10.2307/2589465

Richards, Joan L., 'Augustus De Morgan, the History of Mathematics, and the Foundations of Algebra', *Isis*, 78 (1987), 7–30. https://doi.org/10.1086/354328

—, *Generations of Reason: A Family's Search for Meaning in Post-Newtonian England* (New Haven: Yale University Press, 2021). https://doi.org/10.12987/yale/9780300255492. 001.0001

Robinson, Henry Crabb, *Diary, Reminiscences, and Correspondence of Henry Crabb Robinson*, ed. by Thomas Sadler, 3 vols. (Boston: Fields, Osgood, 1869).

Rouse Ball, Walter William, 'Augustus De Morgan', *The Mathematical Gazette*, 8 (1915–16), 42–45.

Rouse Ball, Walter William, and John A. Venn, ed., *Admissions to Trinity College, Cambridge*, vol. 4 (London: Macmillan, 1911).

Smith, David Eugene, *Rara Arithmetica: A Catalogue of the Arithmetics Written before the Year MDCI, with a Description of Those in the Library of George Arthur Plimpton of New York* (Boston: Ginn, 1908).

Smith, Gordon C., *The Boole-De Morgan Correspondence 1842–1864* (Oxford: Clarendon Press, 1982).

Stirling, Anna M. W., *William De Morgan and his Wife* (London: Thornton Butterworth, 1922).

Tuke, Margaret J., *A History of Bedford College for Women 1849–1937* (London: Oxford University Press, 1939).

von Wowern, Johann, *De Polymathia Tractatio* ([Basel]: Officina Frobeniano, 1603).

Fig. 2 Augustus De Morgan pictured in the 1860s. (Public domain, via MacTutor, https://mathshistory.st-andrews.ac.uk/Biographies/De_Morgan/pictdisplay/)

PART I

SCIENTIFIC WORK

Fig. 3 De Morgan's letters, books and personal papers were often adorned with imaginative and unusual cartoons. Here, he applies his idiosyncratic sense of humour to the subject of algebra. (RAS MSS De Morgan 3, reproduced by permission of the Royal Astronomical Society Library and Archives.)

1. De Morgan and Mathematics

Adrian Rice

> De Morgan did not write mathematics;
> He wrote *about* mathematics!
> — J. J. Sylvester[1]

Introduction

Contemporaries of Augustus De Morgan described him as 'one of the most eminent mathematicians and logicians of his time' and 'one of the profoundest and subtlest thinkers of the nineteenth century'.[2] His mathematical publications spanned an impressive array of subjects, including algebra, logic, probability, analysis, differential equations, actuarial mathematics, mathematical education and the history of mathematics, earning him the reputation as one of Victorian Britain's most respected and influential mathematicians.

Over the years, particularly after the First World War, this reputation faded substantially. By 1935, a contributor to *The Listener* lamented: 'If a book were to appear today on the great scientists of the nineteenth century it would be safe to say you would look in vain in its index for the name of Augustus de Morgan.'[3] And although De Morgan is

---

1 George Bruce Halsted, 'De Morgan to Sylvester', *The Monist*, 10 (1900), 188–97 (p. 197).
2 William Stanley Jevons, 'De Morgan, Augustus', *Encyclopædia Britannica*, 9th edn, vol. 8 (Edinburgh: Adam & Charles Black, 1877), pp. 64–67 (p. 64); Henry Enfield Roscoe, *The Life and Experiences of Sir Henry Enfield Roscoe* (London: Macmillan, 1906), p. 25. The vocabulary echoes that of De Morgan's newspaper obituaries.
3 A. S. Russell, 'Augustus De Morgan, A Forgotten Worthy', *The Listener*, 14 (24 Dec. 1935), 1161.

in fact present in the index—and the main text—of E. T. Bell's *Men of Mathematics* of 1937, it is solely for brief supporting roles in chapters on better-known contemporaneous British mathematicians.[4]

If mathematicians remember Augustus De Morgan at all today, it is primarily for the pair of algebraic laws that bear his name, for he proved no major theorem, made no notable mathematical discoveries and published no *magnum opus*. His mathematical achievements appear slim in comparison with those of his peers, such as Arthur Cayley, James Joseph Sylvester, William Rowan Hamilton and George Boole. Indeed, it is hard to identify much in today's mathematics for which De Morgan was responsible. If he is mentioned in recent historical studies of nineteenth-century mathematics, he is regarded as a minor, although charming, figure of minimal consequence to the overall development of mathematics in the period.

Such comparative indifference after seemingly universal approbation prompts the present-day reader to speculate whether his mathematical contributions are undervalued today, or, by contrast, whether such a high former reputation was deserved. This chapter will explore the range of De Morgan's activities as a mathematician to re-assess his former and his current reputation.

De Morgan's Mathematical Output

De Morgan was a voluminous writer. During a period of over forty years from the beginning of his career in 1828, he published a host of books and research papers, as well as countless unsigned articles and reviews in various journals and periodicals.[5] Some of his textbooks, originally intended for his students at University College London, sold in sufficient numbers to warrant multiple editions and translations, particularly his *Elements of Arithmetic* (1830).[6] This pioneered a new style of textbook exposition in being designed to be read by both teacher and pupil (although teachers would no doubt have been the main beneficiaries). De Morgan presented material in a clear, more user-friendly way than previous textbooks had done, beginning with

---

4   E. T. Bell, *Men of Mathematics* (New York: Simon & Schuster, 1937), pp. 354, 384, 387, 434, 438, 440–41.
5   See Chapters 4 and 12 in this volume.
6   1st edn, 1830; 6th edn, 1876; Marathi translation, 1848; Mandarin translation, 1859.

simple motivating examples before introducing rules, demonstrations and more sophisticated problems. He encouraged the use of tangible objects such as pebbles, coins and counters, following Johann Heinrich Pestalozzi's advocacy of environments in which children could learn through activities and exploration, rather than the more traditional approaches.[7]

Appearing 'at a time when perhaps few teachers, as they submitted the rules of the science to their pupils, cared to establish them upon reason and demonstration', the commercial success of *Elements of Arithmetic* reflected its modest influence in the teaching of arithmetic in Britain.[8] As a reviewer in *Nature* observed in 1870: 'The effect of this work was that a rational arithmetic began to be taught generally, and the mere committing of rules to memory took its due subordinate position in the course of instruction.'[9] De Morgan's interest in mathematics education was particularly noticeable in the early 1830s, when he contributed a series of articles to the short-lived *Quarterly Journal of Education*. While his later publications on didactic matters were fewer in number, subsequent articles and book reviews in the *Educational Times* and *The Athenæum* testified to a lifelong interest in mathematical pedagogy.[10]

By the mid-1830s De Morgan's mathematical interests had broadened to include probability theory and its applications, particularly to problems in insurance. The actuarial profession was in its infancy at the time, and mathematicians with a sufficient understanding of probabilistic methods could pursue lucrative careers in the insurance business.[11] To supplement his professorial income, De Morgan served as a freelance actuarial consultant to various insurance companies and 'occupied the first place [as an actuary], though he was not directly associated with any particular office; but his opinion was sought for by professional

---

7   Florence A. Yeldham, *The Teaching of Arithmetic through Four Hundred Years 1535–1935* (London: Harran, 1936).
8   Anon, 'Our Book Shelf', *Nature*, 7 July 1870, 186.
9   Anon, 'Our Book Shelf'.
10 See Chapter 6 in this volume for a discussion of De Morgan's educational writings, and also Sloan Evans Despeaux and Adrian C. Rice, 'Augustus De Morgan's Anonymous Reviews for *The Athenæum*: A Mirror of a Victorian Mathematician', *Historia Mathematica*, 43 (2016), 148–71 (pp. 156–62).
11 Interestingly, it was the pioneering work of De Morgan's great-grandfather James Dodson (c.1705–57) on mortality tables and long-term life-insurance policies that had led to the formation of the Society for Equitable Assurances on Lives and Survivorships (later known as Equitable Life) in London in 1762.

actuaries on all sides, on the more difficult questions connected with the theory of probabilities, as applied to life-contingencies'.[12] He also wrote several works on actuarial subjects. His *Essay on Probabilities and on their Application to Life Contingencies and Insurance Offices* (1838) was the first book of its kind in English and remained highly regarded in insurance literature for well over a generation.

British mathematicians had neglected the theory of probability for some time and De Morgan's work on the topic, albeit small in comparison to his output in other areas, was significant. Its stimulus was probably a need to understand the great French mathematician Pierre-Simon Laplace's seminal *Théorie Analytique des Probabilités*, which De Morgan had been asked to review, and which he regarded as 'by very much the most difficult mathematical work we have ever met with'.[13] He called it

> the Mont Blanc of mathematical analysis; but the mountain has this advantage over the book, that there are guides always ready near the former, whereas the student has been left to his own method of encountering the latter.[14]

More generally, he wrote: 'There are no questions in the whole range of applied mathematics which require such close attention, and in which it is so difficult to escape error, as those which occur in the theory of probabilities'.[15] De Morgan's book-length article on the 'Theory of Probabilities' for the *Encyclopaedia Metropolitana* constituted the first major nineteenth-century English work on probability theory. Although it contained no original results, it was a readable, if technical, elucidation of Laplace's work and, with its thoughtful simplification and clarification of many of the intricate proofs, functioned both linguistically and mathematically as a translation of Laplace's *Théorie*. It furthermore displayed a knowledge and appreciation of what we would now call mathematical statistics, devoting substantial space to data analysis, including a detailed discussion of the method of least squares.

---

12 Arthur Cowper Ranyard, Obituary Notice of Augustus De Morgan, *Monthly Notices of the Royal Astronomical Society*, 32 (1871–72), 112–18 (p. 116).
13 Augustus De Morgan, 'Theory of Probabilities', *Encyclopaedia Metropolitana*, 2 (London: Baldwin & Cradock, 1845), pp. 393–490 (p. 418).
14 [Augustus De Morgan], Review of *Théorie Analytique des Probabilités* (by P.-S. Laplace), *Dublin Review*, 3 (1837), 237–48, 338–54 (p. 347).
15 [Augustus De Morgan], 'Probability, Probabilities, Theory of', in *The Penny Cyclopaedia*, ed. by George Long, 27 vols. (London: C. Knight, 1833–1843), 19 (1841), pp. 24–30 (p. 29).

De Morgan thus showed himself to be one of the few mainstream British advocates of Laplacean probability during the early- to mid-nineteenth century. Moreover, by promoting its practical utility in the nascent field of insurance he helped to establish it as the basis of modern actuarial mathematics.

The early nineteenth century was marked by a concerted effort by British mathematicians to bring their work up to the standard of their more progressive European counterparts. After the unpleasant priority dispute in the early eighteenth century between Isaac Newton and Gottfried Leibniz over the invention of the calculus, British mathematicians had become considerably isolated from continental developments. This had resulted in a deterioration in the overall quality of British mathematics and a startling ignorance of contemporary mathematics overseas. Laplace's notoriously challenging five-volume *Traité de Méchanique Céleste* (1799–1825) jolted British mathematicians out of their complacency and stimulated them to learn and adopt continental techniques:

> The impact of this work on British science cannot be underestimated. It was immediately recognized by many British mathematicians as a masterpiece which crowned Newtonian mechanics and astronomy. ... The achievements of Laplace in these fields were outstanding: an urgent need arose to understand his work.[16]

As a student at Cambridge in the 1820s, De Morgan had been one of the earliest of a new generation of British mathematicians to be exposed to this renewed interest in continental mathematics. His early publications show a keen interest in, and considerable knowledge of, European (particularly French) work in several areas. One of these was the study of functions, infinite series and the theoretical underpinnings of calculus, known today as mathematical analysis.

Modern-day mathematicians more or less universally accept that the whole framework of calculus is based on the notion of a limit. Although Bernard Bolzano and Augustin-Louis Cauchy first successfully formalised this concept in the 1810s and 1820s, the idea was not yet widely accepted in the 1830s. De Morgan's *Elements of Algebra* of 1835

---

16 Niccolò Guicciardini, *The Development of Newtonian Calculus in Britain 1700–1800* (Cambridge: Cambridge University Press, 1989), p. 117.

stands as the first British work to contain a formal definition of a limit, specifying: 'When, under circumstances, or by certain suppositions, we can make A as near as we please to P, ... then P is called the *limit* of A.'[17]

De Morgan followed this book with a 785-page treatise on *The Differential and Integral Calculus* (1842), in which he affirmed his adherence to the new system of Cauchy and showed a keen awareness and appreciation of the work of other European mathematicians. It covered every conceivable area of calculus and remained the most comprehensive British work on the subject for well over a generation. More than one hundred years later one of the greatest twentieth-century British analysts, G. H. Hardy, called it 'the best of the early English text-books on the calculus, contain[ing] much that is still interesting to read and difficult to find in any other book'.[18]

In addition to being the most exhaustive study of the subject to date in English, De Morgan's treatise contained a number of original results in analysis. For example, he introduced a test for the convergence of infinite series of the form

$$\frac{1}{f(1)} + \frac{1}{f(2)} + \frac{1}{f(3)} + \cdots + \frac{1}{f(n)} + \frac{1}{f(n+1)} + \cdots$$

where $f(n)$ is any increasing positive function of n. By introducing the quantity

$$\rho = \lim_{n \to \infty} \frac{nf'(n)}{f(n)}$$

De Morgan determined that the series would diverge if $\rho < 1$ and converge if $\rho > 1$.[19] Moreover, if $\rho = 1$, his method contained the novel feature of an iterated procedure to determine the convergence of more problematic cases.[20]

---

17   Augustus De Morgan, *The Elements of Algebra* (London: Taylor & Walton, 1835), p. 155.
18   G. H. Hardy, *Divergent Series* (Oxford: Clarendon Press, 1949), p. 19.
19   Augustus De Morgan, *The Differential and Integral Calculus* (London: Baldwin & Cradock, 1842), pp. 235–36. See also C. W. Groetsch, 'De Morgan's Series Test', *The College Mathematics Journal*, 47 (2016), 136–37.
20   De Morgan, *Differential and Integral Calculus*, pp. 324–25. An equivalent procedure was discovered independently by the French mathematician Joseph Bertrand. Consequently, the convergence test is now usually called Bertrand's Test.

De Morgan made his most progressive research contribution to this area on the subject of divergent series, one of the thorniest issues in nineteenth-century mathematics as most mainstream mathematicians categorically rejected them. As the Norwegian mathematician Niels Henrik Abel declared: 'Divergent series are the inventions of the devil, and it is a shame to base on them any proposition whatsoever. By using them, one may draw any conclusion he pleases and that is why these series have produced so many fallacies and so many paradoxes.'[21] De Morgan began his 1844 paper 'On Divergent Series':

> I believe it will be generally admitted that the heading of this paper describes the only subject yet remaining, of an elementary character, on which a serious schism exists among mathematicians as to absolute correctness or incorrectness of results.[22]

Mathematicians of the time regarded strange results like

$$1 - 1 + 1 - 1 + 1 - 1 + \cdots = \tfrac{1}{2}$$

and

$$1 + 2 + 4 + 8 + 16 + \cdots = -1$$

as meaningless and therefore useless. But they are now known to be of great value. For example, string theory in physics makes use of the peculiar formula

$$1 + 2 + 3 + 4 + 5 + \cdots = -\tfrac{1}{12}.$$

De Morgan cautioned strongly against rejecting such results simply through a lack of sufficient understanding: 'The history of algebra shows us that nothing is more unsound than the rejection of any method which naturally arises, on account of one or more apparently valid cases in which such method leads to erroneous results. Such cases

---

21 Gaither's *Dictionary of Scientific Quotations*, ed. by Carl C. Gaither and Alma E. Cavazos-Gaither (New York: Springer, 2012), p. 2301.
22 Augustus De Morgan, 'On Divergent Series and Various Points of Analysis Connected with them', *Transactions of the Cambridge Philosophical Society*, 8 (1844), 182–203 (p. 182).

should indeed teach caution, but not rejection.'[23] Although he admitted that many divergent series were not yet sufficiently understood to be of any use, he warned: 'to say that what we cannot use no others ever can ... seems to me a departure from all rules of prudence'.[24]

Looking back over a century later, the mathematician Morris Kline candidly but fairly described De Morgan's paper as 'acute and yet confused'.[25] De Morgan made some very sound points, and his main thesis—do not reject what you do not understand—turned out to be perfectly valid. As he had written in 1840: 'I am fully satisfied that they [divergent series] have an *algebraical* truth wholly independent of arithmetical considerations: but I am also satisfied that this is the most difficult question in mathematics.'[26] But while De Morgan may have been virtually the only British mathematician of the time capable, not only of appreciating the subtleties of the problem, but also of commenting constructively on the issue, what was most remarkable for twentieth-century mathematicians like Kline and Hardy was 'that so acute a reasoner should be able to say so much that is interesting and yet to miss the essential points so completely'.[27]

In some of his related research on differential equations, De Morgan's presentation was influenced by prior work by his elder British contemporaries, Charles Babbage, John Herschel, and especially his erstwhile teacher George Peacock. Their particular technique was more algebraic in spirit than analytic, focusing principally on the clever manipulation of symbols. For example, to solve the differential equation

$$\frac{d^2y}{dx^2} + \frac{dy}{dx} - 6y = 0,$$

they would re-write it in terms of the differential operator, D, to give

$$D^2y + Dy - 6y = 0,$$

---

23 De Morgan, *Differential and Integral Calculus*, p. 566.
24 De Morgan, 'On Divergent Series', p. 183.
25 Morris Kline, *Mathematical Thought from Ancient to Modern Times* (New York: Oxford University Press, 1972), p. 975.
26 Oxford, Bodleian Library, Dep. Lovelace-Byron, Box 170, Letter from Augustus De Morgan to Ada Lovelace, 15 Oct. 1840, f. 19v.
27 Hardy, *Divergent Series*, p. 20.

where $D = d/dx$. The symbols of operation would then be separated from the variable, y, to give

$$(D^2 + D - 6)y = 0 \text{ or } (D - 2)(D + 3)y = 0.$$

Now, since it was known that the solution of $(D - 2)y = 0$ is $y = c_1 e^{2x}$ and the solution of $(D + 3)y = 0$ is $y = c_2 e^{-3x}$, the solution of the original differential equation would thus be

$$y = c_1 e^{2x} + c_2 e^{-3x}.$$

De Morgan used this technique, known as the 'calculus of operations', in a paper of 1844 to analyse differential equations of the form $(D + a)^n y = X$, where a is a constant.[28] In so doing, he generated a formula in terms of the operator D^{-1}:

$$(D + a)^{-1}\{A(D + a)B\} = AB - (D + a)^{-1}(BDA).$$

Given that the function D represented the derivative, its inverse D^{-1} would naturally represent the opposite operation, namely, the integral. Thus, De Morgan's result, with $a = 0$, $A = u$ and $B = v$, was simply equivalent to the well-known formula for integration by parts, or

$$\int u \, dv = uv - \int v \, du.$$

The calculus of operations 'became a cottage industry of British mathematics in the 1840s and 1850s', featuring prominently in the work of several of De Morgan's contemporaries such as Duncan Gregory, Robert Murphy, George Boole and Arthur Cayley.[29] This algebraic inclination bore notable fruit as British mathematicians furthered the agenda initiated by Peacock's *Treatise on Algebra* (1830), in which symbolic algebra was conceived as a generalisation of the symbols and operations of basic arithmetic.

---

28 Augustus De Morgan, 'On the Equation $(D + a)^n y = X$', *Cambridge Mathematical Journal*, 4 (1844), 60–62.

29 Victor J. Katz and Karen Hunger Parshall, *Taming the Unknown: A History of Algebra from Antiquity to the Early Twentieth Century* (Princeton: Princeton University Press, 2014), p. 405. See also Elaine Koppelman, 'The Calculus of Operations and the Rise of Abstract Algebra', *Archive for History of Exact Sciences*, 8 (1971–72), 155–242.

In a series of four papers, 'On the Foundation of Algebra', and the monograph *Trigonometry and Double Algebra*, De Morgan pushed algebra further towards abstraction by focusing less on the meanings of the individual symbols and more on the laws under which they operated.[30] Examples of such laws included the associative law for multiplication, which states that, for any three numbers, a, b and c,

$$a(bc) = (ab)c$$

and the commutative law,

$$ab = ba.$$

By focusing on these fundamental laws, or axioms, De Morgan helped advance a trend towards axiomatising algebra that gained pace as the nineteenth century progressed. For De Morgan, the algebraic symbols a, b and c could represent numbers, but did not have to, provided that the laws were applied correctly:

> In abandoning the meanings of symbols, we also abandon those of the words which describe them. Thus *addition* is to be, for the present, a sound void of sense. It is a mode of combination represented by +; when + receives its meaning, so also will the word *addition*. ... If any one were to assert that + and − might mean reward and punishment, and A, B, C, &c. might stand for virtues and vices, the reader might believe him, or contradict him, as he pleases—but not out of *this* chapter.[31]

At this time algebra was still largely restricted to solving equations using the basic laws of arithmetic. The most general kind of solutions were complex numbers of the form $z = a + bi$, where a and b were real numbers and $i^2 = -1$. Complex numbers could be represented geometrically in two dimensions; De Morgan termed this 'Double Algebra'. But could such a representation be extended to three-dimensional space? To answer this,

---

30 Augustus De Morgan, 'On the Foundation of Algebra', *Transactions of the Cambridge Philosophical Society*, 7 (1842), 173–187; 'On the Foundation of Algebra, no. II', *Transactions of the Cambridge Philosophical Society*, 7 (1842), 287–300; 'On the Foundation of Algebra, no. III', *Transactions of the Cambridge Philosophical Society*, 8 (1849), 139–42; 'On the Foundation of Algebra, no. IV, on Triple Algebra', *Transactions of the Cambridge Philosophical Society*, 8 (1844), 241–54; *Trigonometry and Double Algebra* (London: Taylor, Walton, & Maberly, 1849).

31 Augustus De Morgan, *Trigonometry*, p. 101.

De Morgan invented a variety of so-called 'Triple Algebras' where each number had the form $z = a + bi + cj$, where $i \neq j$ and $i^2 = j^2 = -1$. But while adding such number triples together was easy, multiplication proved more of a challenge. De Morgan found that, given two triples $z_1 = a_1 + b_1 i + c_1 j$ and $z_2 = a_2 + b_2 i + c_2 j$, their product was the rather messy

$$(a_1 a_2 - b_1 b_2 - c_1 c_2) + (a_1 b_2 + b_1 a_2)i + (a_1 c_2 + c_1 a_2)j + (b_1 c_2 + c_1 b_2)ij.$$

This was clearly not a triple, so something needed to be done about the rogue ij-term at the end. Arbitrarily letting $ij = -1$, De Morgan solved this problem, but he then found that

$$i(ij) = i(-1) = -i$$

but

$$(ii)j = (i^2)j = (-1)j = -j.$$

This meant that, in general,

$$a(bc) \neq (ab)c$$

meaning that multiplication in De Morgan's triple algebra was not associative.[32] He had thus provided the first published example of what mathematicians now call a non-associative algebra.

De Morgan's research on algebra influenced William Rowan Hamilton's ground-breaking 1843 discovery of quaternions, a four-dimensional group of complex numbers, and the first known algebra to be non-commutative with regard to multiplication. Hamilton acknowledged:

> Among the circumstances which assisted to prevent me from losing sight of the general subject ... was probably the publication of Professor De Morgan's first paper on the Foundation of Algebra, of which he sent me a copy in 1841.[33]

This 'liberation' of algebra from the previously unassailable laws of arithmetic was a key development in nineteenth-century mathematics.

---

32 Augustus De Morgan, 'On the Foundation of Algebra, No. IV, on Triple Algebra', *Transactions of the Cambridge Philosophical Society*, 8 (1844), 241–54 (pp. 249–51).
33 William Rowan Hamilton, *Preface to Lectures on Quaternions* (Dublin: Hodges & Smith, 1853), p. 41.

It opened the floodgates for the creation of newer and ever more unorthodox algebraic systems, including matrices, octonions and vectors. By De Morgan's death in 1871, algebra was well on its way to becoming the generalised abstract subject it remains today, with new systems of algebras being created and axioms formalised. De Morgan's contributions to the subject thus stimulated the growth of abstraction and catalysed a remarkable algebraic discovery.

As if this were not enough, De Morgan had a lifelong fascination with the history of mathematics, which he regarded as being extremely useful in mathematical research and teaching, as well as intrinsically interesting, and historical writings accounted for one-sixth of his published output.[34] In the nineteenth century he was an acknowledged authority, as renowned for his historical work as for his mathematics, and described as 'perhaps more deeply read in the philosophy and history of mathematics than any of his contemporaries'.[35] His interests were wide-ranging and his knowledge was extensive: medieval English mathematical authors, calendar reckoning, and the history of arithmetic are just a few of the diverse topics on which he published. Like William Whewell, David Brewster, and other contemporary historians of science, De Morgan in his publications on the history of mathematics generally emphasised primary and archival sources, and aimed to reconstruct as accurate a picture of past events as possible from the available evidence. But in stark contrast to his peers, De Morgan rejected the 'great man' view of history. Instead, he attempted 'to understand his mathematical predecessors not merely as intellectual forefathers but as human beings'.[36] At a time when Newton and Euclid dominated historical studies of mathematics, De Morgan believed that 'names which are now unknown to general fame are essential to a sufficient view of history'[37] and wrote:

---

34 Adrian Rice, 'Augustus De Morgan: Historian of Science', *History of Science*, 34 (1996), 201–40 (p. 201).

35 Walter William Rouse Ball, *A History of the Study of Mathematics at Cambridge* (Cambridge: Cambridge University Press, 1889), p. 133.

36 Joan L. Richards, 'Augustus De Morgan, the History of Mathematics, and the Foundations of Algebra', *Isis*, 78 (1987), 7–30 (p.17).

37 Augustus De Morgan, 'The Progress of the Doctrine of the Earth's Motion between the Times of Copernicus and Galileo, being Notes on the Ante-Galilean Copernicans', *Companion to the Almanac for 1855*, 5–25 (p. 21).

> It would be much too strong a simile to compare the man whose name is in the mouths of all to the engineer who lays the match to the train, and startles the world by an explosion, while no one asks who bored the rock or laid the powder. But though such a comparison would err in degree, it will serve to remind us that, in every great achievement of human intellect of which it falls within the power of history to see the antecedents, we notice a gradual preparation, which is seldom adequately described.[38]

De Morgan is perhaps best remembered for his research into logic.[39] He published two books and a series of five papers 'On the Syllogism' between 1846 and 1863. The famous 'De Morgan's Laws' appeared in his third paper (1858). Today, they are a staple of any introductory undergraduate course in logic, where they might appear as

$$\neg(a \vee b) \equiv \neg a \wedge \neg b \text{ and } \neg(a \wedge b) \equiv \neg a \vee \neg b,$$

and in algebra, in which they could take the form

$$(A \cup B)' = A' \cap B' \text{ and } (A \cap B)' = A' \cup B'.$$

De Morgan, however, expressed them verbally: 'The contrary of an aggregate is the compound of the contraries of the aggregants: the contrary of a compound is the aggregate of the contraries of the components.'[40]

His most original contribution to the whole subject came in his fourth paper of 1860, in which he provided an analysis of the logic of relations that substantially increased the scope of the subject.[41] In particular, he categorised relations as 'identical', 'convertible' and 'transitive' (which

---

38 Augustus De Morgan, 'On Some Points in the History of Arithmetic', *Companion to the Almanac for 1851*, 5–18 (p. 14).
39 See Chapter 2 of this volume.
40 Augustus De Morgan, 'On the Syllogism, No. III and on Logic in General', *Transactions of the Cambridge Philosophical Society*, 10 (1858), 173–320 (p. 208).
41 Augustus De Morgan, *On the Syllogism, and Other Logical Writings*, ed. by Peter Heath (London: Routledge & Kegan Paul, 1966), p. xx; Daniel D. Merrill, *Augustus De Morgan and the Logic of Relations* (Dordrecht: Kluwer Academic Publishers, 1992); Maria Panteki, 'French "logique" and British "logic": On the Origins of Augustus De Morgan's Early Logical Inquiries, 1805–1835', *Historia Mathematica*, 30 (2003), 278–340 (p. 280); Michael E. Hobart and Joan L. Richards, 'De Morgan's Logic', in *Handbook of the History of Logic. Volume 4: British Logic in the Nineteenth Century*, ed. by Dov M. Gabbay and John Woods (Amsterdam: North-Holland, 2008), pp. 283–329.

we would nowadays call reflexive, symmetric and transitive), and proved a variety of results about abstract relations. Today, relations are central to pure mathematics, and De Morgan's ground-breaking paper of 1860 is what initiated this important area, creating a whole new field of study for mathematicians, and making the subsequent development of subjects such as equivalence relations, adjacency relations and partially ordered sets possible.

He was also one of the few mathematicians of his time to realise the importance of logic to mathematics, and vice versa, attempting to bring mathematical ideas into his logic by introducing a numerically precise method of 'quantifying the predicate' and constructing a symbolic notation in which all reasoning could be carried out. This innovative approach directly inspired his friend George Boole to 'resume the almost forgotten thread of former enquiries' and publish his own version, ultimately leading to the creation of Boolean algebra and the birth of modern symbolic logic.[42]

De Morgan as Professor of Mathematics

In addition to being a prominent mathematical writer, De Morgan was a highly influential teacher, holding the chair of mathematics at University College London for almost his entire career. He inspired his students with a love and fascination for the subject and convinced even those who took it no further of the beauty and allure of mathematics. The historian Thomas Hodgkin recalled him from his experience as a student at UCL in the late 1840s thus:

> Towering up intellectually above all his fellows, as I now look back upon him, rises the grand form of the mathematician, Augustus De Morgan, known, I suppose to each succeeding generation of his pupils as 'Gussy'. A stout and tall figure, a stiff rather waddling walk, a high white cravat and stick-up collars in which the square chin is buried, a full but well chiselled face, very short-sighted eyes peering forth through gold-rimmed spectacles; but above all

---

42 George Boole, *The Mathematical Analysis of Logic* (Cambridge: Macmillan, 1847), p. 1. See also: Luis María Laíta, 'Influences on Boole's Logic: The Controversy between William Hamilton and Augustus De Morgan', *Annals of Science*, 36 (1979), 45–65; and Anna-Sophie Heinemann, *Quantifikation des Prädikats und numerisch definiter Syllogismus. Die Kontroverse zwischen Augustus De Morgan und Sir William Hamilton: Formale Logik zwischen Algebra und Syllogistik* (Münster: Mentis, 2015).

such a superb dome-like forehead, as could only belong to one of
the kings of thought: that is my remembrance of De Morgan, and
I feel in looking back upon his personality that his is one of the
grandest figures that I have known.[43]

The lawyer James Bourne Benson affirmed that 'De Morgan [was] looked
upon with awe' by the undergraduates of his day.[44] By his retirement in
1867, De Morgan's name and the quality of his instruction had established
UCL as the prime source for advanced mathematical tuition in London.[45]
From each of the four decades following De Morgan's 1828 inauguration
as UCL's founding professor of mathematics, graduates distinguished
in their chosen fields praised their former mentor fulsomely as an
'eccentric but brilliant teacher' whose lectures were stimulating, often
inspiring, and far from easy.[46] Such students included the future Master
of the Rolls, Herbert Cozens-Hardy; educational reformer Sir Philip
Magnus; and constitutional authority Walter Bagehot.[47] Two students
from the 1840s, Isaac Todhunter and E. J. Routh, spread De Morgan's
pedagogical influence, going on to Cambridge, then the epicentre
of mathematical education in Britain, and dominating the teaching
of mathematics there for half a century: the former wrote a highly
successful series of textbooks, while the latter became one of the most
successful mathematical coaches in its history.

The eminent mathematician James Joseph Sylvester, famous for
Sylvester's law of inertia and for introducing the word 'matrix' into
mathematics, was one of De Morgan's earliest students. Although he

---

43 Thomas Hodgkin, 'University College, London, Fifty Years Ago', *The Northerner*, 1 (1901), 75.
44 UCL MS (1921), Mem. 1B/3: Materials for the history of UCL: James Bourne Benson, 'Some Recollections of University College in the Sixties', f. 3.
45 Adrian Rice, 'What Makes a Great Mathematics Teacher? The Case of Augustus De Morgan', *The American Mathematical Monthly*, 106 (1999), 534–52; Adrian Rice, 'Mathematics in the Metropolis: A Survey of Victorian London', *Historia Mathematica*, 23 (1996), 376-417.
46 David Eugene Smith, *History of Mathematics* (Boston: Ginn, 1923), vol. 1, p. 462.
47 In addition to his college students, De Morgan also took private pupils, the most famous being Ada King, Countess of Lovelace and the only legitimate child of Lord Byron, who famously worked with Charles Babbage on what is widely regarded as the first computer program. De Morgan tutored Lovelace on algebra, functional equations and calculus in 1840 and 1841. See Christopher Hollings, Ursula Martin and Adrian Rice, 'The Lovelace–De Morgan Mathematical Correspondence: A Critical Re-Appraisal', *Historia Mathematica*, 44 (2017), 202–31, and *Ada Lovelace: The Making of a Computer Scientist* (Oxford: Bodleian Library Publishing, 2018).

only studied at UCL for a few months, Sylvester remained proud of his association with De Morgan, the teacher 'whose pupil I may boast to have been'.[48] The economist and logician Stanley Jevons, a student from the 1850s, described him in the *Encyclopaedia Britannica* as 'unrivalled' as a teacher of mathematics, whose 'writings, however excellent, give little idea of the perspicuity and elegance of his *viva voce* expositions, which never failed to fix the attention of all who were worthy of hearing him'.[49] The astrophysicist Arthur Cowper Ranyard corroborated:

> He had a method of interesting his hearers in the subjects on which he lectured, and of making them love mathematics for its own sake, to which few other men have ever attained.[50]

Ranyard collaborated with De Morgan's son George, himself a promising mathematician, to form the London Mathematical Society (LMS) in 1865. This body originated as little more than a mathematics club for De Morgan's current and former students, before quickly growing into Britain's *de facto* national learned society for mathematics.[51] The last major event of De Morgan's mathematical career and perhaps the most tangible by-product of his success as a teacher at UCL was his inaugural presidency of the LMS, and his opening address became the first paper to be published in the Society's *Proceedings*.

De Morgan's Philosophy of Mathematics

De Morgan was, first and foremost, a teacher.[52] Pedagogical concerns and didactic perspectives inform a significant proportion of his mathematical writings, as is apparent from his attention to epistemological issues, clarity of expression and educational utility. De Morgan wrote much of his scholarly work to be precise, intelligible and, above all, instructive, both for his own undergraduate students and for those readers who might be engaged in higher studies or mathematical research.

---

48   E. T. Bell, *Men of Mathematics*, p. 384.
49   Jevons, 'De Morgan', p. 65.
50   Ranyard, *Obituary of De Morgan*, p. 115.
51   Adrian C. Rice, Robin J. Wilson, and J. Helen Gardner, 'From Student Club to National Society: The Founding of the London Mathematical Society in 1865', *Historia Mathematica*, 22 (1995), 402–21.
52   See Rice, 'What Makes a Great Mathematics Teacher?'

Admittedly De Morgan did not always achieve this aim. When laying the foundations of his subject in his inaugural lecture at UCL, for example, he did little more than appeal to idealised extensions of 'self-evident' concepts derived from experience, with striking lack of clarity for a mathematician with a reputation for logical precision:

> From the appearances of the material world, certain distinct notions are gathered, which though their prototypes have no real existence in nature, are the clearest and most definite which our minds contain. Thus, a straight line needs no definition, nor will the mention of it leave the least doubt as to what is meant in the mind of any person present.[53]

Such vagueness was typical of contemporary English scientific writers. In common with many of his countrymen, De Morgan's philosophy of mathematics was rooted in, and heavily influenced by, the English empiricist tradition of John Locke.[54] As G. H. Hardy later commented when surveying De Morgan's work in analysis: 'He talks much excellent sense, but the habits of the time are too strong for him: logician though he is, he cannot, or will not, give definitions.'[55]

De Morgan's rules for operating on his (non-defined) concepts are based on similarly empirically derived axioms, or 'necessary truths', such as 'two straight lines cannot enclose a space'.[56] Despite a largely positivist outlook, his methodology was not entirely straightforward. For example, in 1841 he was the first English mathematician to give an explicit formulation of the fundamental axioms required for an

---

53 UCL Special Collections: Augustus De Morgan, 'An Introductory Lecture delivered at the Opening of the Mathematical Classes in the University of London, Novr. 5th, 1828', MS ADD 3, ff. 14–15. He was similarly vague when writing about the study of numbers: 'The first ideas of arithmetic, as well as those of other sciences, are derived from early observation. How they come into the mind it is unnecessary to inquire; nor is it possible to define what we mean by number and quantity. They are terms so simple, that is, the ideas which they stand for are so completely the first ideas of our mind, that it is impossible to find others more simple, by which we may explain them' (Augustus De Morgan, *On the Study and Difficulties of Mathematics* (London: Baldwin & Cradock, 1831), p. 4).
54 Marie-José Durand-Richard, 'Genèse de l'algèbre en Angleterre: une influence possible de J. Locke', *Revue d'Histoire des Sciences*, 43 (1990), 129–80.
55 Hardy, *Divergent Series*, p. 19.
56 [Augustus De Morgan], 'Mathematics', *The Penny Cyclopaedia*, 15 (1839), 11–14 (p. 12).

algebraic system.⁵⁷ However, having taken such an apparently modern step, he deduced nothing from these axioms. In this paper and more generally, De Morgan's mathematics, although presented as a series of logical arguments, was in reality a pragmatic blend of rigorous deductions, inductive generalisations, philosophical rumination and unsubstantiated intuition.

In her study of De Morgan's algebra, Helena Pycior divided his algebraic work into three distinct periods, or 'stages': an initial 'traditional' stage in which his algebra was grounded on self-evident first principles and motivated by allusions to real-world examples; a second 'abstract' approach, inspired by Peacock's more formal symbol-based methodology; and a final 'ambivalent' stage in which, free to invent new algebraic systems, 'he concentrated on developing a truly meaningful algebra'.⁵⁸ For this reason, she correctly pointed out that 'De Morgan's attitude towards algebra and symbolical algebra in particular changed to such an extent that it is impossible to ascribe to him any single view on the subject'.⁵⁹

Widening this theme, Joan Richards brought in the important ingredient of De Morgan's expertise in the history of mathematics as an additional factor in the shaping of his mathematical philosophy. Just as Pycior emphasized the influence of Peacock on De Morgan's algebra, Richards also underscored the stimulus provided by Whewell with regard to the use of history as a scientific tool. De Morgan's historically inspired awareness that mathematics does not necessarily develop in a linear, orderly manner had profound consequences for his approach to the subject: 'For De Morgan the historical evolution of mathematical ideas provided important insights into the essential nature of the mathematics, which could not be counted on to fit neatly into logical or formal frameworks'.⁶⁰ It also influenced his advice for others engaged in mathematical study—as he wrote in 1859:

---

57 Augustus De Morgan, 'On the Foundation of Algebra, no. II', pp. 287–88. In doing so, he unknowingly gave the first axiomatic definition of a field (minus the law of associativity) – Katz and Parshall, *Taming the Unknown*, p. 403.
58 Helena M. Pycior, 'Augustus De Morgan's Algebraic Work: The Three Stages', *Isis*, 74 (1983), 211–26 (p. 211).
59 Pycior, p. 211.
60 Richards, p. 8.

> Even in geometry and algebra, there is no method of discovery: the rule is, Imitate those who have succeeded, by patiently thinking out, as they did, the method of succeeding. You may be aided by observation of your predecessors: they may give useful hints, but not digested and infallible rules.[61]

Although De Morgan harboured reservations about certain aspects of Whewell's philosophy, he agreed thoroughly with Whewell's 'progressive' epistemological viewpoint.[62] In a review in *The Athenæum* in 1860, he gave the now famous example of the Four-Colour Theorem, which he believed to rely on the principle that 'four areas cannot each have common boundary with all the other three without inclosure'.[63] This principle, he said, though far from obvious was not only incapable of proof, but had also never been noticed by mathematicians before: 'Our knowledge of necessary truth, then, may be progressive'.[64]

Whewell's historically motivated study of the philosophy of science not only convinced De Morgan of the accumulative nature of scientific knowledge, but reinforced his conviction that an essential ingredient for scientific (and therefore mathematical) progress is the study of its history. As he said at the first meeting of the London Mathematical Society in 1865:

> It is astonishing how strangely mathematicians talk of the Mathematics, because they do not know the history of their subject. By asserting what they conceive to be facts they distort its history in this manner. There is in the idea of every one some particular sequence of propositions, which he has in his own mind, and he imagines that that sequence exists in history; that his own order is the historical order in which the propositions have been successively evolved. The mathematician needs to know what the course of invention has been in the different branches of Mathematics ... If he be to have his own researches guided in the way which will best lead him to success, he must have seen the

---

61 [Augustus De Morgan], Review of William Whewell's *Novum organum renovatum*, *The Athenæum*, 1628 (8 January 1859), 42–44 (p. 43).
62 For a detailed discussion of De Morgan's and Whewell's contrasting 'meta-scientific' views, see Chapter 5 in this volume.
63 [Augustus De Morgan], Review of William Whewell's *The Philosophy of Discovery*, *The Athenæum*, 1694 (14 April 1860), 501–03 (p. 502).
64 [De Morgan], Review of Whewell, 1860, p. 502.

curious ways in which the lower proposition has constantly been evolved from the higher.[65]

Throughout his career De Morgan was fascinated with language and its conversion into effective symbolic notation, an extension of his conviction of the importance of precise expression in mathematics. This view also emerges in his 1865 lecture:

> If we do not attend to extension of language, we are shut in and confined by it. Of this Euclid is a good example. He was stunted by want of extension. When we come to study language in connection with Logic, we find a great many things which would hardly have been expected, and by which we may learn how we may best extend the meanings of our terms.[66]

His example was that it is not immediately obvious that the words 'of' and 'but' may be construed as logical opposites. Taking the phrases 'All of men' as meaning 'All men', and 'All but men' as 'Everything except men', he showed that since the first phrase is the opposite of the second, the words 'of' and 'but' can be seen as negations of each other. Thus, in his words, 'we begin for the first time to have a rational power of extending the meanings of words'.[67]

The theme of language was also present in his inaugural lecture of 1828. Citing Locke's *Essay Concerning Human Understanding*, De Morgan contrasted the fluidity of terminology in regular language with the relative precision of mathematical vocabulary.[68] During his career, his fascination with language intersected with his interest in, and development of, symbolic notation. Indeed, in his book-length *Treatise on the Calculus of Functions* (1836), he devoted a discursive section to the evolution of algebraic notation in which he presented the new symbolism as the outcome of an abstraction process with a long history.[69] His research into logic was also furthered by his employment of algebraic

---

65 Augustus De Morgan, 'Speech of Professor De Morgan, President, At the First Meeting of the Society, January 16th, 1865', *Proceedings of the London Mathematical Society*, 1st ser., 1 (1865), 1–9 (p. 6).
66 De Morgan, 'Speech', p. 8.
67 De Morgan, 'Speech', p. 8.
68 De Morgan, 'Introductory Lecture', f. 16.
69 Augustus De Morgan, *A Treatise on the Calculus of Functions* (London: Baldwin & Cradock, 1836), p. 62. [Later published in *Encyclopaedia Metropolitana*, Pure Sciences, 2 (1845), 305–92.]

symbols to facilitate logical inferences previously represented by words.[70] As he wrote, the formal manipulation of mathematical symbols should always lead to truth, but it was the linguistic interpretation of that truth that presented the greater challenge: '[T]here is every reason to hope that the symbols are always right, even though the views of their explanation may require correction.'[71]

De Morgan's philosophy of mathematics, then, was continually evolving. But three motivating features remained constant throughout: the pedagogically-inspired desire for clarity and precision; the fascination with language and notation, prompted by this need for accurate expression; and the belief that true insights into the nature of mathematics were obtainable via the study of its history.

Conclusion

Augustus De Morgan's mathematical legacy lies in four principal achievements. Firstly, his work on algebra in the 1840s was an important catalyst for the discovery of quaternions, one of the great mathematical innovations of the nineteenth century. Moreover, his work furthered the axiomatisation and abstraction of the subject, making him one of the forefathers of modern abstract algebra. Secondly, his research into logic resulted not only in the birth of modern symbolic logic and, indirectly, Boolean algebra, but also in the creation of a totally new area of mathematics, and arguably his most original contribution to the subject, the study of relations. Thirdly, he established UCL as one of the leading centres for the study of mathematics in Britain. Finally, his influential support of the London Mathematical Society at its foundation helped provide a model by which new mathematical ideas could be communicated, extended and preserved—a model that lasts to this day.

De Morgan's high mathematical reputation in the nineteenth century rested primarily on his tremendous expository skill, via both his published works and college teaching. In addition to his undergraduates, he influenced contemporaneous mathematical research. The great variety of his publication venues, from learned

---

70 See Hobart and Richards, 'De Morgan's Logic'.
71 [Augustus De Morgan], 'Algebra', in *Supplement to the Penny Cyclopaedia*, ed. by George Long (London: Charles Knight, 1845), vol. 1, pp. 74–78 (p. 78).

society journals to magazines to encyclopaedias, ensured a wide and diverse audience for his writings. These ranged from textbooks and papers on mathematical pedagogy to lengthy expositions of probability and analysis that introduced sophisticated recent European methods to a British audience, and from original investigations in the history of mathematics to notable contributions to the development of both abstract algebra and mathematical logic.

Yet the very reasons for De Morgan's fame and reputation in and shortly after his lifetime simultaneously explain his relative obscurity today. A teacher's influence decays rapidly in the absence of its immediate beneficiaries. By the 1920s and 1930s, De Morgan's former students were dead, and he had ceased to be a living memory. Although he was a great mathematical writer, he was more of a supporting character than a main protagonist in ground-breaking mathematics, and more of an expositor than an originator. He was one of the most learned and erudite scholars of his time, but no single work that he wrote stands as his masterpiece. Instead, he scattered his erudition in a host of once-popular books and compendia that have become hard to find and in long-defunct esoteric journals. Consequently, the task of reading, let alone appreciating, the entirety of De Morgan's output is next to impossible, so many and varied were his areas of expertise. Perhaps the American historian of mathematics David Eugene Smith put it best when he said of De Morgan in 1923: 'Had he been able to confine himself to one line, he might have been a much greater though a less interesting man.'[72]

Bibliography

Anon, 'Our Book Shelf', *Nature*, 7 July 1870, 186.

Bell, Eric Temple, *Men of Mathematics* (New York: Simon & Schuster, 1937).

Benson, James Bourne, 'Some Recollections of University College in the Sixties', MS (1921), UCL Special Collections, Materials for the history of UCL, Mem. 1B/3.

Boole, George, *The Mathematical Analysis of Logic* (Cambridge: Macmillan, 1847).

---

72 Smith, *History of Mathematics*, p. 462.

De Morgan, Augustus, 'An Introductory Lecture delivered at the Opening of the Mathematical Classes in the University of London, Novr. 5th, 1828', UCL Special Collections, MS ADD 3.

—, *On the Study and Difficulties of Mathematics* (London: Baldwin & Cradock, 1831).

—, *The Elements of Algebra* (London: Taylor & Walton, 1835).

—, *A Treatise on the Calculus of Functions* (London: Baldwin & Cradock, 1836). [Later published in *Encyclopaedia Metropolitana*, Pure Sciences, 2 (1845), 305–92.]

—, *A Treatise on the Theory of Probabilities* (London: Baldwin & Cradock, 1837). [Later published in *Encyclopaedia Metropolitana*, Pure Sciences, 2 (1845), 393–490.]

[—], Review of *Théorie Analytique des Probabilités* (by P.-S. Laplace), *Dublin Review*, 3 (1837), 237–48, 338–54.

—, 'Mathematics', in *The Penny Cyclopaedia*, ed. by George Long, 15 (London: Charles Knight, 1839), pp. 11–14.

—, 'On the Foundation of Algebra II', *Transactions of the Cambridge Philosophical Society*, 7 (1841), 287–300.

—, 'Probability, Probabilities, Theory of', in *The Penny Cyclopaedia*, ed. by George Long, 19 (London: Charles Knight, 1841), pp. 24–30.

—, *The Differential and Integral Calculus* (London: Baldwin & Cradock, 1842).

—, 'On Divergent Series and Various Points of Analysis Connected with Them', *Transactions of the Cambridge Philosophical Society*, 8 (1844), 182–203.

—, 'On the equation $(D + a)^n y = X$', *Cambridge Mathematical Journal*, 4 (1844), 60–62.

—, 'On the Foundation of Algebra, No. IV, on Triple Algebra', *Transactions of the Cambridge Philosophical Society*, 8 (1844), 241–54.

—, 'Algebra', in *Supplement to the Penny Cyclopaedia*, ed. by George Long (London: Charles Knight, 1845), vol. 1, pp. 74–78.

—, *Trigonometry and Double Algebra* (London: Taylor, Walton, & Maberly, 1849).

—, 'On Some Points in the History of Arithmetic', *Companion to the Almanac for 1851*, 5–18.

—, 'The Progress of the Doctrine of the Earth's Motion between the Times of Copernicus and Galileo, being Notes on the Ante-Galilean Copernicans', *Companion to the Almanac for 1855*, 5–25.

—, 'On the Syllogism, No. III and on Logic in General', *Transactions of the Cambridge Philosophical Society*, 10 (1858), 173–320.

[—], Review of William Whewell's *Novum organum renovatum*, *The Athenæum*, 1628 (8 January 1859), 42–44.

[—], Review of William Whewell's *The Philosophy of Discovery*, *The Athenæum*, 1694 (14 April 1860), 501–03.

—, 'Speech of Professor De Morgan, President, At the First Meeting of the Society, January 16th, 1865', *Proceedings of the London Mathematical Society*, 1st ser., 1 (1865), 1–9.

—, *On the Syllogism, and Other Logical Writings*, ed. by Peter Heath (London: Routledge & Kegan Paul, 1966).

Despeaux, Sloan Evans, and Adrian C. Rice, 'Augustus De Morgan's Anonymous Reviews for *The Athenæum*: A Mirror of a Victorian Mathematician', *Historia Mathematica*, 43 (2016), 148–71. https://doi.org/10.1016/j.hm.2015.09.001

Durand-Richard, Marie-José, 'Genèse de l'algèbre en Angleterre: une influence possible de J. Locke', *Revue d'Histoire des Sciences*, 43 (1990), 129–80.

Gaither, Carl C., and Alma E. Cavazos-Gaither, ed., *Gaither's Dictionary of Scientific Quotations* (New York: Springer, 2012).

Groetsch, C. W., 'De Morgan's Series Test', *The College Mathematics Journal*, 47 (2016), 136–37. https://doi.org/10.4169/college.math.j.47.2.136

Guicciardini, Niccolò, *The Development of Newtonian Calculus in Britain 1700-1800* (Cambridge: Cambridge University Press, 1989).

Halsted, George Bruce, 'De Morgan to Sylvester', *The Monist*, 10 (1900), 188–97.

Hamilton, William Rowan, *Lectures on Quaternions* (Dublin: Hodges & Smith, 1853).

Hardy, G. H., *Divergent Series* (Oxford: Clarendon Press, 1949).

Heinemann, Anna-Sophie, *Quantifikation des Prädikats und Numerisch Definiter Syllogismus. Die Kontroverse zwischen Augustus De Morgan und Sir William Hamilton: Formale Logik zwischen Algebra und Syllogistik* (Münster: Mentis, 2015).

Hobart, Michael E., and Joan L. Richards, 'De Morgan's Logic', in *Handbook of the History of Logic. Volume 4: British Logic in the Nineteenth Century*, ed. by Dov M. Gabbay and John Woods (Amsterdam: North-Holland, 2008), pp. 283–329. https://doi.org/10.1016/s1874-5857(08)80010-6

Hodgkin, Thomas, 'University College, London, Fifty Years Ago', *The Northerner*, 1 (1901), 75.

Hollings, Christopher, Ursula Martin and Adrian Rice, 'The Lovelace–De Morgan Mathematical Correspondence: A Critical Re-appraisal', *Historia Mathematica*, 44 (2017), 202–31. https://doi.org/10.1016/j.hm.2017.04.001

—, *Ada Lovelace: The Making of a Computer Scientist* (Oxford: Bodleian Library Publishing, 2018).

Jevons, William Stanley, 'De Morgan, Augustus', *Encyclopædia Britannica*, 9th edn, vol. 8 (Edinburgh: Adam & Charles Black, 1877), pp. 64–67.

Katz, Victor J., and Karen Hunger Parshall, *Taming the Unknown: A History of Algebra from Antiquity to the Early Twentieth Century* (Princeton: Princeton University Press, 2014). https://doi.org/10.23943/princeton/9780691149059.001.0001

Kline, Morris, *Mathematical Thought From Ancient to Modern Times* (New York: Oxford University Press, 1972).

Koppelman, Elaine, 'The Calculus of Operations and the Rise of Abstract Algebra', *Archive for History of Exact Sciences*, 8 (1971–72), 155–242.

Laíta, Luis María, 'Influences on Boole's Logic: The Controversy between William Hamilton and Augustus De Morgan', *Annals of Science*, 36 (1979), 45–65.

Merrill, Daniel D., *Augustus De Morgan and the Logic of Relations* (Dordrecht: Kluwer Academic Publishers, 1992). https://doi.org/10.1007/978-94-009-2047-7

Panteki, Maria, 'French "logique" and British "logic": On the Origins of Augustus De Morgan's Early Logical Inquiries, 1805–1835', *Historia Mathematica*, 30 (2003), 278–340. https://doi.org/10.1016/s0315-0860(03)00025-9

Pycior, Helena M. 'Augustus De Morgan's Algebraic Work: The Three Stages', *Isis*, 74 (1983), 211–26. https://doi.org/10.1086/353244

Ranyard, Arthur Cowper, Obituary Notice of Augustus De Morgan, *Monthly Notices of the Royal Astronomical Society*, 32 (1871–72), 112–18.

Rice, Adrian, 'Augustus De Morgan: Historian of Science', *History of Science*, 34 (1996), 201–40. https://doi.org/10.1177/007327539603400203

—, 'Mathematics in the Metropolis: A Survey of Victorian London', *Historia Mathematica*, 23 (1996), 376–417. https://doi.org/10.1006/hmat.1996.0039

—, 'What Makes a Great Mathematics Teacher? The Case of Augustus De Morgan', *The American Mathematical Monthly*, 106 (1999), 534–52. https://doi.org/10.2307/2589465

Rice, Adrian C., Robin J. Wilson, and J. Helen Gardner, 'From Student Club to National Society: The Founding of the London Mathematical Society in 1865', *Historia Mathematica*, 22 (1995), 402–21. https://doi.org/10.1006/hmat.1995.1032

Richards, Joan L., 'Augustus De Morgan, the History of Mathematics, and the Foundations of Algebra', *Isis*, 78 (1987), 7–30. https://doi.org/10.1086/354328

Roscoe, Henry Enfield, *The Life and Experiences of Sir Henry Enfield Roscoe* (London: Macmillan, 1906).

Rouse Ball, Walter William, *A History of the Study of Mathematics at Cambridge* (Cambridge: Cambridge University Press, 1889).

Russell, A. S., 'Augustus De Morgan, a Forgotten Worthy', *The Listener*, 14 (24 Dec. 1935), 1161.

Smith, David Eugene, *History of Mathematics*, 2 vols. (Boston: Ginn, 1923).

Yeldham, Florence A., *The Teaching of Arithmetic through Four Hundred Years 1535–1935* (London: Harran, 1936).

Fig. 4 De Morgan's artistic flair and keen eye for design were reflected in his 'Zodiac of Syllogism', an attractive arrangement of various logical arguments exhibited in his distinctive symbolic notation, surrounding his personal monogram, which featured the letters ADM arranged in a symmetric formation. This emblem was subsequently used on the reverse of the London Mathematical Society's De Morgan Medal. (MS ADD 7, reproduced by permission of UCL Library Services, Special Collections.)

2. De Morgan and Logic

Anna-Sophie Heinemann

> Logic, the only science which is admitted to have
> made no improvements in century after century,
> is the only one which has grown no symbols.
>
> — Augustus De Morgan[1]

Introduction

Although most logicians of the present day are familiar with the propositional laws regarding conjunction, disjunction and negation that have come to bear Augustus De Morgan's name, little is known about his original work on logic. Historiographers of logic notoriously refer to him as a contemporary to George Boole, but of a more traditional mindset.[2] Clarence Irving Lewis, for example, stated in 1918 that his 'methods and symbolism ally him rather more with his predecessors than with Boole and those who follow'.[3] Eighty years later, Ivor Grattan-Guinness similarly asserted in his influential *Search for Mathematical Roots* that he 'worked largely within the syllogistic

---

1   Augustus De Morgan, 'On the Syllogism, No. III, and on Logic in General', *Transactions of the Cambridge Philosophical Society*, 10 (1858), 173–230 (p. 184).
2   George Boole is usually seen as the founding father of symbolic logic in a modern sense. For a critical assessment of this claim, see, for example, Volker Peckhaus, 'Was Boole Really the "Father" of Modern Logic?', in *A Boole Anthology. Recent and Classical Studies in the Logic of George Boole*, ed. by James Gasser (Dordrecht: Springer, 2000), pp. 271–85.
3   Clarence I. Lewis, *A Survey of Symbolic Logic* (Berkeley: University of California Press, 1918), p. 38.

tradition'.⁴ Authoritative assessments of this kind have rarely been questioned. Therefore, De Morgan's contributions to the logical literature of his times are not usually discussed very extensively.⁵

It is true that De Morgan's approach to logic is primordially rooted in traditional syllogistic logic. Nonetheless, it is also true that De Morgan's logic provides for certain novelties which imply some fundamental revisions of the syllogistic tradition. Apart from De Morgan's logic of relations, which has been widely recognised as a seminal contribution to the development of modern logic,⁶ his theory of what he named the 'abstract copula' as an indication of relations to be defined by logical properties such as reflexivity, symmetry and transitivity should certainly be counted among those innovations.⁷ The present chapter, however, will focus on how De Morgan departed from traditional syllogistic logic and

---

4   Ivor Grattan-Guinness, *The Search for Mathematical Roots 1870–1940. Logics, Set Theories and the Foundations of Mathematics from Cantor through Russell to Gödel* (Princeton University Press, 2000), p. 27.
5   There are of course exceptions. For example, there is an extensive discussion of De Morgan's logic in Maria Panteki's Ph.D. thesis, *Relationships between Algebra, Differential Equations and Logic in England: 1800–1860* (Ph.D. Diss., Middlesex University, London, 1991), pp. 407–92. De Morgan's logic of relations has been dealt with in Daniel D. Merrill, *Augustus De Morgan and the Logic of Relations* (Dordrecht: Springer, 1990). In 2008, Michael Hobart and Joan L. Richards published the illuminating overview 'De Morgan's Logic', in *Handbook of the History of Logic*, vol. 4, ed. by Dov Gabbay and John Woods (Amsterdam: North Holland, 2008), pp. 283–329. I myself devoted more than half of a monograph to De Morgan's logic in *Quantifikation des Prädikats und numerisch definiter Syllogismus. Die Kontroverse zwischen Augustus De Morgan und Sir William Hamilton: Formale Logik zwischen Algebra und Syllogistik* (Münster: mentis, 2015), especially pp. 105–260.
6   Again, see Merrill, *Logic of Relations*, or, for more historical context, Benjamin S. Hawkins Jr., 'De Morgan, Victorian Syllogistic and Relational Logic', *Modern Logic*, 5 (1995), 131–66.
7   De Morgan developed his notion of an 'abstract copula' in 'On the Symbols of Logic, the Theory of the Syllogism, and in Particular of the Copula, and the Application of the Theory of Probabilities to Some Questions of Evidence', *Transactions of the Cambridge Philosophical Society*, 9 (1850), 79–127 (pp. 104–14). De Morgan's abstract copula allows for logical inferences which depart from traditional syllogisms of forms such as 'S is M, M is P, therefore S is P' in that they do not require a middle term (M) in order to derive a conclusion connecting the extremes (S and P). An example often referred to is: 'Every horse is an animal, therefore every head of a horse is a head of an animal'. The present chapter, however, will be concerned with De Morgan's modifications of syllogistic schemes which do have middle terms. While De Morgan elaborated on his logic of relations and the abstract copula from the late 1850s onwards, his modifications of syllogistic schemes with middle terms are situated in his earlier work on logic published in the 1840s and early 1850s. Throughout the present chapter, we will focus on De Morgan's logical writings from this period.

thereby revoked the notion of logical quantity of his times. We will show that De Morgan was serious about 'quantification'[8] in logic: he thought of logical quantity as resulting from an operation of enumerating members of a given set of instances of a term. In other words, De Morgan anticipated a modern sense of quantifying over a domain. However, the originality of De Morgan's stance has hardly ever been honoured.[9]

One of the reasons for this omission may lie in a certain complexity of De Morgan's writings due to which Lewis, for instance, judged De Morgan's articles 'ill-arranged and interspersed with inapposite discussion'.[10] Again, Grattan-Guinness echoed that De Morgan 'was not a clear-thinking philosopher'.[11] There are indeed some passages in De Morgan's writings which appear unclear and confusingly abundant with technical details, the productiveness of which is not always evident. The goal of the present chapter will be to sketch out De Morgan's approach to quantification without reproducing too many of De Morgan's technicalities. To this purpose, we will first summarise

---

[8] Genetically speaking, De Morgan may have adopted the term 'quantification' from Sir William Hamilton, his opponent in the debate over the so-called quantification of the predicate. Hamilton had coined the term 'quantification of the predicate' with regard to a clarification of the propositional forms acknowledged in traditional syllogistic logic, as summarised in the second section of the present chapter. In traditional syllogistic logic, propositions are classified according to the quantity of their subject term. For example, 'All A is B' is universal as to the term A. Hamilton, however, would distinguish between 'All A is all B' and 'All A is some B' in order to 'quantify' the predicate term B. After a personal correspondence on syllogistic logic, Hamilton came to the conclusion that De Morgan had plagiarised his own thought in the paper published in 1847, which we will discuss in the main part of this chapter. The historical course of the debate between De Morgan and Hamilton is outlined in Peter Heath, 'Editor's Introduction', in *Augustus De Morgan: On the Syllogism and Other Logical Writings*, ed. by Peter Heath (London: Routledge & Kegan Paul, 1966), pp. vii–xxxi (pp. xi–xxiv). An overview is also given in Luis María Laíta, 'Influences on Boole's Logic: The Controversy between William Hamilton and Augustus De Morgan', *Annals of Science*, 36 (1979), 45–65 (pp. 51–60). A detailed reconstruction can be found in my *Quantifikation*, pp. 23–58.

[9] Daniel Bonevac, 'A History of Quantification', in *Handbook of the History of Logic*, vol. 11, ed. by Dov Gabbay, John Woods and Francis J. Pelletier (Amsterdam: North Holland, 2012), pp. 63–126, for example, omits reference to De Morgan altogether, except for a casual remark stating that De Morgan adopted Hamilton's scheme of quantification (p. 94). A closer look both at De Morgan's and at Hamilton's writings, however, would have revealed that this cannot possibly be the case. I have tried to substantiate this claim in *Quantifikation*, especially pp. 21–22, 39–41, 52–58.

[10] Lewis, *Survey*, p. 38, fn. 1.

[11] Grattan-Guinness, *Mathematical Roots*, p. 27.

some basic traits of traditional syllogistic logic as a point of departure. Subsequently, we will explain some of De Morgan's modifications by reference to two of his systems of syllogistic inference.

Point of Departure: Traditional Syllogistic Logic

Traditional accounts of syllogistic logic admit of four propositional forms.[12] These are compounds of a subject term, S, and a predicate term, P, to be distinguished by a quantitative specification of the subject term, as indicated by 'all' or 'some', and by a qualitative specification of the copula, to be expressed by 'is' or 'is not'. For short, the letters A, E, I and O stand for

A: All S is P,

E: All S is not P (i.e. No S is P),

I: Some S is P,

O: Some S is not P.

As a propositional form, A is universal and affirmative, E is universal and negative, I is particular and affirmative, and O is particular and

---

[12] For the purposes of the present chapter, the name 'syllogistic logic' should be taken to denote a version of Aristotelian logic handed down to nineteenth-century Britain through early modern writers, most prominently Henry Aldrich. His *Artis Logicae Compendium*, first published in 1691, saw multiple editions and translations into English, as well as abridged and annotated versions for the use of schools up to the year 1900. Other early modern authors to be named are Edward Brerewood, Richard Crackanthorpe, Robert Sanderson, John Wallis and Isaac Watts. Around the middle of the 1820s, a significant revival of interest in syllogistic logic was prompted by Richard Whately's article for the *Encyclopaedia Metropolitana*, first published as a monograph in 1826: *Elements of Logic. Comprising the Substance of the Article in the Encyclopaedia Metropolitana; with Additions, &c.* (London: Mawman, 1826). The effects of this revival are discussed in Chapter 5 of this volume. An overview of early nineteenth century British logic is given in James W. Allard, 'Early Nineteenth-Century Logic', in *The Oxford Handbook of British Philosophy in the Nineteenth Century*, ed. by W. J. Mander (Oxford: Oxford University Press, 2014), pp. 25–43. Calvin Lee Jongsma discussed Whately's role in his Ph.D. dissertation, *Richard Whately and the Revival of Syllogistic Logic in Great Britain in the Early Nineteenth Century* (Ph.D. Diss., University of Toronto, 1982).

The account in this section is based on an 1821 abridged and annotated edition, which is likely to mirror the standard logic of De Morgan's times: Henry Aldrich, *Artis Logicae Rudimenta from the Text of Aldrich. With Illustrative Observations on Each Section*, 2nd edn (Oxford: Baxter, 1821).

negative.[13] In other words, each of the four letters is used to sum up two specifications, each of which may be of two kinds: A proposition may be assigned universal or particular quantity, while it may be of affirmative or negative quality.

On the traditional account, a proposition's quantity is determined according to the quantity of the subject term alone.[14] A proposition of the form A, for example, is universal because it states that all of the subject term S belongs to P, the predicate. A proposition of the form E is also universal since it states that all of S does not belong to P, i.e., that none of S belongs to P. In both cases, the subject term is said to be distributed,[15] which means that the proposition makes a claim about every member of the class denoted by the subject term. In the case of I and O, not all S, but only some of S is stated to belong to P or not to belong to P, respectively. Accordingly, I- and O-propositions are not universal, but particular, and their subject terms are not distributed.[16]

For the purposes of traditional syllogistic logic, no explicit mention of the quantity of P is necessary. In the case of affirmative propositions A and I, it is to be understood that S belongs to P, but does not necessarily exhaust it. For example, all humans are mortal, but it is not the case that all mortals are human. However, it may be that all humans are rational animals and all rational animals are humans. Therefore, the quantity of the predicate remains indefinite in the sense of being unspecified. Hence in terms of traditional syllogistic logic, P is not distributed in affirmative propositions. Negative propositions E and O, however, imply that S does not belong to P. Since this means that S is apart from all of P, the quantity of the predicate is definite. Accordingly, in negative propositions, P is always distributed.[17] In short:

(i) In propositions of the form A, only the subject term is distributed.

(ii) In propositions of the form E, both the subject term and the predicate term are distributed.

---

13   Aldrich, *Rudimenta*, p. 86.
14   Aldrich, *Rudimenta*, p. 79.
15   Aldrich, *Rudimenta*, p. 86.
16   Aldrich, *Rudimenta*, p. 86.
17   Aldrich, *Rudimenta*, pp. 86–87.

(iii) In propositions of the form *I*, neither the subject nor the predicate term is distributed.

(iv) In propositions of the form *O*, only the predicate term is distributed.

These rules have certain implications for the validity of syllogisms. A syllogism is defined as a combination of two propositions serving as premises such that a third proposition, the conclusion, is to be inferred. One common scheme is '*S* is *M*, *M* is *P*, therefore *S* is *P*'.

Syllogistic inferences, then, are possible if and only if the premises share one term, the so-called middle term (*M*), in a way allowing for a specification of the relation between the remaining two components. These are usually called the 'extremes' (*S* and *P*). But according to the traditional account, connecting the extremes is not possible if the middle term is not distributed in at least one of the premises,[18] or if both premises are negative,[19] or if both premises are particular.[20] In other words, the basic guidelines of traditional syllogistic logic are:

(I) The middle term must be distributed in at least one of the premises.

(II) At least one of the premises must be affirmative.

(III) At most one of the premises must be particular.

In the remainder of this chapter, we will discuss how De Morgan's approach to logic departs from (I), (II) and (III) just given. We will point out that the reason De Morgan's logic allows for such departures lies in

---

18 Aldrich, *Rudimenta*, pp. 121–22.
19 Aldrich, *Rudimenta*, p. 125.
20 Aldrich, *Rudimenta*, p. 125. The quoted edition of Aldrich has twelve rules altogether. They can be summarised as follows: 1.) A syllogism must involve three terms. 2.) A syllogism must consist of three propositions. 3.) The middle term must not be ambiguous. 4.) If the middle term is not distributed, no conclusion is possible. 5.) The middle term must be distributed in at least one of the premises to allow for a conclusion. 6.) If one of the other terms is not distributed in the premises, it cannot be distributed in the conclusion. 7.) If both premises are negative, no conclusion is possible. 8.) If one premise is negative, the conclusion must be negative. 9.) If the conclusion is negative, it must be that one of the premises is negative. 10.) If both premises are particular, no conclusion is possible. 11.) If one of the premises is particular, the conclusion must be particular. 12.) If the conclusion is particular, it is not the case that one of the premises is necessarily particular. (Aldrich, *Rudimenta*, pp. 116–132).

the fact that he dismissed the traditional understanding of propositions as codified in rules (i), (ii), (iii) and (iv).

Syllogistic Logic in the 'Language of Numeration of Instances'

According to De Morgan, his logical writings speak a 'language of numeration of instances'.[21] This means that De Morgan's logic is not so much about conceptual spheres, i.e., the meanings of the terms chosen as subject and as predicate, but about sets of instances denoted by these terms. In other words, the relations of inclusion and exclusion between subject terms and predicate terms are to be interpreted extensionally. They pertain to sets of instances of terms, portions of which may map onto each other if there are pairwise coincidences of certain instances of each set. Coincidences of this kind lie in that a given member of a set is an instance both of the subject and the predicate term. For example, each particular member of the set of individuals denoted by the term 'human' is at the same time a member of the set of individuals denoted by the term 'mortal'. In other words, the very same individual is human and mortal at the same time. However, there are instances of the term 'mortal' which do not map onto any of the instances of the term 'human'.

It is a substantial consequence of De Morgan's approach that propositions of the traditional forms of *A, E, I* and *O* may be re-stated by reference to the complements of those portions of sets of instances of terms which are referred to in the original statement. But reference to complements requires a counting of the instances included and those excluded in a term's extension. Of course, this requirement cannot be met in principle if the total number of instances is indefinite. Therefore, the notion of a term's being definite or distributed must be reconsidered. As we will see, De Morgan's reconsiderations allow for revisions of the rules (i), (ii), (iii) and (iv) of traditional syllogistic logic, as summarised in the previous section of this chapter. Consequently, the basic guidelines (I), (II) and (III) become negotiable.

In what follows, we will address De Morgan's revisions of traditional syllogistic logic in his 'system of contraries' and his 'numerically definite

---

21   De Morgan, 'Symbols', p. 96.

system'.[22] De Morgan spelled out his system of contraries in his first substantial paper on logic, 'On the Structure of the Syllogism', published in 1847.[23] He re-stated it in a more systematic fashion in the body of 'On the Symbols of Logic', published in 1850.[24] The numerically definite system was first suggested in an 'Addition' to De Morgan's 'Syllogism' paper of 1847.[25] De Morgan discussed it at length in his monograph on *Formal Logic*, published in 1847.[26]

Notably, the system of contraries dispenses with rule (I), i.e. that the middle term must be distributed in at least one of the premises. At least hypothetically, it also undermines rule (II), i.e., that one premise at least must be affirmative. The numerically definite system additionally revokes rule (III), i.e., that the premises may not both be particular.

The 'System of Contraries': Terms and Contraries

A predicate, De Morgan said in 1847, is basically a 'term' or 'name' which should be understood as a 'word' which may legitimately be applied to

---

22 De Morgan introduced both terms retrospectively in 1850 ('Symbols', p. 101, p. 102 and p. 79, respectively.)
23 Augustus De Morgan, 'On the Structure of the Syllogism, and on the Application of the Theory of Probabilities to Questions of Argument and Authority', *Transactions of the Cambridge Philosophical Society*, 8 (1847), 379–408. The paper was included in the volume of the Transactions for 1847, which, however, did not appear until 1849 (as can be seen in Chapter 12 of this volume). It consists of two parts, the main text and an 'Addition' (pp. 406–08). The main text is dated 3 October 1846, and it was read before the Society on 9 November. The 'Addition', however, is dated 27 February 1847. In the course of the De Morgan-Hamilton debate on the 'quantification of the predicate' (see fn. 8), Hamilton tried to substantiate his claim that De Morgan had plagiarised his own innovations in the time elapsed between acceptance of the paper by the Society and the submission of the 'Addition'. However, Hamilton apparently did so without having seen any of the two parts: He accused De Morgan of plagiarism in a letter dated 13 March 1847, while on 27 March he confirmed that he had not yet received the preprints which De Morgan had announced to him on 16 March. The correspondence is reproduced in William Hamilton, *A Letter to Augustus De Morgan, Esq. Of Trinity College, Cambridge, Professor of Mathematics in University College, London, on His Claim to an Independent Re-Discovery of a New Principle in the Theory of Syllogism. Subjoined, the Whole Previous Correspondence, an A Postscript in Answer to Professor De Morgan's 'Statement'* (Edinburgh: Maclachlan & Stewart, 1847), p. 26.
24 It was only in 1850 that De Morgan referred to it by the name of 'system of contraries' ('Symbols', p. 101, p. 102).
25 De Morgan, 'Structure', pp. 406–08.
26 Augustus De Morgan, *Formal Logic, or: The Calculus of Inference, Necessary and Probable* (London: Taylor & Walton, 1847), pp. 141–70. Again, De Morgan introduced the name 'numerically definite system' only in 1850 ('Symbols', p. 79).

any instance in a 'collection of objects of thought'. As a general rule, attributions of this kind are encoded in affirmative propositions of the form 'S is P', or, as De Morgan preferred to put it, 'X is Y'. However, according to De Morgan, it is not the case that in negative propositions of the kind 'X is not Y', Y is to be taken as the predicate, connected to X via a negative copula. On De Morgan's account, 'X is not Y' should instead be read as 'X is non-Y', the predicate 'non-Y' being affirmatively connected to the subject term. Hence 'non-Y' may be taken as the complement of Y, or in De Morgan's terminology, Y's 'contrary', i.e., 'y'.[27]

At first glance, the versed logician might object that attributing y to any subject will be contradictory, not contrary to attributing Y. To be contradictory would mean that y should refer to everything that is not Y, to the effect that an attribution of y and an attribution of Y could neither be true nor false at the same time. Like a pair of contradictory assertions, two contrary attributions cannot both be true. However, it is possible that both are false at the same time. De Morgan thought of contraries as opposed to given terms within a restricted frame of reference. He labelled the frame of reference in question the 'universe of a proposition, or of a name'.[28] Taken together, a term and its contrary are apt to exhaust the given universe. In De Morgan's words, 'every thing in the universe is either X or x'.[29] But on a larger scale, there may be things which belong neither to X nor x. De Morgan's example is that if the universe is that of humans (or citizens), 'Briton' and 'foreigner' are contrary to each other.[30] There is no question of stones, trees, books and the like, which, viewed on absolute terms, are of course also non-Britons, but just as well non-foreigners. Within the given universe, however, foreigners may be referred to as non-British, or Britons as non-foreigners. Hence in relation to a given universe, terms may count as each other's complements without producing contradictory assertions on an absolute scale.

The 'System of Contraries': Propositions

It is obvious from De Morgan's example just quoted that context determines which is to be taken as the positive term and which as the

---

27 De Morgan, 'Structure', p. 379.
28 De Morgan, 'Structure', p. 380.
29 De Morgan, 'Structure', p. 380.
30 De Morgan, 'Structure', p. 380.

negative, i.e., the contrary. But therefore, assertions about each of them will be re-statable by reference to the other. This is why De Morgan claimed that as soon as contraries are systematically considered, distinctions between affirmative and negative as well as between universal and particular propositions turn out to be 'accidents of language, at least for logical purposes'.[31] For it is possible, as De Morgan suggested, that an expression which denotes a certain set of objects may be rendered in another language only as a negative correlative of another, while a third language may provide no name for the whole set of objects in question at all.[32] But therefore, translations of assertions from the first into the second language would require an apparent change of a proposition's quality. Translations from one of them into the third language, however, would call for particular propositions where the first and the second employ universals. However, according to De Morgan, this does not imply a difference as to logical structure and import. It is in this sense that he arrived at the conclusion that

> in truth, every proposition distributes, wholly or partially, among the individuals of the predicate, or of its contrary; making one particular or universal, according as the other is universal or particular.[33]

For example, since all humans are mortal, but not all mortals are human, the proposition 'all humans are mortal' distributes partially among the individuals denoted by the term 'mortal'. On the other hand, it is evident that all humans are excluded from the contrary term 'non-mortal'. Therefore, the proposition 'all humans are not non-mortals' distributes wholly among the individuals of the contrary of 'mortal'. Moreover, De Morgan's account seems to imply that 'all humans are mortal, but not all mortals are human' implies that there are some non-humans which are not mortal. It is in this sense that he explained in 1850: 'Again, "Every X is Y" denies of some xs that they are Ys: for Ys must not fill the universe.' Similarly, he claimed that by '"some Xs are Ys" I deny something of every x: namely, that any one of them is one of those Ys'.[34]

---

31 De Morgan, 'Structure', p. 380.
32 De Morgan, 'Structure', p. 380.
33 De Morgan, 'Structure', p. 382.
34 De Morgan, 'Symbols', p. 92.

Consequently, a given proposition can always be transformed into an equivalent expression if on substitution of terms by their contraries, correlated variations of quantity and quality are taken into account. A catalogue of transformation rules was first suggested in De Morgan's 1847 'Syllogism' paper.[35] An extended version was offered in De Morgan's 1850 article.[36] We will return to De Morgan's transformation rules in the following subsection, as soon as a short exposition of De Morgan's notational systems has been given with a special eye to the relativity of quantity and quality.

Table 2.1 includes both versions of De Morgan's notation for the four traditional forms of A, E, I and O, interpreted extensionally.

Table 2.1 De Morgan's notational systems.

Traditional syllogistic logic	De Morgan's notation of 1847[37]	Interpretation as of 1847[38]	De Morgan's notation of 1850[39]	Interpretation as of 1850[40]
A	$X)Y$	Every X is Y.	$X))Y$	Every X is [some] Y.
E	$X.Y$	No X is Y.	$X).(Y$	No X is [any] Y.
I	XY	Some X is Y.	$X(\ )Y$	Some Xs are [some] Ys.
O	$X:Y$	Some X is not Y.	$X(.(Y$	Some Xs are not [any] Ys.

The interpretations given for the 1847 version correspond to De Morgan's own. The interpretations given for the 1850 notation, however, add quantifiers for the predicate term. This addition to the verbal circumscription is warranted by De Morgan's systematic use of parentheses in his notational system: according to De Morgan's paper of 1850, universal quantity is indicated by a bracket to suggest a circle around the term sign such that it 'would be inclosed if the oval

---

35   De Morgan, 'Structure', p. 381.
36   De Morgan, 'Symbols', p. 91.
37   De Morgan, 'Structure', p. 381.
38   De Morgan, 'Structure', p. 381.
39   De Morgan, 'Symbols', p. 91.
40   De Morgan, 'Symbols', p. 91, quantifiers for the predicate term added.

were completed',[41] as in 'X)'. Accordingly, the bracket in 'X(' might be interpreted as the remnant of an intersection of two circles which cuts out a portion of X's scope.

It is evident that De Morgan's notation for propositions has symbolic quantifiers for both the subject and the predicate term. If we remind ourselves that interpretations must conform to De Morgan's principle of making his logical systems speak a 'language of numeration of instances',[42] more detailed ways of verbal circumscription suggest themselves. De Morgan's A could be read as 'for every member of the set of objects denoted by X, there is a member of the set of objects denoted by Y'. His E would be 'for every member of the set of objects denoted by X, there is not a member of the set of objects denoted by Y, or, X and Y denote mutually exclusive sets of objects'. De Morgan's I could be interpreted as 'for at least one of the members of the set of objects denoted by X, there is a member of the set of objects denoted by Y'. Finally, De Morgan's O would be 'for at least one of the members of the set of objects denoted by X, there is not a member of the set of objects denoted by Y'.

We can now see more clearly the way De Morgan conceived of the relations between subject terms and predicates as overlaps between sets of instances of terms, i.e., of objects. The following quote gives evidence that these relations imply quantification in the sense of enumeration:

> The Xs being distinguished as X_1, X_2, X_3 &c., the universal "Every X is Y" affirms that X_1 is Y, and that X_2 is Y, and that X_3 is Y, *et caetera* [while] the particular "some Xs are not Ys" only declares that either X_1 is not Y, or that X_2 is not Y, or that X_3 is not Y, *aut caetera*.[43]

In other words, 'the universal speaks conjunctively, the particular disjunctively, of the same set'.[44] Again, it should be emphasised that on De Morgan's account, this goes for both the subject and the predicate term. An affirmative universal proposition, for example, speaks conjunctively of its subject term, but it speaks disjunctively of its predicate since, traditionally speaking, the latter is not distributed.

---

41   De Morgan, 'Symbols', p. 86.
42   De Morgan, 'Symbols', p. 96.
43   De Morgan, 'Symbols', pp. 81–82.
44   De Morgan, 'Symbols', p. 81.

In 1847, De Morgan makes the four forms indicated in the table yield eight variants altogether: $X)Y$, $Y)X$, $X.Y$, $Y.X$, XY, YX, $X:Y$ and $Y:X$. The eight variants reduce to six since according to De Morgan, $X.Y$ is equivalent to $Y.X$ and XY is equivalent to YX as to their logical import. However, on De Morgan's account, substitution of terms by contraries additionally yields $x)y$, $x.y$, xy, $x:y$.[45] De Morgan took $x)y$ to be equivalent to the conversion of $X)Y$, i.e., to $Y)X$; similarly, he held that $x:y$ is equivalent to a converted $X:Y$, i.e., $Y:X$. However, $x.y$ and xy seem to have no equivalents. For short, De Morgan labelled these forms e and i, respectively. Following De Morgan's own interpretation, i states that X and Y are not contraries and therefore do not exhaust a given universe. In other words, there are objects in the universe which are neither X nor Y. But, according to De Morgan, e is the negation of xy. Hence in De Morgan's interpretation, e asserts that it is false that there are objects in the universe which are neither X nor Y. However, it does not preclude that there are objects which are both.[46] As an interpretation of De Morgan's 1850 notation, i might be read as 'some non-Xs are some non-Ys', or, 'for at least one member of the set that is the complement of all instances of X, there is at least one member of the set that is the complement of all instances of Y'. For e, however, the interpretation could be 'all non-Xs are not among any of the non-Ys', or, 'the complement of all instances of X and the complement of all instances of Y are mutually exclusive'.

Table 2.2 indicates De Morgan's full inventory of 'fundamental propositions'. Again, both versions of De Morgan's notation are compared.

---

45 De Morgan, 'Structure', p. 382. It is unclear why in 1847, De Morgan did not consider inverted variants in the case of contraries. Granting that $y.x$ and yx are dispensable in the sense of being equivalent to $x.y$ and xy, one should still assume that on systematic variation, $x)y$ and $y:x$ should also be taken into account. The exposition which De Morgan offered in 1850 is much more systematic in this respect.
46 De Morgan, 'Structure', p. 382.

Table 2.2: De Morgan's fundamental propositions.

	Notation of 1847[47]	Notation of 1850[48]
A	$X)Y$	$X))Y$
a	$Y)X = x)y$	$x))y$
E	$X.Y$	$X).(Y$
e	$x.y$	$x).(y$
I	XY	$X(\,)Y$
i	xy	$x(\,)y$
O	$X:Y$	$X(.(Y$
o	$Y:X = x:y$	$x(.(y$

Clearly, the fundamentals of De Morgan's logic go beyond traditional syllogistic logic in providing twice as many propositional forms. But moreover, we should remind ourselves that on De Morgan's account, any negative proposition should be capable of being transformed into an affirmative on substitution of terms by contraries. Table 2.3 contains a systematic catalogue of transformations:[49]

Table 2.3 Transformations in De Morgan's system of contraries.

Notation of 1850
$X))Y = X).(y = x((y = x(.)Y$
$x))y = x).(Y = X((Y = X(.)y$
$X).(Y = X))y = x(.)Y = x((Y$
$x).(y = x))Y = X(.)Y = X((y$
$X()Y = X(.(y = x)(y = x).)Y$

---

47 De Morgan, 'Structure', p. 381.
48 De Morgan, 'Symbols', p. 91.
49 De Morgan did not himself develop the full catalogue (cf. 'Symbols', p. 91). He did, however, give transformation rules which allow for its completion. I have tried for a more detailed discussion of these rules in '"Horrent with Mysterious Spiculae": Augustus De Morgan's Logic Notation of 1850 as a "Calculus of Opposite Relations"', *History and Philosophy of Logic*, 39 (2018), 29–52.

$x()y = x(.(Y = X)(Y = X).)y$
$X(.(Y = X()y = x).)y = x)(Y$
$x(.(y = x()Y = X)(y = X).)Y$

The 'System of Contraries': Syllogisms

Departing from his fundamental propositions, De Morgan claimed to derive all forms of syllogistic inference which are valid on the traditional account. However, De Morgan's extended syllogistic logic also provides for two forms of inference which cannot be accounted for in the traditional system. Notably, they involve the new variants e and i, as introduced above.

In his first 'Syllogism' paper, De Morgan claimed to have derived the inference schemes i_{AA} and I_{ee}, i.e., a syllogism which infers an i-conclusion from two affirmative universal premises, and a syllogism which infers an affirmative particular conclusion from two e-premises. Both violate the rules of traditional syllogistic logic as summarised in the second section of the present chapter.

In i_{AA}, both premises are affirmative and universal at least as to their subject term. In terms of traditional syllogistic logic, their subject terms are distributed. However, according to presumption (i), their predicate terms cannot be. The reason is that in affirmative propositions, nothing is said about whether the predicate term is exhausted by the subject term (extensionally or intensionally). Granting that the middle term would be in predicate position (as in 'All S is M, all P is M, therefore ...'), no conclusion would be possible since it is not distributed in either of the premises and nothing can be said about an intersection between S and P via M if rule (I) holds. For example, if all men are cheese-eaters and all mice are cheese-eaters, no conclusion can be drawn about a relationship between men and mice if the traditional account is granted. De Morgan, however, derived a conclusion which does not make a statement about an intersection between S and P, but about how the complements of S and P may relate to each other since i refers to contraries only.

To contextualise our example, we may quote from De Morgan's correspondence with Sir William Hamilton. 'This is an old trap for a beginner', De Morgan said,

> A man eats cheese,
> A mouse eats cheese,
> Therefore...
> The beginner who falls into the trap says, "a man is a mouse," and his teacher shows him, as he thinks, that no inference can be drawn. But there is an inference, namely, that there are things which are neither men nor mice, namely all which do not eat cheese.[50]

Of course, the only way of making sense of this example is to presuppose that De Morgan had in mind a generic interpretation of 'a man' and 'a mouse'—and that vegan lifestyles had not yet been invented for humans (nor for mice). Granting these limitations, however, i_{AA} is an exception to rule (I) inasmuch as it allows for the middle term not being distributed.

In I_{ee}, a particular affirmative conclusion is inferred from two negative universals. Again, traditional syllogistic logic precludes inferences from two negative premises according to rule (II). In De Morgan's case, however, both premises do not concern the terms of the conclusion, but their contraries. In other words, the premises speak of the complements of what is denoted by the terms that the conclusion speaks of. The case of I_{ee} is a bit less perspicuous than i_{AA}. Adapting the example quoted above, the premises could look like

1. All non-humans are not among any of the non-cheese-eaters (i.e., the set of non-humans and the set of non-cheese-eaters are mutually exclusive).

2. All non-mice are not among any of the non-cheese-eaters (i.e., the set of non-mice and the set of non-cheese-eaters are mutually exclusive).

Of course, it would be blatantly false to conclude 'Some humans are mice'. But it seems that this is due to empirical, not logical reasons. Remember that on De Morgan's account, e states that it is false that there are objects in the universe which are neither X nor Y but does not preclude that there are objects which are both.[51] In other words, e leaves open if the terms whose contraries it speaks of are themselves contrary to each other. Hence the terms whose contraries are connected in an e-proposition do not necessarily exhaust the universe in question.

---

50   Quoted in Hamilton, *A Letter to Augustus De Morgan*, p. 23.
51   De Morgan, 'Structure', p. 382.

Accordingly, it must remain an open question whether the sets denoted by the terms are indeed disjoint. Therefore, there is at least a hypothetical conclusion to the possibility of an overlap. In the case at issue, this would mean that the universe of cheese-eaters could include more than men and mice and that a separate criterion would be required to test whether the set of men and the set of mice are mutually exclusive. Hence I_{ee} at least hypothetically undermines rule (II) since it allows for the premises both being negative.

The 'Numerically Definite System'

The previous section served to show how De Morgan introduced two novel inference schemes which violate the most prominent guidelines of traditional syllogistic logic as summarised in the second section: On the one hand, i_{AA} is an exception to rule (I) inasmuch as it allows for the middle term not being distributed. On the other, I_{ee} at least hypothetically undermines rule (II) since it allows for the premises both being negative. However, there is a third principle not yet touched upon in De Morgan's system of contraries, namely rule (III), which precludes inferences from two particular premises. In an 'Addition' to his first 'Syllogism' paper,[52] however, De Morgan outlined some considerations that imply the very possibility of dispensing with rule (III). He then extended upon these considerations in *Formal Logic*, published in the same year.

De Morgan's numerically definite system shares one essential presupposition with his system of contraries, namely that any term or contrary 'distributes among the individuals' which it denotes. However, within the context of a proposition, it may do so 'wholly or partially'.[53] Accordingly, a universal proposition such as 'Every X is Y', De Morgan said, 'is distributively true, when by "Every X" we mean each one X: so that the proposition is "The first X is Y, and the second X is Y, and the third X is Y, &c."'[54]

This approach conforms to De Morgan's principle that the instances contained in a given universe must be countable at least in principle. De Morgan's extension presently discussed, however, requires that they

---
52 De Morgan, 'Structure', pp. 406–408.
53 De Morgan, 'Structure', p. 382.
54 De Morgan, *Formal Logic*, p. 144.

must be numerically specified. A particular proposition such as 'Some Xs are Ys' should then be spelled out as 'Every one of *a* specified Xs is one or other of *b* specified Ys.' A negative particular, on the other hand, would read 'No one of *a* specified Xs is any of *b* specified Ys'.[55]

This approach implies a sense of predication—and therefore, of logical inference—as based on one-to-one-mappings of instances of terms. Therefore, De Morgan labelled it the 'numerically definite system',[56] as based on the notion of 'definite particulars'.[57] Numerically definite inferences, then, should be derivable from premises such as 'if there be 100 Ys and we can say that each of 50 Xs is one or other of 80 Ys, and that no one of 20 Zs is any one of 60 Ys'.[58]

In fact, it is not immediately evident where premises of this kind lead to. De Morgan's *Formal Logic* provides a very detailed technical apparatus including case-by-case analyses for specified numbers of Xs being greater or smaller than the specified numbers of Ys or Zs. However, for reasons of both space and clarity, we will omit further references to these discussions here. Nevertheless, there is one aspect of De Morgan's numerically definite system that we will take up for the very reason that it facilitates dispensing with rule (III) of traditional syllogistic logic, as stated in the second section of the present chapter. This aspect is De Morgan's 'ultratotal quantification' of the middle term. Maybe its clearest statement in *Formal Logic* is as follows:

> We cannot show that Xs are Zs by comparison of both with a third name, unless we can assign a number of instances of that third name, *more than filled up* by Xs and Zs: that is to say, such that the very least number of Xs and Zs which it can contain are together more in number than there are separate places to put them in. ... Accordingly, so many Xs at least must be Zs as there are units in the number by which the Xs and Zs to be placed, together exceed the number of places for them.[59]

In this context, De Morgan distinguished two combinations of premises that allow for inferences, namely a combination of two affirmative

---

55 De Morgan, 'Structure', p. 406.
56 De Morgan, 'Symbols', p. 79.
57 De Morgan, *Formal Logic*, p. 15.
58 De Morgan, 'Structure', p. 406.
59 De Morgan, *Formal Logic*, p. 154.

propositions on the one hand, and a combination of one affirmative and one negative proposition on the other. In both cases, Y serves as the middle term and a limited universe is given. The first combination is that

1. A specified number m out of a total of ξXs maps onto the same number of instances of Y.
2. A specified number n out of a total of ηYs maps onto the same number of instances of Z.

The second case is a combination of one affirmative and one negative premise, namely that

1. A specified number m out of a total of ξXs maps onto the same number of instances of Y.
2. A specified number n out of a total of ζZs does not map onto a specified number s out of the total of ηYs.

Hence according to De Morgan, given

ξ: total number of Xs,

η: total number of Ys,

ζ: total number of Zs,

ν: total number of instances in the universe,[60]

the combinations of premises are

$$mXY + nYZ,$$

$$mXY + nZ : sY.[61]$$

In the first case, neither m nor n exceed the total number of instances of the middle term, η. In other words, both the specified number of Xs and the specified number of Zs each fall short of the total number of Ys. However, if their conjunction does exceed the total number η of Ys, it is possible to infer that there is an overlap between the specified scope of X and the specified scope of Z. This overlap must then contain as many elements as lie between η and the sum of m and n:

---

60 De Morgan, *Formal Logic*, pp. 143–44.
61 De Morgan, *Formal Logic*, p. 145.

$$mXY + nYZ = (m + n - \eta)\, XZ.^{62}$$

For the second combination of premises, De Morgan considered a case-by-case-analysis for the sum of m and s exceeding η on the one hand, and the sum of n and s exceeding η on the other. According to De Morgan, if the sum of m and s is larger than η, the excess elements do not map onto any of the nZs. In case n and s exceed η, the conclusion is that the specified mXs do not map onto any of the excess elements:

for $m + s > \eta$, $\quad mXY + nZ{:}sY = (m + s - \eta)X{:}nZ$
for $n + s > \eta$, $\quad mXY + nZ{:}sY = mX{:}(n + s - \eta)Z.^{63}$

Since in all the cases just discussed, the numbers m, n and s are specified selections out of a total number, they may all be classified as inferences from particular premises in which the middle term is not distributed amongst the total number of individuals denoted by it. Therefore, we may infer that De Morgan's numerically definite system is apt to allow for conclusions both from two particular premises and from pairs of premises that lack a distributed middle. Hence they undermine the guidelines (I) and (III) of traditional syllogistic logic, as given in the second section of this chapter. However, note that the numerically definite system does not allow for inferences from two negative premises, which means that it does not dispense with rule (II).

---

62   De Morgan, *Formal Logic*, p. 145. The thought suggests itself that the size of the overlap, which depends on the numerically specified extensions of the terms, determines how probable it is for one particular individual denoted by X to be identical with an instance of Z. It is interesting that much earlier than engaging with logic, De Morgan went beyond pure mathematics in publishing *An Essay on Probabilities, and Their Application to Life Contingencies and Insurance Offices* (London: Longman, Brown, Green & Longmans, 1838). A related hypothesis would be that De Morgan's numerically definite approach in syllogistic logic stems from his involvement with applied probability. Section V of his first 'Syllogism' paper as well as Chapters IX and X of *Formal Logic* point in a similar direction. For a discussion of relations between De Morgan's stances in probability and in syllogistic logic, see Adrian Rice, '"Everybody Makes Errors": The Intersection of De Morgan's Logic and Probability, 1837–1847', *History and Philosophy of Logic*, 24 (2003), 289–305, especially pp. 293–96.
63   De Morgan, *Formal Logic*, pp. 145–46. De Morgan himself did not make use of >. Therefore, we add the case-by-case-analysis above according to his verbal descriptions.

Summary

In the course of the present chapter, we have endeavoured to give an overview of De Morgan's early thought on logic. We pointed out that it is rooted in the syllogistic tradition handed down to nineteenth-century Britain through various editions of early modern works. Unlike De Morgan's logic of relations and his notion of an abstract copula, his syllogistic logic never did turn out to be particularly trendsetting. As mentioned in our introduction, his attempts at casting syllogistic logic into a more technical form are usually regarded as inferior, especially when compared to the achievements of George Boole. However, we have tried to show that De Morgan's syllogistic logic does provide for certain novelties and that they relate to his approach to quantification.

In his first 'Syllogism' paper, De Morgan introduced the notion of 'contraries' of terms within a given 'universe'. According to this, a term and its contrary exhaust the given universe. This implies that the instances denoted by each of them are countable at least in principle. The same holds for the total number of instances contained in the universe. Granting these assumptions, De Morgan arrived at the conclusion that assertions about terms and contraries turn out to be re-statable by reference to the other if on substitution of terms by their contraries, correlated variations of propositions' quantity and quality are taken into account. On this basis, De Morgan introduced two novel inference schemes which violate the principles of traditional syllogistic logic, as summarised in our second section: The scheme i_{AA} is an exception to rule (I) inasmuch as it allows for the middle term not being distributed. The scheme I_{ee}, however, at least hypothetically undermines rule (II) since it allows for the premises both being negative. In short, De Morgan's 'system of contraries' dispenses with the guidelines (I) and (II), but keeps rule (III), which demands a distributed middle term.

In an 'Addition' to his 'Syllogism' paper and in his monograph on *Formal Logic*, however, De Morgan developed a 'numerically definite' system, which allows for inferences from pairs of premises lacking a distributed middle. It requires that the instances of terms within a given universe must not only be countable in principle, but numerically specified. If this requirement is met, the system allows for violations of guidelines (I) and (III), but it keeps rule (II), according to which no inferences can be drawn from two negative premises.

De Morgan's numerically definite system appears to be consistent with his claim that all works on logic speak a 'language of numeration of instances'. However, while some of its specific assumptions were items of controversy even in his own times,[64] logicians of the present day usually judge it a dead-end in the history of modern formal logic.[65]

We may conclude that even if De Morgan's early logical systems have not themselves been very influential, both give evidence of De Morgan's sense of logical quantification, which amounts to conjunction or disjunction of definite or at least specifiable numbers of instances both in universal and particular cases. It is this sense of quantification which has survived in modern formal logic when it comes to quantifying over domains.

Bibliography

Aldrich, Henry, *Artis Logicae Rudimenta from the Text of Aldrich. With Illustrative Observations on Each Section*, 2nd edn (Oxford: Baxter, 1821).

Allard, James W., 'Early Nineteenth-Century Logic', in *The Oxford Handbook of British Philosophy in the Nineteenth Century*, ed. by W. J. Mander (Oxford: Oxford University Press, 2014), pp. 25–43. https://doi.org/10.1093/oxfordhb/9780199594474.013.001

Bonevac, Daniel, 'A History of Quantification', in *Handbook of the History of Logic*, vol. 11, ed. by Dov Gabbay, John Woods and Francis J. Pelletier

---

[64] See, for example, James Broun, 'The Supreme Logical Formule [sic]', *The Athenæum*, 1025, 19 June 1847, 645–646. De Morgan himself did not elaborate on the numerically definite system in his later writings. There is, however, a paper by Boole entitled 'On Propositions Numerically Definite', which was posthumously read to the Cambridge Philosophical Society by De Morgan in 1868 and published in 1871 (George Boole, 'On Propositions Numerically Definite. By the late George Boole, F.R.S., Professor of Mathematics in Queen's College. Communicated by A. De Morgan, Esq.', *Transactions of the Cambridge Philosophical Society*, 11 (1871), 396–411).

[65] Ian Pratt-Hartmann, for example, holds that 'no finite collection of syllogism-like rules, broadly conceived, is sound and complete for the numerical syllogistic' ('No Syllogisms for the Numerical Syllogism', in *Languages: From Formal to Natural. Essays Dedicated to Nissim Francez on the Occasion of his 65th Birthday*, ed. by Orna Grumberg, Michael Kaminski et al. (Berlin, Heidelberg: Springer, 2009), pp. 192–203 (p. 192); cf. 'The Syllogistic With Unity', *Journal of Philosophical Logic*, 42 (2013), 391–407 (p. 391).

(Amsterdam: North Holland, 2012), pp. 63–126. https://doi.org/10.1016/b978-0-444-52937-4.50002-2

Boole, George, 'On Propositions Numerically Definite. By the late George Boole, F.R.S., Professor of Mathematics in Queen's College. Communicated by A. De Morgan, Esq.', *Transactions of the Cambridge Philosophical Society*, 11 (1871), 396–411.

Broun, James, 'The Supreme Logical Formule [sic]', *The Athenæum*, 1025, 19 June 1847, 645–46.

De Morgan, Augustus, *An Essay on Probabilities, and Their Application to Life Contingencies and Insurance Offices* (London: Longman, Brown, Green & Longmans, 1838).

—, 'On the Structure of the Syllogism, and on the Application of the Theory of Probabilities to Questions of Argument and Authority', *Transactions of the Cambridge Philosophical Society*, 8 (1847), 379–408.

—, *Formal Logic, or: The Calculus of Inference, Necessary and Probable* (London: Taylor & Walton, 1847).

—, 'On the Symbols of Logic, the Theory of the Syllogism, and in Particular of the Copula, and the Application of the Theory of Probabilities to Some Questions of Evidence', *Transactions of the Cambridge Philosophical Society*, 9 (1850), 79–127.

—, 'On the Syllogism, No. III, and on Logic in General', *Transactions of the Cambridge Philosophical Society*, 10 (1858), 173–230.

Grattan-Guinness, Ivor, *The Search for Mathematical Roots 1870–1940. Logics, Set Theories and the Foundations of Mathematics from Cantor through Russell to Gödel* (Oxford: Princeton University Press, 2000). https://doi.org/10.1515/9781400824045

Hamilton, William, *A Letter to Augustus De Morgan, Esq. Of Trinity College, Cambridge, Professor of Mathematics in University College, London, on His Claim to an Independent Re-Discovery of a New Principle in the Theory of Syllogism. Subjoined, the Whole Previous Correspondence, and A Postscript in Answer to Professor De Morgan's 'Statement'* (Edinburgh: Maclachlan & Stewart, 1847).

Hawkins, Benjamin S. Jr, 'De Morgan, Victorian Syllogistic and Relational Logic', *Modern Logic*, 5 (1995), 131–66.

Heath, Peter, 'Editor's Introduction', in *Augustus De Morgan: On the Syllogism and Other Logical Writings*, ed. by Peter Heath (London: Routledge & Kegan Paul, 1966), pp. vii–xxxi.

Heinemann, Anna-Sophie, *Quantifikation des Prädikats und numerisch definiter Syllogismus. Die Kontroverse zwischen Augustus De Morgan und Sir William Hamilton: Formale Logik zwischen Algebra und Syllogistik* (Münster: mentis, 2015).

—, '"Horrent with Mysterious Spiculae": Augustus De Morgan's Logic Notation of 1850 as a "Calculus of Opposite Relations"', *History and Philosophy of Logic*, 39 (2018), 29–52. https://doi.org/10.1080/01445340.2017.1319593

Hobart, Michael E., and Joan L. Richards, 'De Morgan's Logic', in *Handbook of the History of Logic*, vol. 4, ed. by Dov Gabbay and John Woods (Amsterdam: North Holland, 2008), pp. 283–329. https://doi.org/10.1016/s1874-5857(08)80010-6

Jongsma, Calvin Lee, *Richard Whately and the Revival of Syllogistic Logic in Great Britain in the Early Nineteenth Century* (Ph.D. Diss., University of Toronto, 1982).

Laíta, Luis María, 'Influences on Boole's Logic: The Controversy between William Hamilton and Augustus De Morgan', *Annals of Science*, 36 (1979), 45–65. https://doi.org/10.1080/00033797900200121

Lewis, Clarence Irving, *A Survey of Symbolic Logic* (Berkeley: University of California Press, 1918). https://doi.org/10.1525/9780520398252

Merrill, Daniel D., *Augustus De Morgan and the Logic of Relations* (Dordrecht: Springer, 1990). https://doi.org/10.1007/978-94-009-2047-7

Panteki, Maria, *Relationships between Algebra, Differential Equations and Logic in England: 1800–1860* (Ph.D. Diss., Middlesex University, London, 1991).

Peckhaus, Volker, 'Was Boole Really the "Father" of Modern Logic?', in *A Boole Anthology. Recent and Classical Studies in the Logic of George Boole*, ed. by James Gasser (Dordrecht: Springer, 2000), pp. 271–85. https://doi.org/10.1007/978-94-015-9385-4_15

Pratt-Hartmann, Ian, 'No Syllogisms for the Numerical Syllogism', in *Languages: From Formal to Natural. Essays Dedicated to Nissim Francez on the Occasion of his 65th Birthday*, ed. by Orna Grumberg, Michael Kaminski et al. (Berlin, Heidelberg: Springer, 2009), pp. 192–203. https://doi.org/10.1007/978-3-642-01748-3_13

—, 'The Syllogistic With Unity', *Journal of Philosophical Logic*, 42 (2013), 391–407. https://doi.org/10.1007/s10992-012-9229-3

Rice, Adrian, '"Everybody Makes Errors": The Intersection of De Morgan's Logic and Probability, 1837–1847', *History and Philosophy of Logic*, 24 (2003), 289–305. https://doi.org/10.1080/01445340310001599579

Whately, Richard, *Elements of Logic. Comprising the Substance of the Article in the Encyclopaedia Metropolitana; with Additions, &c.* (London: Mawman, 1826).

Fig. 5 De Morgan's personal copy of a volume containing his twelve biographies of eminent scientists, originally published in *The Gallery of Portraits: with Memoirs* (1833–37), features several witty and whimsical drawings, including this cartoon of 'Saturn and his Ring'—further evidence of his playful and somewhat eccentric sense of humour. (RAS MSS De Morgan 3, reproduced by permission of the Royal Astronomical Society Library and Archives.)

3. Augustus De Morgan, Astronomy and Almanacs

Daniel Belteki

> Astronomy signifies the *laws of the stars* . . .
> If we except general terms, such as *science*, there
> is perhaps no single word which implies so many
> and different employments of the human intellect.
>
> — Augustus De Morgan[1]

Introduction

How can an individual contribute to astronomy? Is it only by making observations of celestial bodies, or are there other means? Augustus De Morgan's contributions to astronomy raise precisely these questions. De Morgan never identified himself as an astronomer, and blindness in one eye rendered him unable to make reliable observations with astronomical instruments.[2] Yet he participated actively in the astronomical community during the mid-nineteenth century, becoming involved as Secretary of the Royal Astronomical Society in major events and controversies that shaped both British and international astronomical practice during the 1840s, and making himself through his

---

1   Augustus De Morgan, 'Astronomy', *Penny Cyclopaedia*, vol. 2 (London: Charles Knight, 1834), pp. 529–38 (p. 529).
2   Sophia Elizabeth De Morgan, *Memoir of Augustus De Morgan* (London: Longmans, Green, 1882), p. 5.

writings an authoritative 'expounder and historian' of astronomy and its instruments.[3]

Therefore, any examination of De Morgan's contributions to astronomy is best achieved not by counting the number of planets, comets or stars he discovered, but by analysing how he shaped the fabric of the astronomical community during the middle of the nineteenth century, and how he wove new and forgotten threads into the history of the field. This chapter revisits the origins of De Morgan's interest in astronomy and his close relationships with leading astronomers of the nineteenth century. It discusses his activities as a writer, arguing that while he raised awareness of history's forgotten and overlooked astronomers, his publications also reaffirmed the contemporary and historical boundaries of the astronomical community. Finally, it examines De Morgan's writings about calendrical reforms and an apparent paradox regarding the determination of the date of Easter, to demonstrate how he combined his interests in antiquarianism, ecclesiastical and legal history with his knowledge of mathematics and astronomy to participate in a debate of interest to the wider public.

Early Interest in Astronomy

Augustus De Morgan's interest in astronomy arose through his studies in mathematics. During the early part of the nineteenth century, astronomers began to place an increasing emphasis on the use of mathematics to solve astronomical problems. For instance, observations made by previous astronomers were recalculated on the basis of revised astronomical values with new mathematical techniques.[4] The discovery of the planet Neptune was seen as a culmination of the achievements of this new approach.[5] When De Morgan entered Trinity College, Cambridge

---

3   S.E. De Morgan, *Memoir*, p. 50.
4   David Aubin, Charlotte Bigg, and H. Otto Sibum, eds, *The Heavens on Earth: Observatories and Astronomy in Nineteenth-Century Science and Culture* (Durham and London: Duke University Press, 2010).
5   For an overview of the controversies surrounding the discovery of Neptune, see Robert W. Smith, 'The Cambridge Network in Action: The Discovery of Neptune', *Isis*, 80:3 (1989), 395–422; Nicholas Kollerstrom, 'An Hiatus in History: The British Claim for Neptune's Co-Prediction, 1845–1846: Part 1', *History of Science*, 44 (2006), 1–28.

in 1823, he was surrounded by men of science widely advocating such an approach. He counted among his teachers at least three such figures in George Peacock, William Whewell and George Airy. Peacock was a founding member of the Analytical Society devoted to reforming mathematics at Cambridge.[6] Whewell became the Master of Trinity College and remains known as a polymath due to his contributions to various fields of science.[7] Airy was appointed Astronomer Royal in 1835 and remained the director of the Royal Observatory at Greenwich until 1881.[8]

Although there is no clear evidence of De Morgan's engagement in astronomy during his years in Cambridge, his wife, Sophia Elizabeth De Morgan, recalled his exceptional knowledge of Eastern astronomy at the time of their first meeting in 1827.[9] The first major milestone in De Morgan's involvement in astronomical matters occurred in 1828, when he was elected a Fellow of the Royal Astronomical Society.[10] The Society had been founded only eight years earlier by individuals who were interested in the applications of the mathematical methods used in astronomy to matters of business. Its founding members promoted an 'astronomical book-keeping' reliant on the use of mathematics as it could be found in the offices of accountants and insurance companies.[11] This approach to astronomy through mathematics suited De Morgan. The community also welcomed De Morgan's mathematical investigations, as they were linked to the profit-seeking motive of the Society's members. Such interests were exemplified by the Society's successful efforts to shape the *Nautical Almanac*, a key publication for the purposes of navigation at sea (for example, for the shipping of goods) and for providing astronomical data to astronomers.

---

6   Kevin Lambert, 'A Natural History of Mathematics: George Peacock and the Making of English Algebra', *Isis*, 104 (2013), 278–302.
7   Richard Yeo, *Defining Science: William Whewell, Natural Knowledge and Public Debate in Early Victorian Britain* (Cambridge: Cambridge University Press, 2003).
8   Allan Chapman, 'Science and the Public Good: George Biddell Airy (1801–92) and the Concept of a Scientific Civil Servant', in *Science, Politics and the Public Good: Essays in Honour of Margaret Gowing*, ed. by Nicolaas A. Rupke (Basingstoke: Macmillan, 1988), pp. 36–62.
9   S.E. De Morgan, *Memoir*, p. 21.
10   At the time, it was still known as the Astronomical Society of London. The Society did not receive its royal charter until 1831.
11   William J. Ashworth, 'The Calculating Eye: Baily, Herschel, Babbage and the Business of Astronomy', *British Journal for the History of Science*, 27 (1994), 409–41.

Members of the Society included eminent men of science such as Francis Baily and John Herschel. Baily would later influence the development of De Morgan's interest in the history of astronomy.[12] Herschel, as well as becoming a lifelong friend and correspondent, would later recommend De Morgan for the presidency of the Society.[13] De Morgan's close acquaintances from his Cambridge years were also members of the Society. Richard Sheepshanks (a Trinity College graduate, a patron of the Cambridge Observatory and the son of a wealthy textile manufacturer) served as its Secretary. Airy was its president and received the Gold Medal of the Society for his various scientific achievements. Sophia De Morgan characterised the Airy-De Morgan-Sheepshanks triangle as an 'intimate friendship'.[14] The three men and their families frequently congregated at the Sheepshanks residence and spent the afternoons playing music together. Such encounters were initially easy to organise as Sheepshanks lived near De Morgan.[15] Recalling these visits, Sophia De Morgan wrote: 'All were fond of music, and Mrs. Airy's and her sister's ballads, sung with a spirit that gave them a character equal to Wilson's,[16] were sometimes accompanied by Mr. De Morgan's flute, and are still among my pleasantest remembrances'.[17]

---

12   Rebekah Higgitt, *Recreating Newton: Newtonian Biography and the Making of Nineteenth-Century History of Science* (London and New York: Routledge, 2015).
13   For an overview of John Herschel's life, see Stephen Case, *Making Stars Physical: The Astronomy of Sir John Herschel* (Pittsburgh: University of Pittsburgh Press, 2018).
14   S.E. De Morgan, *Memoir*, p. 48.
15   Unfortunately, Sheepshanks' personal correspondence is scattered around archival collections in small numbers. Sheepshanks is the most prolific correspondent in RAS MSS De Morgan at the Royal Astronomical Society, with 68 letters, 1842-1852 (RAS MSS De Morgan 1; subjects covered include, alongside matters discussed in this chapter, whether or not Maria Mitchell should be elected as an honorary member of the Society). Letters exchanged between Airy and De Morgan survive in the Royal Greenwich Observatory Archives at Cambridge University Library (see Chapter 11 of this volume). Another key source is De Morgan's correspondence with Herschel, held in the archives of the Royal Society (also discussed in Chapter 11). Together, these letters provide a window into the dynamic between core members of the Royal Astronomical Society.
16   This is probably a reference to John Wilson's *Cheerful Ayres or Ballads: First Composed for One Single Voice and Since Set for Three Voices*, first published in 1660.
17   S.E. De Morgan, *Memoir*, p. 47.

De Morgan as the Secretary of the Royal Astronomical Society

De Morgan was elected a member of the Council of the Royal Astronomical Society in 1830, and the following year he became its Honorary Secretary.[18] This position had been created in 1824 to assist the work of the Society's Secretary in the increased number of clerical duties.[19] As Honorary Secretary, De Morgan drew up documents relating to the Society's operations, arranged meetings, helped to edit the Society's two journals (*Monthly Notices of the Royal Astronomical Society* and *Memoirs of the Royal Astronomical Society*), corresponded with members, and edited—and in some cases also wrote—obituaries of its deceased fellows. Dreyer and Turner in their history of the Society even state that the detailed summaries of papers published in the *Monthly Notices* became the publication's characteristic feature through De Morgan's efforts.[20] In addition, his frequent interactions with the members of the Society enabled him to demonstrate his mathematical and tutoring skills. For example, the astronomer George Bishop (who would become President of the Society in 1857) even took lessons in algebra from De Morgan.[21]

As an active member of the Society, De Morgan became involved in various debates that rippled through the astronomical community during the mid-nineteenth century. An example of this was the infamous *Troughton & Simms* v. *South* court case, a legal battle and subsequent controversy which historian Michael Hoskin later labelled the 'Astronomers at War' saga.[22] It concerned the performance of a

---

18   S.E. De Morgan, *Memoir*, pp. 41–42. For De Morgan's correspondence in connection with this role, see London, Royal Astronomical Society, RAS Letters 1831–1866, De Morgan. These are mainly letters about forthcoming meetings and publications and are addressed to the Society's Assistant Secretary. Letters from De Morgan to other astronomers are included in other RAS MSS series, such as correspondence relating to the asteroid discoveries of John Russell Hind in RAS MSS Hind.

19   John Louis Emil Dreyer and Herbert Hall Turner, eds, *History of the Royal Astronomical Society 1820–1920* (London: Royal Astronomical Society and Wheldon & Wesley, 1923), p. 44.

20   Dreyer and Turner, p. 79.

21   S.E. De Morgan, *Memoir*, p. 49.

22   Michael Hoskin, 'Astronomers at War: South vs Sheepshanks', *Journal for the History of Astronomy*, 20 (1989), 175–212; Michael Hoskin, 'More on "South v.

telescope constructed by the instrument makers Troughton & Simms for the wealthy astronomer James South. South considered the performance of the telescope subpar, while the instrument makers argued that South did not allow the construction to be finished. The astronomical community (including members of the Royal Astronomical Society) was divided in its support for the two sides. Airy and Sheepshanks supported the instrument makers, while Charles Babbage came to the support of South. South lost the ensuing legal battle, which included a back-and-forth of letters and opinion pieces published in newspapers. The final decision in favour of Troughton & Simms did not calm the sensibilities of the losing side, and both Babbage and South continued their attacks in the ensuing years.[23]

In the decades-long conflict, De Morgan sided with his intimate friends, Airy and Sheepshanks. His association with them made him a target for the wrath of their opponents, most notably when South publicly demanded to know on what basis De Morgan had been elected a Fellow of the Society in the first place.[24] South explicitly asked the Assistant Secretary of the Society to see the letters of recommendation that had testified to De Morgan's contributions to the field and his suitability to be a member of the Society. This was a serious and potentially threatening development. As De Morgan neither made astronomical observations nor published scientific articles in the Society's journals, South was attacking De Morgan from a very delicate angle and questioning both the legitimacy of his role and the evaluations of the astronomers who had supported his election. Luckily for De Morgan, other members of the Society rallied round to dismiss the request, and it had no effect on his involvement within the Society nor with the astronomical community at large. Indeed, what South's futile attack ultimately demonstrated was how deeply De Morgan was embedded within the Society's core group, to the extent that accusations by his enemies failed to affect his reputation within it.

Sheepshanks"', *Journal for the History of Astronomy*, 22 (1991), 174–79; Anita McConnell, 'Astronomers at War: The Viewpoint of Troughton & Simms', *Journal for the History of Astronomy*, 25 (1994), 219–35.

23 Doron David Swade, *Calculation and Tabulation in the Nineteenth Century: Airy versus Babbage* (Unpublished Ph.D. Diss., University College London, 2003).

24 S.E. De Morgan, *Memoir*, pp. 63-64.

In addition to participating in debates, De Morgan was able through his active role in the Society to witness the impact of major astronomical discoveries at close quarters. One of these was the discovery of Neptune, which provoked a controversy about the circumstances surrounding the breakthrough.[25] Both Urbain Le Verrier and John Couch Adams worked on the challenge of predicting the path of a new planet. Le Verrier's calculations and predictions were verified in 1846. The British astronomical community later showed that Adams had sent similar predictions to Airy to be verified by observations but that Adams had failed to respond to Airy's follow-up letter, which had led to delays in the search for the planet. The ensuing debates pitted claims of national, personal and scientific interests against each other. De Morgan contributed to the discussion with two articles in the *Athenæum*, a weekly magazine aimed at a middle-class audience with a growing appetite for scientific news,[26] to which his friends, like Airy, also regularly contributed.

In De Morgan's first piece to the magazine about the new planet, he summarised a recent meeting of the Society, during which Airy had presented the chronology of his correspondence with Adams. Even at this early stage De Morgan predicted that the controversy surrounding the discovery would be discussed by future historians of science. In addition, he claimed that England missed out on the discovery because 'the mathematicians of this country had not faith enough in their own science'.[27] In his next article he defended Airy's scepticism about the possibility of a new planet on account of Adams's lack of response.[28] De Morgan further argued that as soon as Le Verrier communicated similar findings to Airy, the Astronomer Royal initiated the search for the planet precisely because of Adams's previous communications. De Morgan also defended the actions of James Challis (the director of the Cambridge University Observatory, who aided Adams's investigations)

---

25 See Smith, 'The Cambridge Network in Action'; Kollerstrom, 'An Hiatus in History'; Allan Chapman, 'Private Research and Public Duty: George Biddell Airy and the Search for Neptune', *Journal for the History of Astronomy*, 19 (1988), 121–39.

26 Susan Holland and Steven Miller, 'Science in the Early *Athenæum*: A Mirror of Crystallization', *Public Understanding of Science*, 6 (1997), 111–30.

27 Augustus De Morgan, 'The New Planet', *The Athenæum*, 21 November 1846, p. 1191.

28 Augustus De Morgan, 'The New Planet', *The Athenæum*, 5 December 1846, pp. 1245–46.

by claiming that Challis was in no position to give up his other duties and to devote his entire attention to Adams's claims. The article then lashed out at François Arago (the director of the Paris Observatory), who proposed naming the new planet Le Verrier without waiting for the Royal Astronomical Society to present the historical circumstances of the discovery. According to De Morgan, this demonstrated that Arago's judgement was 'subjected to his distorting mirror of national bias'.[29]

As Secretary of the Society, De Morgan's activity extended beyond participation in the debate to mediation of the discussions surrounding the award of the Society's Gold Medal for the discovery of Neptune.[30] Members faced a conundrum that arose from trying to acknowledge the contributions of both Le Verrier and Adams, even though Le Verrier could claim to be the first who made his discovery public. At a Council meeting of the Society in February 1847, in the absence of the required three-to-one majority during voting relating to the medal, no agreement was reached. Babbage (a supporter of South and a critic of the De Morgan-Airy-Sheepshanks triangle) was one of the majority who supported awarding the medal to Le Verrier only. He summarised the events in a letter sent to *The Times*,[31] claiming that there was a two-to-one majority in support of awarding the medal to Le Verrier: ten votes in support and five against (the five against including Airy). As a result, a motion by Airy was adopted after the vote, which called for an extraordinary meeting to discuss awarding two or more medals. Babbage submitted a letter to the extraordinary meeting (as he was unable to attend), which supported the Gold Medal being awarded to Le Verrier and an extraordinary medal awarded to Adams. However, his letter was not read out at the meeting. Somewhat surprisingly, Babbage's suggestion was not radical. Even Sheepshanks, despite his previous clashes with Babbage, supported a similar approach: Le Verrier should be awarded a medal first in the usual manner, and Adams could be awarded a medal decided on by a special meeting. In contrast, Airy argued that if no

---

29   Augustus De Morgan, 'The New Planet', *The Athenæum*, 5 December 1846, pp. 1245–46.
30   For a summary of the debates surrounding the awards, see S.E. De Morgan, *Memoir*, pp. 132–36.
31   Charles Babbage, 'The Planet Neptune and the Royal Astronomical Society's Medal', *The Times*, 15 March 1847, p. 5.

medals were awarded this time, then it would be impossible to award any medal in the future.

Ultimately though, it was De Morgan whose proposed solution was adopted. He stated that the established procedure for awarding a medal was to obtain a three-to-one majority at the relevant meeting of the Council. The consequence of failing to reach this threshold, he argued, should not be the creation of a new by-law (i.e. awarding extra medals), but rather a decision to refrain from awarding any medal. On the basis of this argument, the Society refused to award any medals, and decided to acknowledge the contributions of Le Verrier and Adams through testimonials instead.

That De Morgan's views directly influenced the steps taken by the Society reflected his integral role within it. It may be seen as a natural consequence that, parallel to the discovery of Neptune, discussions arose about the possibility of electing him as the President of the Society: discussions which clearly demonstrate that, despite not being a 'practical astronomer', he was held in high esteem and was seen as a fitting leader of the astronomical community. His refusal to take on the role shows that he continued to view himself as a non-practising astronomer, albeit as an active participant within the community. In a letter to another member of the Society, Captain William Smyth, De Morgan argued that only a 'practical astronomer' was suitable to become the president of the Society: 'the President must be a man of brass—a micrometer-monger, a telescope-twiddler, a star-stringer, a planet-poker, and a nebula-nabber'.[32] Similarly, in a letter to John Herschel, De Morgan described himself as 'a person who has never promoted astronomy otherwise than as promoting mathematics is indirectly doing so'.[33] At the same time, his refusal of the Presidency was motivated by his interest in promoting Herschel to the same role: he stated that he would only take on the role of Vice-President or Secretary if Herschel were willing to become President.

De Morgan also directed some of his prodigious energy to the Society's library. He volunteered his expertise in bibliography to assist with the arrangement and cataloguing of the Society's hitherto 'literally inaccessible' library, working with the Assistant Secretary, James Epps,

---

32   S.E. De Morgan, *Memoir*, pp. 153–54.
33   S.E. De Morgan, *Memoir*, p. 155.

on an eighty-five-page catalogue published in 1838.[34] And when the old Spitalfields Mathematical Society was dissolved in 1845, he co-signed the report on the absorption of its members and its library into the Royal Astronomical Society,[35] and supervised appraisal of incoming texts, as acknowledged by Assistant Secretary John Williams in a report of 1848.[36]

Given De Morgan's uncompromising stance on issues of importance to him, it is perhaps appropriate that his long period of service on the Society's Council came to an end over a matter of principle. In 1861, the wealthy amateur astronomer and philanthropist John Lee was elected President of the Society, and although De Morgan was not opposed to Lee's election per se, the manner in which he perceived it had been conducted, which departed somewhat from the usual conventions, was to him distasteful. And despite the fact that this same election saw him elected to the position of Vice-President, he declined to serve, promptly resigning as a member of the Council on which he had served for over three decades.[37]

The Astronomical Publications of Augustus De Morgan

De Morgan's publications on astronomy can be categorised into three distinct areas: articles on the history of astronomy and astronomers; reviews of published books on astronomy; and contributions to almanacs, particularly on matters relating to calendrical reckoning and the determination of the date of Easter.

History of Astronomy

In the first category, De Morgan's editorship of obituaries of deceased fellows of the Royal Astronomical Society, combined with his antiquarian interests, led him to write extensively about the lives of astronomers. In addition, his interest in the underdogs and forgotten contributors

---

34 *Catalogue of the Library of the Royal Astronomical Society* (London: printed by James Moyes, 1838). See Dreyer and Turner, p. 64.
35 London, Royal Astronomical Society, RAS Papers 37.
36 London, Royal Astronomical Society, RAS Papers 45.
37 De Morgan's lengthy letter of resignation is reproduced in full in the RAS minutes: see London, Royal Astronomical Society, RAS Papers 2.2 (Council minutes for March 1861). See also S.E. De Morgan, *Memoir*, pp. 272–77.

to the field led him to examine the lives of lesser-known individuals. Such texts were usually published as articles in the *Penny Cyclopaedia*, as contributions to biographical collections, and as obituaries in the *Monthly Notices of the Royal Astronomical Society*.

De Morgan's earliest writings about astronomy were written for the *Companion* to the *British Almanac*. These discussed the nature of eclipses (1832), comets (1833, 1835), and the moon's orbit (1834).[38] His desire to contribute to the dissemination of knowledge and to the education of the public is further demonstrated by his explanation of the *Maps of the Stars*.[39] Titled *An Explanation of the Gnomonic Projection of the Sphere* and published in 1836, this book devoted an entire chapter to the historical analysis of gnomonic projections, i.e. charts that depict the great circles of a sphere as straight lines.[40] De Morgan's work as the Secretary of the Royal Astronomical Society editing the detailed obituaries in *The Monthly Notices of the Royal Astronomical Society*, already noted, helped to shape his biographical research skills and widened his knowledge of the lives of astronomers. These obituaries still serve as essential starting points for historians of astronomy. De Morgan's most popular contributions relating to astronomy, however, were his large number of entries about astronomers and astronomical concepts for the *Penny Cyclopaedia*.[41] The entries bear witness to De Morgan's skills in historical research. They include entries on the various Astronomers Royal at the Royal Observatory, Greenwich (James Bradley, John Flamsteed, Edmond Halley, Nevil Maskelyne and John Pond). De Morgan also wrote the entry on the celebrated discoverer of the planet Uranus, William Herschel, and his entry on Jeremiah Horrocks contributed

---

38   These appeared in *The British Almanac of the Society for the Diffusion of Useful Knowledge* for the years 1832 to 1835. The relevant sections can be found in the part of the *Almanac* titled *The Companion to the Almanac; or Year-Book of General Information*.
39   Also published by the Society for the Diffusion of Useful Knowledge.
40   Augustus De Morgan, *An Explanation of the Gnomonic Projection of the Sphere* (London: Baldwin & Cradock, 1836). Great circles are the largest circles that can be drawn on a sphere. Such projections usually result in circular charts centred around a single point where the great circles intersect. Furthermore, only one hemisphere is depicted on such charts.
41   The *Penny Cyclopaedia* was also a publication of the Society for the Diffusion of Useful Knowledge. See Chapter 4 of this volume.

to efforts to raise the historical reputation of this relatively unknown seventeenth-century astronomer.[42]

Another strand of De Morgan's biographical research activities consisted of his contributions to *The Gallery of Portraits: with Memoirs*, published by Charles Knight between 1833 and 1837.[43] De Morgan wrote twelve 'portraits' in total about astronomers, mathematicians, and instrument makers for the publication: James Bradley, Jean-Baptiste Delambre, René Descartes, John Dollond, Leonard Euler, Edmond Halley, John Harrison, William Herschel, Joseph-Louis Lagrange, Pierre-Simon Laplace, Gottfried Leibniz and Nevil Maskelyne. The Royal Astronomical Society retains a collection of De Morgan's contributions bound together in a single volume, annotated by De Morgan, with additional information and corrections relating to the various mini-biographies.[44] While the style of most of the portraits in Knight's *Gallery* is dry, De Morgan's manuscript includes various cartoons illustrating events from the written accounts, which are characteristic of his sense of humour. The annotations exhibit his playful writing style, which is best showcased by his play on the word 'Uranus' ('you're an ass').

De Morgan's biographical research led him to analyse the life and work of Isaac Newton, following early-nineteenth-century efforts to examine Newton's life more critically than had previously been the case.[45] Francis Baily had pioneered this approach with a collection of texts about the life of John Flamsteed and his quarrel with Newton. A key point of the debate was Flamsteed's unwillingness to fund the publication of his observations, which meant limited access to the astronomical data he gathered. Although Newton helped to procure assistance from the Crown and the Royal Society to publish the results, he also altered the final publication in places without Flamsteed's consent. By showcasing part of the injustice done to Flamsteed, Baily's collection of texts was unflattering toward Newton. De Morgan wrote several accounts of

---

42 For a full list of his contributions to the *Penny Cyclopaedia*, see S.E. De Morgan, *Memoir*, pp. 407–14.
43 Sian Prosser, 'From the Collections: Illustrating Scientific Lives', *Astronomy & Geophysics*, 59 (2018), 4.11.
44 London, Royal Astronomical Society, MSS De Morgan 3.
45 For a more detailed discussion of De Morgan's contribution to the study of the life of Newton, see Adrian Rice, 'Augustus De Morgan: Historian of Science', *History of Science*, 34 (1996), 201–40; Higgitt, *Recreating Newton*.

Newton's life that followed Baily's critical approach, possibly influenced in his views (as Sophia De Morgan claimed) by his friendship with Baily.[46] He also aimed to demonstrate that there was a clear distinction between the image of Newton that survived and the Newton who lived at the time. This did not mean a devaluation of Newton's skills, talent, and contributions; instead, De Morgan argued that just because a person's work was highly valued, it did not immediately follow that his character was flawless.

As demonstrated by William Whewell's famous *History of the Inductive Sciences*, historical research based almost exclusively on secondary sources was considered perfectly acceptable at this time. But De Morgan and Baily shared an approach to historical research that emphasised a reliance on primary rather than secondary sources, thereby breaking from the established ways of examining the lives of astronomers and the history of astronomy. As Rebekah Higgitt has demonstrated, De Morgan was able to practise his writing in an intellectual community and among a circle of friends who shared the same approach towards collecting and consulting original sources.[47] Although De Morgan was not an ardent collector of letters and manuscripts, this antiquarian spirit was partly responsible for his accumulating a large collection of books.[48] Another factor was his willingness to write book reviews.

Athenæum Book Reviews

It is difficult to overstate the importance of *The Athenæum* in the nineteenth-century periodical world: established in 1828, it was the century's best-selling weekly, and its reviews were influential, renowned for their disinterestedness.[49] De Morgan was an avid book reviewer there, publishing anonymously almost one thousand reviews in *The Athenæum* over many years.[50] Many of the books he reviewed

---

46   Adrian Rice, 'Vindicating Leibniz in the Calculus Priority Dispute: The Role of Augustus De Morgan', in *The History of the History of Mathematics*, ed. by Benjamin Wardhaugh (Oxford: Lang, 2012), pp. 89–114.

47   Higgitt, *Recreating Newton*, pp. 106–10 and 116–24.

48   See Chapter 10 of this volume.

49   See Leslie A. Marchand, *The Athenæum: A Mirror of Victorian Culture* (Chapel Hill: University of North Carolina Press, 1941).

50   For an overview of the variety of subjects that De Morgan reviewed, see Sloan Evans Despeaux and Adrian C. Rice, 'Augustus De Morgan's Anonymous Reviews

were astronomical, and De Morgan used his reviews to reflect on the field and on its developments. Alongside major works within the field, the books reviewed included privately published volumes written by individuals unknown within the established astronomical community, thereby reflecting De Morgan's fascination with the contributions of 'underdogs' and individuals at the boundaries of the field. Mistakes in publications allowed him to exercise his playful writing style for his own enjoyment, and reviews became springboards for De Morgan to reflect upon more contemporary issues by criticising or praising the actions of astronomers that led to the publication of their books. His judgements of books ranged from scathing criticism, through simple summary, to genuine praise. A good example of the sharp edge of his critical style is the last line of his review of *A Theory of the Structure of the Sidereal Heavens*: 'as we cannot argue either for or against pure speculation, we stop here, wishing the author had expended his time, money, and very neat copper plates in something more likely to do him and others good.'[51]

De Morgan rarely engaged in detailed criticism of the content of the books, especially when reviewing observations published by observatories. In a review of the observations made at the Toronto Observatory he noted: 'it is not the province of our journal to enter upon the details of such a work'.[52] Instead, he gave a short description of the larger magnetic project in which the Observatory participated. Similarly, when the first volume of observations made at the Naval Observatory at Washington was published, De Morgan marked it as a historical moment in the development of astronomy: 'This is the first large volume of observations, that we have ever seen, emanating from a fixed observatory in the United States.'[53] Thereby, De Morgan used his reviews not only to reflect on the contents of books but also on the contexts of their production.

---

for *The Athenæum*: A Mirror of a Victorian Mathematician', *Historia Mathematica*, 43 (2016), 148–71. A selection of his reviews from *The Athenæum* was later reproduced in Augustus De Morgan, *A Budget of Paradoxes* (London: Longmans, Green, 1872).

51 [Augustus De Morgan], Review of *A Theory of the Structure of the Sidereal Heavens*, *The Athenæum* (25 March 1843), p. 284.

52 [Augustus De Morgan], Review of *Toronto Magnetical and Meteorological Observations*, Vol. I. 1840–42, *The Athenæum* (5 April 1845), p. 332.

53 [Augustus De Morgan], Review of *Astronomical Observations made at the Naval Observatory*, Washington, *The Athenæum* (9 January 1847), p. 45.

As a historian of astronomy, De Morgan also found pleasure in noting mistakes and misconceptions about Newton in astronomical publications. In a review of *Astrology As It Is, Not As It Has Been Represented*, he remarked on the author's claim that Newton was an astrologer: 'this we never heard before, and we never found any trace of it in his writings. We hope the author will tell us how he makes this out'.[54] In another review, he mentioned the mistaken belief that 'Newton [had suppressed] the manuscript of the Principia for many years, lest the savans [of the Royal Society] should be offended', and noted the involvement of Halley and others in the publication of Newton's work.[55] And while his review of *The Wonders of Astronomy* was on the whole laudatory, he noted that '[i]t is startling to see that the law of gravitation was "revealed to Newton by the fall of an apple."'[56] These examples indicate how book reviews served partly as an outlet for De Morgan to engage in historical commentaries about the lives of scientific practitioners such as Newton.

The above-mentioned reviews were relatively short and were included within the 'Our Literary Table' section of the magazine devoted to brief reviews. In addition, De Morgan wrote substantial reviews of major publications within the field of astronomy. The focus of these reviews, several columns in length, was rarely the works themselves. A salient example was a review of John Herschel's astronomical observations made at the Cape of Good Hope from 1834 to 1838.[57] Herschel's project was to survey and catalogue the double stars and nebulae visible from the Cape. De Morgan's review began with an overview of Herschel's initial aims and how the project was an expansion of his father's (i.e. William Herschel's) research. It then explained how the younger Herschel had not intended to publish the observations separately but that the Duke of Northumberland had offered to fund their individual publication. De Morgan also gave a brief overview of the content and noted the minute

---

54 [Augustus De Morgan], Review of *Astrology as it is, not as it has been represented*, The Athenæum (14 February 1857), p. 213.
55 [Augustus De Morgan], Review of *The Solar System as it is, not as it is represented*, The Athenæum (18 July 1857), pp. 908–09.
56 [Augustus De Morgan], Review of *The Wonders of Astronomy*, The Athenæum (26 December 1846), p. 1324.
57 [Augustus De Morgan], Review of *Astronomical Observations...*, The Athenæum (21 August 1847), pp. 885–86.

descriptions of nebulae found in the volumes. He found the importance of the publication in its being a 'mass of observations, deductions, and results as has rarely appeared at one time from one individual'. In addition, he praised the 'undivided labour of twelve years' that yielded the work. In brief, the review functioned as an encomium in order to communicate to the readers the accomplishments of an astronomer. We see the same technique in De Morgan's review of Ormsby Mitchel's *The Planetary and Stellar Worlds*. The review began by praising Mitchel for being an excellent popular writer on astronomy, and for creating a work that has 'intrinsic merit [... in] the freshness of its illustrations and [...] newness of its language'.[58] The rest of the review retold the story of the establishment of the Cincinnati Observatory. De Morgan noted Mitchel's involvement in its founding, his visit to European observatories, subsequent financial troubles, and the fire that burnt down the Observatory, which resulted in Mitchel's transformation into an itinerant lecturer. De Morgan used this story to build up a half-joking and half-serious proposal that the Cincinnati Observatory should be renamed the Mitchel Observatory. These two reviews demonstrate how De Morgan used his lengthier articles to contextualise the circumstances of their production. In this light, his longer book reviews also served as commentaries on contemporary developments within the astronomical community.

Between 1849 and 1856, there is a gap in De Morgan's reviews on astronomy for *The Athenæum*, a lacuna arising at least partly (1850–1854) from a disagreement between De Morgan and the journal's editor from 1846 to 1853, Thomas Kibble Hervey. He returned in full force in 1856 with a long review of François Arago's *Popular Astronomy*. De Morgan laid out his somewhat sceptical opinion of Arago's achievements in the first paragraphs: 'there are men among the living and the dead who ought to stand far above Arago, but who have never attained any reputation even remotely approaching to the brilliancy and the universality of that obtained by him.'[59] This remark set the tone for the rest of the review. It examined Arago's 'social public character' as well

---

58 [Augustus De Morgan], Review of *The Planetary and Stellar Worlds*, *The Athenæum* (21 October 1848), pp. 1051–52.
59 [Augustus De Morgan,] Review of *Popular Astronomy*, *The Athenæum* (5 January 1856), pp. 5–6.

as his 'faculty of illustration, both in speaking and writing' as essential components for his rise to fame. De Morgan criticised Arago's book for making erroneous claims about history and about the discovery of Neptune. De Morgan's reaction was unsurprising, given that Arago had sought to exclude John Couch Adams from the claims for the discovery of the planet. Besides the Neptune controversy, De Morgan's friendship with the disgraced Italian mathematician, historian and bibliophile Guglielmo Libri must have also influenced the review, as Arago and Libri were 'implacable enemies'.[60] Nonetheless, the review considered the book to be a useful popular account of astronomy, stopping short at 'being fit to decide nice controversies from original research'.

The change in tone of the longer reviews from praise to criticism was also present in a piece that discussed a pamphlet by James South, in which South continued his attacks against many members of the British astronomical community.[61] The beginning of De Morgan's review brought the readers up to date with the events and introduced the main characters of the 'Astronomers at War' saga described above (Troughton & Simms, Sheepshanks, Airy, Babbage and South). It characterised South as a person who alienated all his friends by his conspiracy theories. It also offered an overview of how his previous attacks were rebutted. The rest of the review countered the allegations made in the pamphlet, which largely centred around an admission by Sheepshanks that during his younger years he had smuggled a foreign instrument into the country by engraving the name of an English instrument maker on it. De Morgan dismissed any criticism of such an act on the grounds that it was 'a thing frequently done' and that it took place in Sheepshanks' youth. Through this commentary we see De Morgan once again using his reviews less to reflect upon the content of the publications than to contextualise their productions and to express his own views on the relevant debates that they concerned.

In summary, De Morgan's reviews of astronomical books within the pages of *The Athenæum* show us three important points about his

---

60   Rebekah Higgitt, '"Newton dépossédé!" The British Response to the Pascal Forgeries of 1867', *British Journal for the History of Science*, 36 (2003), 437–53 (pp. 446–47).

61   [Augustus De Morgan], Review of *A Letter to the Fellows of the Royal and the Royal Astronomical Societies*, *The Athenæum* (26 April 1856), pp. 513–15.

involvement with astronomy. First, they reflect his interest in astronomy as a historian, as a communicator of recent developments, and also as an educator (in his criticism of poorly written textbooks and publications). Second, De Morgan used his reviews to communicate his opinion about astronomical controversies. Third, the reviews distinguished 'useful' books from books that contained many mistakes, and works that furthered astronomical knowledge from those that were purely speculative. Thereby, even if indirectly, he gave the impression of serving as a gatekeeper to the astronomical community. As Despeaux and Rice argued, 'De Morgan also used *The Athenæum* as a place to debunk fallacious claims made in mathematics and science'.[62] The extent to which his reviews influenced the views of the readers of *The Athenæum* and the members of the astronomical community remains difficult to measure. However, the publication provided the best possible outlet for De Morgan to reach 'the growing middle classes, an audience which had a growing thirst for science'.[63]

Calendrical Reckoning and the Date of Easter

De Morgan's work on calendars combined his varied interests in history, mathematics, astronomy, legal matters and theology. He exhibited this best in his analysis of the confusion relating to the date of Easter Day in 1845. He examined the history of the development of calendars, the mathematical calculations upon which they were based, the astronomical principles underpinning their construction and the history of such calendars being incorporated into ecclesiastical and state legislation.

The core of the debate arose in relation to the ecclesiastical calendar for the year 1845, which denoted 23 March 1845 as Easter Day. This resulted in the appearance of an apparent paradox in the calculation of Easter and confusion about the rules for determining the exact day upon which it should fall.[64] The general rule for finding Easter states that if the full moon that follows 21 March falls upon a Sunday, then Easter Sunday is the one following it. Yet in 1845, Easter Sunday was denoted as falling on a Sunday immediately after 21 March, which was also a

---

62 Despeaux and Rice, p. 162.
63 Holland and Miller, p. 112.
64 A. De Morgan, *Budget*, pp. 217–30.

full moon. Within the British context there were also further problems, as British law prescribed that the determination of Easter Day rested on the tables and rules provided by the church. Thereby, a mistake in the determination was not only an ecclesiastical issue, but also a legal one.

The problem of establishing the exact day of Easter was neither a new nor an unknown problem, nor was it confined to the annals of history. Mathematicians and astronomers had undertaken the task of solving the challenges posed by the question as recently as the beginning of the nineteenth century. For example, the German mathematician Carl Friedrich Gauss and the French mathematician-astronomer Jean-Baptiste Delambre had each developed their own algorithms for calculating the exact day of Easter.[65] Moreover, the same problem with the calendar had arisen in 1818. Although discussions about the correctness of the calendar had taken place in that year, De Morgan later called them 'useless'.[66]

De Morgan wrote about the issue for four different publications. His first statement appeared as a letter in *The Athenæum*.[67] The next was a detailed examination of the problem, which appeared in the *Companion to the British Almanac*.[68] This was followed by a second paper one year later in the same publication, which discussed the history of the earliest printed almanacs.[69] And, within his *Budget of Paradoxes*, he wrote a shorter account of the debate.[70] This was a re-edited version of his method for finding the date of Easter, which appeared in the *Book of Common Prayer* as published by the Ecclesiastical History Society in 1849.[71] The variety of publication venues once again reflects how De Morgan's varied interests were combined in the Easter question. *The Athenæum* was the publication where he acted as a gate-keeper to astronomical, mathematical and scientific knowledge, and in which

---

65 See Reinhold Bien, 'Gauss and Beyond: The Making of Easter Algorithms', *Archive for History of Exact Sciences*, 58 (2004), 439–52.
66 Augustus De Morgan, 'Easter-Day, 1845', *The Athenæum*, 13 July 1844, p. 646.
67 Augustus De Morgan, 'Easter-Day, 1845', p. 646.
68 Augustus De Morgan, 'On the Ecclesiastical Calendar', *Companion to the Almanac for 1845*, 1–36.
69 Augustus De Morgan, 'On the Earliest Printed Almanacs', *Companion to the Almanac for 1846*, 1–31.
70 A. De Morgan, *Budget*, pp. 217–30.
71 *The Book of Common Prayer*, vol. 1. (London: Ecclesiastical History Society, 1849), pp. 57–64.

his reviews and articles served as reflections on contemporary political matters. The inclusion of the detailed analysis in the *Companion of the British Almanac* showed it as a mathematical and historical problem that needed to be explained to the public. Finally, the inclusion of a summary of his writings in the *Book of Common Prayer* showcased his interest in ecclesiastical history.

De Morgan's answer to the Easter Day problem rested on his historical analysis of texts. He argued that the terms 'moon' and 'lunations' within the ecclesiastical documents did not relate to actual astronomical objects. Instead, he distinguished between the 'moon of the heavens' and the 'moon of the calendar'.[72] The 'moon of the calendar' was a 'mean' or 'fictitious moon', which closely resembled the movement of the real body, but never precisely replicated it.[73] Similarly, the full moon referenced in ecclesiastical texts to define Easter Day was derived from the mean or fictitious moon rather than the movement of the real body.[74] De Morgan claimed that the lack of clarification about these distinctions was one of the chief causes of the apparent Easter Day paradox. In addition, there was the problem of nomenclature relating to the full moon. The term gradually replaced the wording used by the original makers of the rule for determining Easter: 'fourteenth day of the moon'.[75] The problem was complicated further in Britain, as an Act of Parliament adopted (with a few changes) the definitions provided by the Roman Catholic Church for determining Easter Day. In particular, the Act did not clarify that the term 'moon' in its text refers not to the real body but to the mean moon. The same Act similarly used the term 'full moon' without any explanation that it refers to 'the fourteenth day from the day of the new moon inclusive'.[76]

After explaining the history and the sources of common misconceptions about determining the day of Easter, De Morgan provided the reader with a step-by-step guide to find the exact day of Easter for any given year. For the sake of completeness, he not only provided it for the Gregorian calendar but also for the Julian calendar to illustrate the

---

72   A. De Morgan, 'Easter-Day, 1845'.
73   This same fictitious motion was reflected by clock time during De Morgan's life. A. De Morgan, *Budget*, pp. 217, 221.
74   A. De Morgan, *Budget*, pp. 217, 221.
75   De Morgan, 'Easter-Day, 1845'.
76   A. De Morgan, 'On the Ecclesiastical Calendar', p. 3.

different results and methods required for the two calendrical systems. Ultimately, the problem of finding Easter never left De Morgan's mind, and he eventually published a separate booklet on how to ascertain Easter Day.[77] This work, *The Book of Almanacs*, allowed its users to convert days between the Gregorian and Julian calendars between the years 1582 and 2000. More importantly, it enabled its readers to find out the days of full moon and new moon in both calendars. The preface also stated that the publication was intended for the use of almanac constructors rather than for the general public. Thus, it was a publication created for a specialist audience as opposed to his other writings about the date of Easter Day, which were communicated in a form accessible to a more general audience. In brief, De Morgan's involvement with the Easter Day question demonstrated how he applied his skills to everyday problems that incorporated aspects of various subjects that he enjoyed and in which he actively engaged.

A Mathematician Among Astronomers

De Morgan matched the intellectual astronomical spirit of the times. His interest in the subject arose through his love of mathematics, and in the Royal Astronomical Society he found a community of astronomers who promoted 'astronomical book-keeping' and thereby acknowledged the value of De Morgan's mathematical skills. Sheer administrative hard work as Secretary of the Society and dutiful editorship of its publications combined with his high intellect to bolster his legacy.

Moving beyond the realm of the Society, this chapter has also shed light on De Morgan's activities as a historian of astronomy and as a reviewer of astronomical books. As a historian of astronomy, he was among the leaders in the field, while as a prolific reviewer of astronomical books in a major periodical, he guided thought. De Morgan combined his interest in astronomy with his love of mathematics and of history to analyse contemporary debates relating to the determination of the date of Easter. At the very beginning, we asked how someone could contribute to astronomy without making a single telescopic observation—and this

---

77   Augustus De Morgan, *The Book of Almanacs* (London: Taylor, Walton, & Maberly, 1851).

chapter provides an answer. For De Morgan demonstrated that such contributions were possible by weaving his academic interests into the activities and publications of practising astronomers, promoting and publicising the discipline to a general audience, while at the same time upholding and maintaining the fabric of the astronomical community.[78]

Bibliography

Ashworth, William J., 'The Calculating Eye: Baily, Herschel, Babbage and the Business of Astronomy', *British Journal for the History of Science*, 27 (1994), 409–41. https://doi.org/10.1017/s0007087400032428

Aubin, David, and C. Bigg and H. O. Sibum, eds, *The Heavens on Earth: Observatories and Astronomy in Nineteenth-Century Science and Culture* (Durham and London: Duke University Press, 2010). https://doi.org/10.1215/9780822392507

Babbage, Charles, 'The Planet Neptune and the Royal Astronomical Society's Medal', *The Times*, 15 March 1847, p. 5.

Bien, Reinhold, 'Gauss and Beyond: The Making of Easter Algorithms', *Archive for History of Exact Sciences*, 58 (2004), 439–52. https://doi.org/10.1007/s00407-004-0078-5

The Book of Common Prayer, vol. 1. (London: Ecclesiastical History Society, 1849).

Case, Stephen, *Making Stars Physical: The Astronomy of Sir John Herschel* (Pittsburgh: University of Pittsburgh Press, 2018). https://doi.org/10.2307/j.ctv7n0c4q

Chapman, Allan, 'Science and the Public Good: George Biddell Airy (1801–92) and the Concept of a Scientific Civil Servant', in *Science, Politics and the Public Good: Essays in Honour of Margaret Gowing*, ed. by Nicolaas A. Rupke (Basingstoke and London: Macmillan, 1988), pp. 36–62.

—, 'Private Research and Public duty: George Biddell Airy and the Search for Neptune', *Journal for the History of Astronomy*, 19 (1988), 121–39.

De Morgan, Augustus, 'Astronomy', *Penny Cyclopaedia*, vol. 2 (London: Charles Knight, 1834), pp. 529–38.

—, *An Explanation of the Gnomonic Projection of the Sphere* (London: Baldwin & Cradock, 1836).

---

78 Thanks are due to Sian Prosser and Jane Trodd for contributing archival information and references to this chapter.

[—], Review of *A Theory of the Structure of the Sidereal Heavens*, *The Athenæum*, 25 March 1843, p. 284.

—, 'Easter-Day, 1845', *The Athenæum*, 13 July 1844, p. 646.

—, 'On the Ecclesiastical Calendar', *Companion to the Almanac for 1845*, 1–36.

[—], Review of *Toronto Magnetical and Meteorological Observations*, Vol. I. 1840–42, *The Athenæum*, 5 April 1845, p. 332.

—, 'On the Earliest Printed Almanacs', *Companion to the Almanac for 1846*, 1–31.

—, 'The New Planet', *The Athenæum*, 21 November 1846, p. 1191.

—, 'The New Planet', *The Athenæum*, 5 December 1846, pp. 1245–56.

[—], Review of *The Wonders of Astronomy*, *The Athenæum*, 26 December 1846, p. 1324.

[—], Review of *Astronomical Observations made at the Naval Observatory, Washington*, *The Athenæum*, 9 January 1847, p. 45.

[—], Review of *Astronomical Observations…*, *The Athenæum*, 21 August 1847, pp. 885–6.

[—], Review of *The Planetary and Stellar Worlds*, *The Athenæum*, 21 October 1848, pp. 1051–52.

—, *The Book of Almanacs* (London: Taylor, Walton, & Maberly, 1851).

[—], Review of *Popular Astronomy*, *The Athenæum*, 5 January 1856, pp. 5–6.

[—], Review of *A Letter to the Fellows of the Royal and the Royal Astronomical Societies*, *The Athenæum*, 26 April 1856, pp. 513–15.

[—], Review of *Astrology as it is, not as it has been represented*, *The Athenæum*, 14 February 1857, p. 213.

[—], Review of *The Solar System as it is, not as it is represented*, *The Athenæum*, 18 July 1857, pp. 908–09.

—, *A Budget of Paradoxes* (London: Longmans, Green, 1872).

De Morgan, Sophia Elizabeth, *Memoir of Augustus De Morgan* (London: Longmans, Green, 1882).

Despeaux, Sloan Evans, and Adrian C. Rice, 'Augustus De Morgan's Anonymous Reviews for *The Athenæum*: A Mirror of a Victorian Mathematician', *Historia Mathematica*, 43 (2016), 148–71. https://doi.org/10.1016/j.hm.2015.09.001

Dreyer, John Louis Emil, and Herbert Hall Turner, eds, *History of the Royal Astronomical Society 1820–1920* (London: Royal Astronomical Society and Wheldon & Wesley, 1923).

Higgitt, Rebekah, '"Newton dépossédé!" The British Response to the Pascal Forgeries of 1867', *British Journal for the History of Science*, 36 (2003), 437–53. https://doi.org/ 10.1017/s0007087403005144

—, *Recreating Newton: Newtonian Biography and the Making of Nineteenth-Century History of Science* (London and New York: Routledge, 2015). https://doi.org/10.4324/ 9781315653068

Holland, Susan, and Steven Miller, 'Science in the Early *Athenæum*: A Mirror of Crystallization', *Public Understanding of Science*, 6 (1997), 111–30. https://doi.org/10.1088/0963-6625/6/2/001

Hoskin, Michael, 'Astronomers at War: South vs Sheepshanks', *Journal for the History of Astronomy*, 20 (1989), 175–212.

—, 'More on "South v. Sheepshanks"', *Journal for the History of Astronomy*, 22 (1991), 174–9.

Kollerstrom, Nicholas, 'An Hiatus in History: The British Claim for Neptune's Co-Prediction, 1845–1846: Part 1', *History of Science*, 44 (2006), 1–28. https://doi.org/10.1177/007327530604400101

Lambert, Kevin, 'A Natural History of Mathematics: George Peacock and the Making of English Algebra', *Isis*, 104 (2013), 278–302. https://doi.org/10.1086/670948

McConnell, Anita, 'Astronomers at War: The Viewpoint of Troughton & Simms', *Journal for the History of Astronomy*, 25 (1994), 219–35.

Marchand, Leslie A., *The Athenæum: A Mirror of Victorian Culture* (Chapel Hill: University of North Carolina Press, 1941).

Prosser, Sian, 'From the Collections: Illustrating Scientific Lives', *Astronomy & Geophysics*, 59:4 (2018), p. 4.11. https://doi.org/10.1093/astrogeo/aty186

Rice, Adrian, 'Augustus De Morgan: Historian of Science', *History of Science*, 34 (1996), 201–40. https://doi.org/10.1177/007327539603400203

—, 'Vindicating Leibniz in the Calculus Priority Dispute: The Role of Augustus De Morgan', in *The History of the History of Mathematics*, ed. by Benjamin Wardhaugh (Oxford: Lang, 2012), pp. 89–114. https://doi.org/10.3726/978-3-0353-0261-5/7

Smith, Robert W., 'The Cambridge Network in Action: The Discovery of Neptune', *Isis*, 80 (1989), 395–422.

Swade, Doron David, 'Calculation and Tabulation in the Nineteenth Century: Airy versus Babbage' (Unpublished Ph.D. Diss., University College London, 2003).

Yeo, Richard, *Defining Science: William Whewell, Natural Knowledge and Public Debate in Early Victorian Britain* (Cambridge: Cambridge University Press, 2003). https://doi.org/10.1017/cbo9780511521515

[illegible struck text] the principles themselves. Identity
I have given my own views of
and difference are things too deep for definition. In
this they are analogues of space and time. We cannot
define either that space is extension, and time is duration,

* I might give many proofs that this arguing
is not finished; I will content myself with one.
The late Professor Spalding, whose learning and acute-
ness will not be questioned, says that the excluded
middle lies deeper than the other two principles; "a
proposition disobeying either of them would be in-
consistent with its data, but yet possible; a
proposition disobeying the third axiom would be
inconceivable." If by a 'possible' proposition be
meant of possible statement, surely it is as conceivable in contradiction of excluded middle as of identity: if 'possibly true' be meant, the sentence
is bewildering.

the abstract notion which is concreted in "a pig is a pig";
and difference is similarly related to "a pig is not a
sheep". Learning cannot go deeper than this, though learning
has tried hard. It has happened that too much
investigation has led to strange results. Hamilton (i. 80) has
actually founded 'A is A' upon an anterior principle. "The
principle of Identity .... is expressed in the formula A is A
.... The principle of Identity is an application of the principle of the absolute equivalence of a whole and all its
parts taken together to the thinking of a thing by the attribution of constituent qualities or characters". No remark
is wanted upon this inversion. The equivalence of a
whole to all its parts taken together comes before, and
on the road to, the principle that the whole is the whole
and each part is itself.
Between those who find a basis for 'X is X' — which I believe to be the basis of eggs is eggs — and those who find occasion to dispense with it, [struck text] I should be in a T.O

Fig. 6 This page from the manuscript of an unpublished paper from 1868 gives an evocative insight into the fervent process of composition, correction and revision that went into De Morgan's publications. Intended as the sixth installment of his series of papers entitled 'On the Syllogism', De Morgan's text is a hodge-podge of cut-and-pasted amendments and insertions. (MS 241, reproduced by permission of Senate House Library, University of London.)

4. De Morgan, Periodicals and Encyclopaedias

Olivier Bruneau

> De Morgan's energy ... was chiefly absorbed by his voluminous writings upon mathematical, philosophical, and antiquarian points.
>
> — Leslie Stephen[1]

Introduction

According to Sally Shuttleworth and Geoffrey Cantor, Britain produced about 125,000 different periodical and newspaper titles during the nineteenth century.[2] Many of these had a very short life span, and few were devoted to science. De Morgan did not publish in all of them—far from it—but, by his own account, he was intensely active between 1830 and 1870, publishing more than 2,200 items of between a few lines and several dozen pages in about twenty newspapers and periodicals, including some of the most important scientific journals of this period.[3] Although De Morgan's research publications focused on

---

1. Leslie Stephen, 'De Morgan, Augustus (1806–1871)', *Dictionary of National Biography*, vol. 14 (London: Smith, Elder, 1888), pp. 331–34 (p. 333).
2. Sally Shuttleworth and Geoffrey Cantor, 'Introduction', in *Science Serialized: Representations of the Sciences in Nineteenth-Century Periodicals*, ed. by Geoffrey Cantor and Sally Shuttleworth (Cambridge, MA: MIT Press, 2004), pp. 1–16. For an overview of nineteenth-century periodical literature, see *Dictionary of Nineteenth-Century Journalism in Great Britain and Ireland*, ed. by Laurel Brake and Marysa Demoor (London: British Library, 2009).
3. For an overview of mathematics periodicals during the Victorian period, see, for instance, Sloan Evans Despeaux, 'A Voice for Mathematics, Victorian Mathematical

mathematics and logic, his book reviews reveal his interest in wide areas of knowledge. His contribution to encyclopaedias, in contrast, was more restricted and limited to four main venues, where he wrote only on his areas of expertise.

This review relies on the biography of Augustus De Morgan published by his widow, Sophia, in 1882 and on G. C. Smith's *The Boole-De Morgan Correspondence*, which appeared a century later.[4] Both authors present substantial lists of De Morgan's articles and other publications. However, these lists are incomplete, because they included neither the numerous book reviews and notes De Morgan wrote for the *Athenæum* nor his fairly short contributions to *Notes & Queries*. Yet he contributed abundantly to both publications, publishing more than 210 notes in *Notes & Queries* between 1850 and 1864, plus nearly one thousand book reviews, around fifty articles in his *Budget of Paradoxes* series, and several other short notes in the *Athenæum* between 1840 and 1870.

This chapter thus aims to present a more holistic overview of the scholarly and journalistic output of Augustus De Morgan.[5]

De Morgan's Involvement in Periodicals: He is Everywhere!

When De Morgan began his academic career at the University of London in 1828, his initial aim was to write textbooks for students and beginners. Thus, as early as March 1827, he offered to write a treatise on acoustics for the recently-formed Society for the Diffusion of Useful Knowledge (SDUK)[6] and published several textbooks under its auspices in subsequent years. His involvement in journal publication began in

---

Journals and Societies', in *Mathematics in Victorian Britain*, ed. by Raymond Flood, Adrian Rice, and Robin Wilson (Oxford: Oxford University Press, 2011), pp. 155–74. De Morgan may have published more than this, for example in the Royal Astronomical Society's *Memoirs* and *Monthly Notices*; anonymous authorship in these and other periodicals renders it impossible to attribute authorship to De Morgan with certainty.

4 Sophia Elizabeth De Morgan, *Memoir of Augustus De Morgan* (London: Longmans, Green, 1882), pp. 401–15; G. C. Smith, ed., *The Boole-De Morgan Correspondence 1842–1864* (Oxford: Clarendon Press, 1982), pp. 141–47.
5 See also Chapter 12 in this volume.
6 University College London, SDUK/24: Letter from De Morgan to Thomas Coates, 30 March 1827.

1830 with his first submissions to two periodicals, both published by the SDUK. The first was the *Companion to the Almanac* for 1831, a compendium of miscellaneous articles of general interest published jointly with the SDUK's annual *British Almanac*. De Morgan's article was a twenty-page introduction to the subject of life assurance (not to be confused with life insurance),[7] containing an account of the principal London assurance companies, an extensive table comparing the premiums charged for various policies by these companies and a discussion of how such premiums were calculated. This article represented De Morgan's first public foray into the realm of actuarial mathematics, in which he was to become an expert. In addition to publishing a variety of further articles on numerous aspects of the subject, he would also work as a freelance actuarial consultant to supplement his income during his academic career. In the *Companion to the Almanac*, he went on to publish one article every year between 1831 and 1857. And although it is difficult to compare the volume of De Morgan's contributions with those of other authors because so many of the articles were anonymous, it is hard to imagine any other contributor surpassing the sheer range of his interests: his twenty-seven articles are learned but accessible treatments of subjects pertaining to astronomy, mathematics and its history, bibliography, the calendar, currencies and actuarial matters.

At roughly the same time, De Morgan contributed the first of many pedagogical articles to the *Quarterly Journal of Education*, a new journal recently set up by the SDUK. In this article, published in the inaugural volume, De Morgan gave the first detailed account in English of the education provided at 'the most celebrated school of instruction for engineers which has ever existed', namely, the École Polytechnique in Paris.[8] Detailing the opening decades of its operation from its foundation in 1794, he presented the school as a model for advanced mathematical and scientific teaching in Great Britain. The *Quarterly Journal of Education* appeared in ten volumes from 1831 to 1835.[9] De Morgan contributed to it throughout the entire period with thirty-three articles divided

---

7 The key difference is that life insurance covers a policyholder for a specific term, whereas life assurance usually provides coverage for the policyholder's entire life.
8 'Polytechnic School of Paris', *Quarterly Journal of Education*, 1 (1831), 57–74 (p. 73).
9 Re-published in facsimile with a new introduction by Christopher Stray, 10 vols. (London: Routledge, 2008).

between original observations on the state of teaching—particularly of mathematics—in various institutions and reviews of works essentially intended for teaching science.[10] As with the *Companion to the Almanac*, the anonymity of most contributions to this journal makes comparisons difficult, but probably the only contributor likely to have rivalled De Morgan's involvement with the *Quarterly Journal* would have been its editor, George Long, formerly professor of Greek and De Morgan's colleague at the University of London.

De Morgan was also an active member of the Cambridge Philosophical Society (founded by some of his former lecturers and friends). It was in its *Transactions* that he published his first scientific research papers and continued to report on his chief mathematical findings, particularly in his series of papers 'On the Foundation of Algebra' and 'On the Syllogism', which had a catalytic effect on contemporary developments in algebra and symbolic logic respectively. He delivered several papers to the Society between 1830 and 1868, and these were published between 1833 and 1871, the year of his death.[11]

Despite its irregular publication schedule, De Morgan considered the *Transactions of the Cambridge Philosophical Society* to be a privileged venue in which to publish mathematics at the time. Yet he also disseminated his scientific work elsewhere. From 1835 until 1852, he published about twenty articles, mostly on mathematics, in the journal now called the *Philosophical Magazine*.[12] He published twenty-one articles in the newly-founded *Cambridge Mathematical Journal*, established by Duncan Gregory, Archibald Smith and Samuel Greatheed, between 1841 (its

---

10 See also Chapter 6 in this volume.
11 De Morgan presented his first paper in the session of November 15, 1830 and the last on October 26, 1868, but due to erratic publication delays, the first published article, corresponding to his first paper, did not appear until vol. 4 (1833), and the last paper was published in vol. 11 (1871). On the history of the Society, see A. Rupert Hall, *The Cambridge Philosophical Society, A History, 1819–1969* (Cambridge: Cambridge Philosophical Society, 1969) and Susannah Gibson, *The Spirit of Inquiry: How One Extraordinary Society Shaped Modern Science* (Oxford: Oxford University Press, 2019).
12 During the period of publication of De Morgan's articles, this periodical changed its title several times, being first *The London and Edinburgh Philosophical Magazine and Journal of Science* and then, after 1840, *The London, Edinburgh and Dublin Philosophical Magazine and Journal of Science*.

second volume) and 1858.[13] He also agreed to publish, at the request of the editors of *The Mathematician*—a periodical that was intended to be the successor of Leybourn's *Mathematical Repository*—some notes in one of the volumes of that journal.[14]

Of all the journals with which De Morgan was involved, it was undoubtedly in the weekly literary and scientific magazine *The Athenæum* that he was most active, contributing nearly one thousand book reviews over a period of more than thirty years (between 1840 and 1870) across a wide range of interests, from mathematics to religion to weights and measures.[15] These all appeared anonymously, but reviewers' names in the editors' file copies reveal the extent of reviewing De Morgan undertook for the journal.[16] In addition to these anonymous reviews, De Morgan also published his famous series of humorous articles under the heading *A Budget of Paradoxes*, together with its *Supplements*, from 1863 until 1867, along with some shorter articles. The *Budget*, a collection of witty articles and reviews about eccentric dabblers in science (and especially mathematics), was subsequently edited and expanded by their author and published by his widow the year after his death.[17]

A notable offshoot from *The Athenæum* began on 3 November 1849 with the publication of the first issue of *Notes & Queries*, a weekly

---

13 This journal also changed names several times, becoming first the *Cambridge and Dublin Mathematical Journal* in 1846 and then the *Quarterly Journal of Pure and Applied Mathematics* from 1855. For a history of this journal, see especially Tony Crilly, 'The *Cambridge Mathematical Journal* and its Descendants: The Linchpin of a Research Community in the Early and Mid-Victorian Age', *Historia Mathematica*, 31 (2004), 407–97.

14 Unfortunately, this journal produced only three volumes. For a history of the *Mathematical Repository*, see most recently, Olivier Bruneau, 'Le Royal Military College, un centre éditorial pour quelles mathématiques?: le Mathematical Repository', in *Circulations mathématiques dans et par les journaux*, ed. by Hélène Gispert, Philippe Nabonnand and Jeanne Peiffer (London: College Publications, 2024).

15 For De Morgan's involvement in *The Athenæum*, see Sloan Evans Despeaux and Adrian C. Rice, 'Augustus De Morgan's Anonymous Reviews for *The Athenæum*: A Mirror of a Victorian Mathematician', *Historia Mathematica*, 43 (2016), 148–71. De Morgan undertook similar review activity for the *Dublin Review*, a Catholic journal published in London, with ten reviews and notes over ten years.

16 These copies, available in the archives of City University, London, formed the basis of an online database: City University, *The Athenæum Projects*, available at: https://athenæum.city.ac.uk/

17 Augustus De Morgan, *A Budget of Paradoxes* (London: Longmans, Green, 1872). See Chapter 7 in this volume.

journal devoted to the exchange of information between scholars and professionals in the fields of literature and history.[18] De Morgan was involved from its second issue until 1864, offering notes and remarks on a wide range of subjects. In all, 214 notes are attributed to De Morgan. Short though these texts are—often only a few lines—they show an intense activity and again demonstrate the breadth of his interests, from the English language to antiquarian books to musical notation.

In 1848, a group of London actuaries founded the Institute of Actuaries in order to promote and better organise the profession, in which De Morgan's involvement has already been noted. They also decided to publish a periodical, *Assurance Magazine*,[19] the first issue of which came out in 1851. Very soon after, in the fourth issue of the first volume, De Morgan published an article on the equivalence of compound interest and simple interest paid when due. By 1868, he had submitted twenty-four articles to this journal, dealing mostly with the mathematical treatment of life contingencies. Its editors also reprinted parts of his *Budget of Paradoxes*, previously published in *The Athenæum*.

Finally, towards the end of his life, De Morgan published several articles in the *Proceedings of the London Mathematical Society*, arising from his privileged position as the Society's first president. He had played a supporting but important role in its foundation, but apart from the inaugural speech he gave at the Society's first meeting on 16 January 1865, his papers in its *Proceedings* are brief and of little particular interest.

When analysing the topics of De Morgan's periodical publications, it makes little sense to include his entire output, as the nearly 1,030 book reviews in various journals distort the picture.[20] If we discount the reviews and focus purely on his research articles and commentaries, De Morgan published nearly two hundred items.[21] Articles on mathematics and logic are the most numerous, numbering eighty-one (i.e. forty-one

---

18 For a historical study of this review, see Patrick Leary, 'A Victorian Virtual Community', *Victorian Review*, 25:2 (2000), 61–79.

19 This later became the *Journal of the Institute of Actuaries*.

20 De Morgan's book reviews can be found in *The Athenæum, Quarterly Journal of Education, Dublin Review, Notes & Queries* and *Assurance Magazine*. As the vast majority of these reviews are in *The Athenæum*, we refer to the article by Despeaux and Rice cited above (fn.15).

21 This figure does not include the 214 notes in *Notes & Queries* owing to their extreme brevity and anecdotal nature; they are clearly not in the same league as research articles or even commentaries on a subject.

per cent).²² Next, thirty-one articles (sixteen per cent) are on history (mainly concerning mathematics and astronomy) and biography.²³ After this come his reflections on matters related to actuarial science (life assurance, mortality tables, etc.) with twenty-one articles (eleven per cent).²⁴ Other sciences follow with sixteen articles (eight per cent), then astronomy, with thirteen articles published (seven per cent).²⁵ De Morgan's interest in education and teaching in general is found in seven per cent of his output and is concentrated towards the beginning of his career. This is principally due to the number of articles he contributed to the *Quarterly Journal of Education* during this period, before his publication of original research gained significant momentum. After 1838, articles on pedagogical matters become far less frequent.²⁶ One subject that recurs throughout De Morgan's career is his advocacy for the decimalisation of the currency, weights and measures used in Great Britain, and he devoted seven articles to this topic (four per cent), principally during the 1850s at the height of public interest in Britain on the question of a decimal coinage.²⁷

---

22 The literature on De Morgan's mathematics and logic is relatively abundant: we refer to the dedicated chapters in this book.
23 On De Morgan as a historian of science, see Adrian Rice, 'Augustus De Morgan: Historian of Science', *History of Science*, 34 (1996), 201–40.
24 On De Morgan's involvement with actuarial science, for instance see Timothy Alborn, 'A Calculating Profession: Victorian Actuaries among the Statisticians', *Science in Context*, 7 (1994), 433–68.
25 It is surprising to see a relatively small number of papers dealing with astronomy in his output, but this does not include his biographies of astronomers or his articles on the history of astronomy. In fact, his interest in this field was significant and he became a member of the Astronomical Society of London (which became the Royal Astronomical Society in 1831) in 1828, holding positions on its council for many years. See Adrian Rice, 'Augustus De Morgan and the Development of University Mathematics in London in the Nineteenth Century' (Ph.D. Diss., Middlesex University, 1997), pp. 101–08, and Chapter 3 in this volume.
26 On De Morgan's work in education, see Chapter 6 in this volume, as well as Abraham Arcavi and Maxim Bruckheimer, 'The Didactical De Morgan: A Selection of Augustus De Morgan's Thoughts on Teaching and Learning Mathematics', *For the Learning of Mathematics*, 9 (1989), 34–39; Adrian Rice, 'What Makes a Great Mathematics Teacher? The Case of Augustus De Morgan', *American Mathematical Monthly*, 106 (1999), 534–52; Adrian Rice, 'Augustus De Morgan and the Development of University Mathematics', pp. 111–21.
27 From its beginning in 1854, De Morgan was involved with the International Association for a Decimal System, which included many of his contemporaries: George Airy, Charles Babbage, John Herschel and George Peacock.

Thus, to summarise, the *Transactions of the Cambridge Philosophical Society* remained De Morgan's preferred venue for publishing lengthy mathematical (or logical) research papers from the beginning of his career until the end of his life. The *Cambridge Mathematical Journal* and its successors played a similar role but over a narrower period, and for shorter papers. These venues were among the only places in which to publish substantial research in mathematics in Britain at this time, aside from the publications of the Royal Society, a body that De Morgan famously refused to join. For his broader interests, De Morgan had a much wider range of non-specialist publications from which to choose. But he clearly enjoyed publishing in *The Athenæum* and, to a lesser extent, *Notes & Queries* to satisfy the breadth of his interests, including mathematical biography, bibliography and more. His motivations for choosing these particular venues are hard to determine precisely, but there is no doubt that his willingness to write so many book reviews for *The Athenæum*—and the editor's willingness to publish them—greatly facilitated the expansion of De Morgan's personal library at no cost to himself.

Although De Morgan played no founding role in any of the journals in which he published, he was privileged to contribute to the earliest issues of several of them, especially those edited by the SDUK. In some instances, as with the *Quarterly Journal of Education*, this arose from his closeness to the publishers. In others, the presence of an article by a well-known and recognised scholar could help to ensure the viability of young journals. This certainly applied to the journals of both the Institute of Actuaries and the London Mathematical Society, the fledgling publications of which both benefitted in no small measure from the imprimatur bestowed by the inclusion of De Morgan's work. De Morgan, then, published extensively in a variety of venues ranging from highly specialised mathematical journals to publications aimed at a very broad audience without a scientific background: he was both a polymath and a communicator at multiple levels.

The British Encyclopaedic Landscape in the First Half of the Nineteenth Century

Although the first English-language encyclopaedia was published in the seventeenth century, the genre began to develop in the eighteenth century, reaching its peak at the turn of the nineteenth century.[28] These reference works, mostly general, comprised several volumes arranged alphabetically, going beyond simple definitions and dealing with a vast range of subjects.[29] They were more or less scholarly, addressing either a broad public or a specific community.[30]

Between 1800 and 1854, the year of publication of the *English Cyclopaedia*, at least twenty-four encyclopaedias were partially or completely published in Great Britain, to which must be added multiple editions of the *Encyclopaedia Britannica* (from the fourth to the eighth) and reissues of earlier dictionaries. John Harris's *Lexicon Technicum* (1704) is considered to be the first attempt to organise scientific knowledge in a work of reference in the English language. Among the British encyclopaedias of the eighteenth century, Ephraim Chambers' *Cyclopaedia* (1728), had an important influence on this type of production in Britain and elsewhere,[31] as did the *Encyclopaedia Britannica*

---

28   For a useful bibliographical review of this period, see S. Padraig Walsh, *Anglo-American General Encyclopaedias: A Historical Bibliography, 1703–1967* (New York and London: Bowker, 1968).

29   Not all encyclopaedias were arranged alphabetically; for example, the *Encyclopaedia Metropolitana* was arranged by genre and then by subject.

30   Studies on encyclopaedias are mainly concentrated on those of the eighteenth century, but see: Richard Yeo, 'Reading Encyclopaedias: Science and the Organization of Knowledge in British Dictionaries of Arts and Sciences, 1730–1850', *Isis*, 82 (1982), 24–49; Richard Yeo, *Encyclopedic Visions, Scientific Dictionaries and Enlightenment Culture* (Cambridge: Cambridge University Press, 2001); *Notable Encyclopaedias of the Seventeenth and Eighteenth Centuries: Nine Predecessors of the* Encyclopédie, ed. by Frank A. Kafker (Oxford: Voltaire Foundation, 1981); *Notable Encyclopaedias of the Late Eighteenth Century: Eleven Successors of the* Encyclopédie, ed. by Frank A. Kafker (Oxford: Voltaire Foundation, 1994); *The Early* Britannica: *The Growth of an Outstanding Encyclopedia*, ed. by Frank A. Kafker and Jeff Loveland (Oxford: Voltaire Foundation, 2009); Peter Burke, *A Social History of Knowledge: From Gutenberg to Diderot* (Cambridge: Polity, 2000); Jeff Loveland, *The European Encyclopaedia, from 1650 to the Twenty-First Century* (Cambridge: Cambridge University Press, 2019).

31   Diderot and D'Alembert's *Encyclopédie* was initially a translation of this work. On the conditions of writing and the genesis of this enterprise, see among others Marie Leca-Tsiomis, *Écrire l'Encyclopédie* (Oxford: Voltaire Foundation, 1999);

(1771), first edited by William Smellie, and Abraham Rees's update of Chambers' *Cyclopaedia*.[32] In the first half of the nineteenth century, several encyclopaedias competed with each other: the various editions of the *Britannica* faced the *Encyclopaedia Londinensis* by John Wilkes (24 volumes, 1810–29), the *Pantologia* by Newton Bosworth, Olinthus Gregory and John Mason Good (12 volumes, 1813) and the *Penny Cyclopaedia* edited by George Long (27 volumes, 1833–1843). When the first sheets of the latter came out in 1833, it was competing with about twenty encyclopaedias of varying sizes. The shortest was Samuel Maunder's two-volume *Treasury of Knowledge and Library of Reference* (1830), and the largest Dionysius Lardner's *Cabinet Cyclopaedia*, with 133 volumes. The others fell into three categories: those between four and six volumes long, those of a dozen volumes and those with more than twenty volumes.

The objectives differed from one encyclopaedia to another. All of them aimed to present and disseminate scientific knowledge for educational purposes, but some, like the first edition of the *Encyclopaedia Britannica*, contained a preliminary discourse that was either not very or not at all technical.[33] The aim could be to build up a small library of mathematics, as with Harris's *Lexicon Technicum*. The publication of some encyclopaedias was justified by the need to replace older and outdated ones: those published in the second half of the eighteenth century were published to replace the multiple editions of Chambers' *Cyclopaedia*, which had changed little, or not at all, since 1728. Diderot and D'Alembert's *Encyclopédie* was also the driving force behind a reaction, and the *Britannica* can in part be seen as a response to this French enterprise.

Finding a new approach made it possible to establish a foothold in this extremely competitive market. For example, the *Encyclopaedia Metropolitana* was based on a division of knowledge and each part is

---

Robert Lewis Collison, *Encyclopaedias: Their History throughout the Ages* (London and New York: Hafner, 1966).

32 The *Encyclopaedia Britannica* remains in existence: it has gone through fifteen print editions and since 2012 it has been digital only. For a history of the early editions, see Kafker and Loveland, *The Early Britannica*, and, more broadly, Herman Kogan, *The Great EB: The Story of the Encyclopaedia Britannica* (Chicago: University of Chicago Press, 1958).

33 Later, as the number of volumes and authors recruited increased, the level of the articles was raised.

treated as a treatise.[34] In the pure sciences, we find not only mathematics but also grammar and morals—which can be seen as a return to an old form of encyclopaedic organisation, as in Gregor Reisch's *Margarita Philosophica* (1504).[35] The *Encyclopaedia Metropolitana* was distinguished by having treatises written by subject experts to quite a high level—in addition to De Morgan's two contributions, treatises by John Herschel, Richard Whately and George Peacock were also highly regarded. These encyclopaedias were often expensive, and hence intended for an educated population.

Of course, no encyclopaedia should be studied in isolation, as encyclopaedias were, at least in part, interrelated. Indeed, borrowings and verbatim copying between two encyclopaedias were frequent.[36] Moreover, encyclopaedias compiled by editors who were not necessarily subject experts might draw on multiple sources—previous encyclopaedias as well as reference works or manuals—and two encyclopaedias could follow the same sources, so it is not uncommon in an article to read verbatim extracts from several manuals. One author could also be involved in several encyclopaedic undertakings: for example, William Wallace wrote the article on fluxions both in the fourth edition of the *Britannica* (1810) and in the *Edinburgh Encyclopaedia* (1815).[37] The fact that some encyclopaedias were written in reaction or opposition to others renders comparison additionally useful. Finally, some titles ran through several editions, which may or may not have

---

34 On the organisation of knowledge in encyclopaedias, see for example Yeo, *Encyclopedic Visions*.
35 The first division contained the so-called formal sciences (universal grammar, logic, mathematics and metaphysics) and real sciences (morals, law and theology). The mathematical part is divided into fifteen treatises (from 50 to over 200 pages).
36 If two texts are identical, this does not necessarily mean that one is a copy of the other: they may be a copy of a third text or there could be an intermediate version. This methodological point is developed in Jeff Loveland, 'Two Partial English-Language Translations of the *Encyclopédie*: The Encyclopaedias of John Barrow and Temple Henry Croker', in *British-French Exchanges in the Eighteenth Century*, ed. by Kathleen Hardesty Doig and Dorothy Medlin (Newcastle: Cambridge Scholars, 2007), pp. 168–187 (p. 176).
37 On the study of fluxion articles in British encyclopaedias and on the construction phenomena of these articles see Olivier Bruneau, 'Encyclopaedias as Markers of Heritage Building: Fluxion Articles in British Encyclopaedias, 1704–1860', *Philosophia Scientiæ*, 26:2 (2022), 67–90.

been updated. For example, between 1771 and 1853, the *Britannica* underwent eight editions and two supplements.[38]

De Morgan and Encyclopaedias: Promoting Mathematics and Astronomy for All

De Morgan was involved in several encyclopaedic ventures. He contributed to the *Encyclopaedia Metropolitana*, Lardner's *Cabinet Cyclopaedia*, the SDUK's *Penny Cyclopaedia*, and the *English Encyclopaedia*, in addition to writing seven entries about mathematicians and astronomers for William Smith's *Dictionary of Greek and Roman Biography and Mythology*.[39] These different encyclopaedias had different objectives and targeted different audiences. The *Encyclopaedia Metropolitana* was intended to be scholarly and was aimed at a literate public wanting an authoritative collection of scientific, legal, moral and theological texts by prominent recognised authors. De Morgan was commissioned to write two texts for the second volume of the section devoted to the so-called pure sciences. The first, a book-length treatise on 'Calculus of Functions', built on the recent work of the Frenchmen Lazare Carnot and Augustin-Louis Cauchy and of De Morgan's friends Charles Babbage, John Herschel and George Peacock. Its eighty-eight densely printed pages, which highlighted the latest advances in the field at the time of its publication in 1836, can be considered as a monograph in its own right. The second text, published in 1837 and of an equivalent size, was a mathematical treatment of probability theory, inspired chiefly by the work of Pierre-Simon Laplace.[40]

---

38   Updates between editions are very uneven and do not necessarily encompass all entries. For example, the mathematics articles in the first three editions are identical, even reproducing typographical errors from the first edition.

39   William Smith, ed., *A Dictionary of Greek and Roman Biography and Mythology*, 3 vols. (London: Taylor & Walton, 1844) and subsequent editions. For Smith and his dictionaries, see Christopher Stray, 'William Smith and his Dictionaries: A Study in Scarlet and Black', in *Classical Books: Scholarship and Publishing in Britain since 1800*, ed. by Christopher Stray, Bulletin of the Institute of Classical Studies, Supplement 101 (London: Institute of Classical Studies, University of London, 2007), pp. 35–54.

40   For a study of this text and the relationship with Laplace, see Adrian Rice, '"Everybody Makes Errors": The Intersection of De Morgan's Logic and Probability, 1837–1847', *History and Philosophy of Logic*, 24 (2003), 289–305; S. L. Zabell, 'De Morgan and Laplace: A Tale of Two Cities', *Journal électronique d'Histoire des Probabilités et de la Statistique*, 8 (2012), 1–29, www.jehps.net.

Dionysius Lardner's *Cabinet Cyclopaedia* (1829–1844) tried to reach a wider public by offering small volumes on individual subjects.[41] Lardner, by then an erstwhile colleague of De Morgan at the university, asked him to write a text on the theory of probability. De Morgan's *Essay on Probabilities and on their Application to Life Contingencies and Insurance Offices* was the 107th volume in the series and was published in 1838. The *Cabinet Cyclopaedia* was much more affordable and more application-oriented than the *Encyclopaedia Metropolitana*. But the appearance of De Morgan's *Essay* prompted a minor controversy between its author and the editors of the *Encyclopaedia Metropolitana*, who, being unaware that the *Essay* differed substantially in content and mathematical sophistication from his earlier book-length treatise on probability in the *Metropolitana*, accused De Morgan of copyright infringement by publishing what 'might be deemed a second edition of the treatise'. Although the matter was eventually settled, the affair led to the end of De Morgan's association with the *Encyclopaedia Metropolitana*, whose editors, he said, had been guilty of 'wasting a good deal of grumbling'.[42]

It was to be the *Penny Cyclopaedia* in which De Morgan invested most of his time.

The *Penny Cyclopaedia*: An Encyclopaedia for the Masses

In 1826, London University (now University College London) had been created in opposition to the Anglican universities of Oxford and Cambridge. One of its founders, then a radical leader in the House of Commons, was Henry Brougham. Highly committed to promoting knowledge to a very wide audience, he created the Society for the Diffusion of Useful Knowledge (SDUK) in the same year, with the aim of providing the working and middle classes with the means to access knowledge, often at modest prices.[43] The two institutions were born of

---

41 For a history of this original encyclopaedic enterprise, see Morse Peckham, 'Dr. Lardner's *Cabinet Cyclopaedia*', *Papers of the Bibliographical Society of America*, 45 (1951), 37–58.
42 S.E. De Morgan, *Memoir*, p. 93.
43 For a study of the SDUK in the publishing world, see Lucy Warwick, 'Print and British Imperialism: The Society for the Diffusion of Useful Knowledge, 1826–46' (Unpublished Ph.D. Diss., Oxford Brookes University, 2016). For a study of the involvement of De Morgan in this society, see Rice, 'Augustus De Morgan and the

the same ideal and, because many of the university's teachers were also members of the SDUK, they cooperated closely. As noted above, De Morgan had offered his services to the SDUK as early as 1827. Following his recruitment to the university in 1828, he became an active member and soon published books and articles under the SDUK aegis.

From January 1833, on the initiative of the SDUK and under the direction of George Long, the founding professor of Greek at London University, Charles Knight published a few pages of a new encyclopaedia every Saturday, priced at one penny. The aptly-named *Penny Cyclopaedia* was thus aimed at the less well-off who could not afford the other, very expensive encyclopaedias. Together with a monthly supplement, the annual cost of a volume amounted to six shillings.[44]

For the first few years the *Penny Cyclopaedia* sold 55,000 copies each week. By the end of the 1830s, however, only 20,000 copies were sold weekly. As the venture neared its conclusion, publication intensified, with the result that poorer readers were less able to keep up financially. Nevertheless, in 1843, the encyclopaedia reached completion in twenty-seven substantial volumes.[45] It advertised itself as having shorter articles under more heads than other encyclopaedias, enabling more rapid retrieval of information.[46] Some contemporary critics saw this plan as providing access to superficial knowledge assembled for convenience. They complained, moreover, that because of the multiplicity of entries which were often very short, readers would be unable to make the link with other acquired knowledge.[47] While most of the entries did not exceed a few columns, some were very long, extending to 86 columns on 'Yorkshire' and 98 columns on 'Rome'. Articles describing animals were also quite long: for example, 46.5 columns on 'Antelope' and 22.5 columns on 'Bear'.

Development of University Mathematics', pp. 108–22; Monica Grobel, 'The Society for the Diffusion of Useful Knowledge, 1826–1846' (Unpublished MA Diss., University of London, 1933).

44 Mark W. Turner, 'Companions, Supplements, and the Proliferation of Print in the 1830s', *Victorian Periodicals Review*, 43 (2010), 119–32 (p. 123).
45 Supplements were published in 1845, 1846, 1851 and 1858.
46 *Penny Cyclopaedia of the Society for the Diffusion of Useful Knowledge*, 27 vols. (London: C. Knight, 1833–1843), vol. 1, p. iii.
47 De Morgan was one of these critics, see De Morgan, *Budget*, pp. 438–48, on p. 442.

De Morgan and the *Penny Cyclopaedia*: A Major Investment

After resigning in 1831 as a university mathematics professor, De Morgan found himself temporarily without salaried employment. Yet he did not remain inactive and compensated for his lack of professorial income partly through heavy involvement in the *Penny Cyclopaedia*.[48] From her husband's copy (apparently no longer extant) and that at the British Museum, Sophia De Morgan was able to reconstruct a list of De Morgan's contributions. Although this list excludes articles on constellations and planets, because of uncertainty regarding their authorship, it nonetheless reveals that De Morgan wrote more than seven hundred articles, approximately one sixth of the entire encyclopaedia.[49]

The list of contributors in the last volume of the *Penny Cyclopaedia* names De Morgan as the author of the articles on astronomy and mathematics. According to Sophia's list, he also covered topics from other areas including physics, scientific instruments and the calendar. Furthermore, he was the main, but not the sole, contributor of mathematical or astronomical articles. For example, Thomas Stephens Davies, mathematical master at the Royal Military Academy in Woolwich, wrote the article on the compass, and perhaps a few others.

In common with most entries in the *Penny Cyclopaedia*, the majority of De Morgan's articles are quite short, with an average length of 1.75 columns (roughly 1,300 words).[50] Thirty per cent are less than fifteen lines long, and half are less than three-quarters of a column. Nevertheless, over one quarter of the articles (twenty-seven per cent) are longer than one page (i.e. more than 1,500 words). Fewer than five per

---

48   Although extant correspondence and documentation relating to De Morgan's involvement with the SDUK contain no details of remuneration, payment for his articles for the *Penny Cyclopaedia* must be assumed in accordance with standard practice. For authorship and remuneration in the periodical press of the time, see Patrick Leary and Andrew Nash, 'Authorship', in *The Cambridge History of the Book in Britain*, 7 vols. (Cambridge: Cambridge University Press, 1999–2019), VI: *1830–1914*, ed. by David McKitterick (2009), pp. 172–213 (pp. 173–74, 178–81).

49   See S.E. De Morgan, *Memoir*, pp. 407–14. George Long's annotated copy of the *Penny Cyclopaedia*, formerly housed in Brighton, has vanished. Attributions can be made from the marked copy in the British Library, 733L.

50   But the standard deviation is 2.9, which means that there is a large disparity between the size of the items. A full page contained two columns.

cent exceed four pages. The longest articles are 'Trigonometrical Survey' (25.5 columns),[51] 'Weights and Measures' (21 columns), 'Astronomy' (18.2 columns), 'Circle, astronomical' (18 columns) and 'Virtual Velocities' (17 columns). Although they are too short to be considered as mini-treatises, they contain a substantial amount of information. In the articles 'Weights and Measures' and 'Astronomy', for example, De Morgan provides a detailed history of each subject and refers more technical or specific questions to other articles.

Forty-four per cent of De Morgan's articles pertain to mathematics or logic; sixteen per cent to each of astronomy and biography. Sixty articles, or eight per cent of the total, are on physics. Articles on weights and measures are also present but in smaller numbers (four per cent), albeit with the second-longest article. Finally come several categories which have few articles, but which nevertheless show De Morgan's interest in these subjects (such as actuarial science, navigation, climatology, architecture, etc.). Over half of the biographical articles (fifty-five per cent) are about mathematicians, with others devoted mainly to astronomers (twenty-seven per cent) and to scientists more generally. The biography of François Viète is more than fifteen columns long: but this is because De Morgan considered previous biographies of him to be in error and because he also added biographical details of Luca Pacioli and Fibonacci. Of the others, De Morgan devoted 6.8 columns to Laplace and 6.5 columns to Bernoulli. Fifty-six per cent of the biographies of mathematicians are less than one column long. Of De Morgan's biographies of astronomers, the longest entry is that about Tycho Brahe with 7.5 columns, the second is Copernicus (6 columns), the third is William Herschel (5.3 columns), then the Cassini dynasty and James Bradley (4 columns each). Physicists account for eight per cent of De Morgan's biographical entries, philosophers for five per cent.[52]

Of course, the category with the most articles is mathematics, within which the dominant subject is geometry, with more than thirty-eight per cent of the articles, followed by algebra (twenty per cent), miscellaneous

---

51 This entry is included in the list provided by Sophia De Morgan, but the list of authors in the last volume of the *Penny Cyclopaedia* names Thomas Galloway as the author.

52 Comparing De Morgan's entries using their lengths in columns rather than the number of articles results in more or less the same proportions.

(i.e. concerning several branches or generalities) with eighteen per cent, arithmetic (thirteen per cent) and analysis (nine per cent). Probability theory is found in only four articles.[53] Two-thirds of the geometry articles are short or very short, being less than one column. With just over ten columns, the article entitled 'Geometry' is the second-longest, behind the article on trochoidal curves (more than fifteen columns).[54] Those concerning algebra are also brief except for two. The first one, 'Viga Ganita' (fourteen columns), presents a survey of the content of this Hindu work on algebra before turning to the topics of Hindu arithmetic and astronomy which had not been covered sufficiently by entries in the previous volumes. The second article, which extends to over fifteen columns, gives an exposition of 'Negative and Impossible Quantities': after stating the main principles and giving some results, De Morgan lists his sources by providing a bibliography of French and British authors, including himself.[55]

Such self-citation by De Morgan in the *Penny Cyclopaedia* was rare, occurring on only three occasions: in the article on 'Negative and Impossible Quantities' above (vol. 16), as well as 'Micrometer' (vol. 15) and 'Probability' (vol. 19). This can perhaps be explained by the fact that De Morgan published his most important writings after the completion of the *Penny Cyclopaedia* in 1843. By contrast, in the *English Cyclopaedia*, which appeared in the 1860s, De Morgan was not only quoted more often by others, but also cited himself in his revised articles.

Only two of the mathematical articles concern specific theorems. One is on Sturm's theorem, which makes it possible to find the number of

---

53   Nevertheless, the article on the theory of probability is still more than eleven columns long. De Morgan starts by giving the main principles and some examples, before stating that it is one of the subjects in which mathematicians are prone to make the most mistakes. He then provides a detailed history of probability and provides bibliographical references, citing the works of Laplace, Condorcet, Lacroix and Poisson, while noting that few English-language works are available, these being almost exclusively his own.

54   Many geometry articles are concerned with particular curves, but there appears to be no obvious reason why the article on trochoidal curves is considerably longer than the others. The engravings of epicycloids in this article were produced by Henry Perigal using a chuck constructed by John Ibbetson, the characteristics of which are given in John Holt Ibbetson, *A Brief Account of Ibbetson's Geometric Chuck, Manufactured by Holtzapffel & Co.* (London: A. Hancock for the author, 1833).

55   The other British authors are Robert Woodhouse, John Warren, Benjamin Gompertz, George Peacock, Davies Gilbert and William Rowan Hamilton; the French ones are Jean-Robert Argand and Augustin-Louis Cauchy.

real roots of an algebraic equation. The first version of this theorem was published in 1829, and it was used very soon afterwards, especially for teaching algebra at the École Polytechnique. Although it was fashionable in France in the 1830s, it is surprising to find an entry on it in a British general encyclopaedia. The only other encyclopaedia of this period to refer to Sturm's theorem was Charles Knight's *English Cyclopaedia*, an updated and expanded version of the *Penny Cyclopaedia* which, along with many other De Morgan-authored articles, carried the piece over verbatim. The other article is on Taylor's theorem, a crucial result in mathematical analysis which expresses a differentiable function in terms of an infinite series. First published by Brook Taylor in 1715, this was the subject of a sixteen-column article which also contained a biography of its author.

Two of De Morgan's articles on mathematics have a special status because they are about books.[56] The first discusses Newton's masterpiece *Philosophiæ Naturalis Principia Mathematica*, and the second the notorious *Commercium Epistolicum*. The former article, on Newton's iconic 1687 work, is over fourteen columns long. In it, De Morgan described the text and commented on it section by section, revealing his interest in Newton's work and its importance more than 150 years after its publication. The latter article is only one and a half columns long, but its significance is far greater. The *Commercium Epistolicum*, published by the Royal Society in 1712, was the highly partisan report used by Newton and his English supporters to disseminate unfounded allegations of plagiarism against the German mathematician Gottfried Leibniz, with respect to the invention of the calculus. De Morgan was subsequently to devote considerable time to detailed research on the calculus priority dispute, revealing previously unknown flaws in Newton's personality in the process, such that this short article served as a prelude to his rehabilitation of Leibniz's mathematical reputation in Britain.

What conclusions can then be drawn from the mass of information contributed by De Morgan to the *Penny Cyclopaedia*? What can explain the extreme length of certain articles? What factors determined the topics on which De Morgan wrote his entries? And what characteristics of De Morgan can be seen through his articles? Firstly, it must be remembered

---

56 The Viga Ganita article is also about a book, but it goes well beyond this with a lengthy account of Hindu astronomy.

that, as a ten-year commercial enterprise, the whole project was constantly having to re-adjust and re-calibrate in response to a variety of circumstances, from the death of contributors to the need to correct mistakes or omissions in previous volumes. Consequently, articles originally destined for inclusion on a certain subject sometimes missed the deadline for their intended volume and had to be incorporated into another (usually related) entry. For example, De Morgan's article on the 'Viga Ganita' was artificially lengthened to accommodate material on Hindu arithmetic and astronomy that should have appeared in an earlier volume. Thus, articles of above-average length were often the result of the merging of multiple topics under the same heading.

Secondly, it is clear that the topics covered in De Morgan's *Penny Cyclopaedia* articles are consistent with his scholarly interests: a significant proportion of the entries concerned subjects which related directly to his teaching, research, prior (and future) publications, professional activities or other academic concerns. And although the subjects of a large number, probably the majority, of his entries would have been commissioned by the editor, there is no doubt that De Morgan's own idiosyncrasies played a not insignificant role in the selection of several topics. Who but De Morgan would have considered a publication as recondite as the *Commercium Epistolicum* to be an appropriate entry for inclusion in an encyclopaedia of general knowledge? But of course this is not surprising given his strong interest, not just in mathematics and science, but in their historical development. Moreover, in common with his later books and research articles, his articles in the *Penny Cyclopaedia* contributed more generally to the dissemination of mathematical and astronomical knowledge via their history.[57]

Finally, one has only to read a few of the articles themselves instantly to recognise De Morgan's voice. His literary style in this format is distinctive, simultaneously both formal and accessible. His pedagogical talent shines through clarity of exposition, which combines with good use of examples and a pithy turn of phrase to make even the driest of subjects quite readable. We see him also as an innovator, for it was in an article in volume 12 that he introduced the term 'induction' to describe a method of proof that is now universally used by mathematicians

---

57 See Rice, 'Augustus De Morgan: Historian of Science'.

across the globe. But above all, the *Penny Cyclopaedia* articles show De Morgan to have been early Victorian Britain's most gifted mathematical expositor. And as such, they enabled him to communicate his love of mathematics and its history to the masses.

Conclusion

Augustus De Morgan wrote prolifically from the start of his career to the end of his life. With more than 1,400 articles, notes, reviews and over seven hundred encyclopaedia entries, he was clearly interested in many subjects, but regularly focused on mathematics, logic and astronomy. His strong interest in history and biography pervades all of his writing, featuring not only in many of his research papers, but also permeating many of his expository articles. The result is a distinctive style of mathematical discourse quite different from most of his contemporaries, in Britain at least.

De Morgan's participation in journals increased the number of his monographs. The most salient example is his posthumous *A Budget of Paradoxes*, which arose from his regular columns in *The Athenæum*. De Morgan's final book, *Newton, his Friend, and his Niece*, a historical investigation of the marital status of Isaac Newton's niece, Catherine Barton, was the extension of an article intended for, and rejected by, the *Companion to the Almanac*.[58]

To the extent that De Morgan was known to the general Victorian public, it would have been primarily as an author of textbooks which were well received and commercially successful, and which displayed his name prominently. Most readers of his more popular expository articles would have been unaware of his authorship, as they were mostly published anonymously. Only among the clique of scientific cognoscenti would his authorship presumably have been a well-kept but open secret, while practising contemporaneous mathematicians knew and respected him for his research papers in prestigious journals.

Today, mathematicians remember De Morgan principally for De Morgan's Laws and historians of science for his *Budget of Paradoxes*, both

---

58 Augustus De Morgan, *Newton, his Friend, and his Niece*, ed. by S.E. De Morgan and A.C. Ranyard (London: Elliot Stock, 1885). For discussion of this, see S.E. De Morgan, *Memoir*, p. 264.

of which have their roots in journal articles.[59] Thus, while much of his writing is now long forgotten, De Morgan's greatest claims to fame can both be traced back to the periodicals of yesteryear.

Bibliography

Alborn, Timothy, 'A Calculating Profession: Victorian Actuaries among the Statisticians', *Science in Context*, 7 (1994), 433–68. https://doi.org/10.1017/S0269889700001770

Arcavi, Abraham and Maxim Bruckheimer, 'The Didactical De Morgan: A Selection of Augustus De Morgan's Thoughts on Teaching and Learning Mathematics', *For the Learning of Mathematics*, 9 (1989), 34–39.

Brake, Laurel, and Marysa Demoor, eds, *Dictionary of Nineteenth-Century Journalism in Great Britain and Ireland* (London: British Library, 2009).

Bruneau, Olivier, 'Encyclopaedias as Markers of Heritage Building: Fluxion Articles in British Encyclopaedias, 1704–1860', *Philosophia Scientiæ*, 26:2 (2022), 67–90. https://doi.org/10.4000/philosophiascientiae.3483

Burke, Peter, *A Social History of Knowledge: From Gutenberg to Diderot* (Cambridge: Polity, 2000).

Collison, Robert Lewis, *Encyclopaedias: Their History Throughout the Ages* (London and New York: Hafner, 1966).

Crilly, Tony, 'The *Cambridge Mathematical Journal* and its Descendants: The Linchpin of a Research Community in the Early and Mid-Victorian Age', *Historia Mathematica*, 31 (2004), 407–97. https://doi.org/10.1016/j.hm.2004.03.001

De Morgan, Augustus, *A Budget of Paradoxes* (London: Longmans, Green, 1872).

Despeaux, Sloan Evans, 'A Voice for Mathematics, Victorian Mathematical Journals and Societies', in *Mathematics in Victorian Britain*, ed. by Raymond Flood, Adrian Rice and Robin Wilson (Oxford: Oxford University Press, 2011), pp. 155–74.

Despeaux, Sloan Evans and Adrian C. Rice, 'Augustus De Morgan's Anonymous Reviews for *The Athenæum*: A Mirror of a Victorian

---

59 De Morgan's Laws appeared, for example, in Augustus De Morgan, 'On the Syllogism III, and on Logic in General', *Transactions of the Cambridge Philosophical Society*, 10 (1858), 173–230 (p. 208). The multiple articles from *The Athenæum* (1863–67) that were later re-published in *A Budget of Paradoxes* (1872) are listed in Chapter 12 of this volume.

Mathematician', *Historia Mathematica*, 43 (2016), 148–71. https://doi.org/10.1016/j.hm.2015.09.001

Gibson, Susannah, *The Spirit of Inquiry: How One Extraordinary Society Shaped Modern Science* (Oxford: Oxford University Press, 2019).

Grobel, Monica, 'The Society for the Diffusion of Useful Knowledge, 1826–1846' (Unpublished MA Diss., University of London, 1933).

Hall, A. Rupert, *The Cambridge Philosophical Society, A History, 1819-1969* (Cambridge: Cambridge Philosophical Society, 1969).

Ibbetson, John Holt, *A Brief Account of Ibbetson's Geometric Chuck, Manufactured by Holtzapffel & Co.* (London: Printed for the author, by A. Hancock, 1833).

Kafker, Frank A. (ed.), *Notable Encyclopaedias of the Seventeenth and Eighteenth Centuries: Nine Predecessors of the* Encyclopédie (Oxford: Voltaire Foundation, 1981).

—, *Notable Encyclopaedias of the Late Eighteenth Century: Eleven Successors of the* Encyclopédie (Oxford: Voltaire Foundation, 1994).

Kafker, Frank A. and Jeff Loveland, eds, *The Early* Britannica*: The Growth of an Outstanding Encyclopaedia* (Oxford: Voltaire Foundation, 2009).

Kogan, Herman, *The Great EB: The Story of the Encyclopaedia Britannica* (Chicago: University of Chicago Press, 1958).

Leary, Patrick, 'A Victorian Virtual Community', *Victorian Review*, 25(2) (2000), 61–79. https://doi.org/10.1353/vcr.2000.0032

Leca-Tsiomis, Marie, *Écrire l'Encyclopédie* (Oxford: Voltaire Foundation, 1999).

Loveland, Jeff, 'Two Partial English-Language Translations of the *Encyclopédie*: The Encyclopaedias of John Barrow and Temple Henry Croker', in *British-French Exchanges in the Eighteenth Century*, ed. by Kathleen Hardesty Doig and Dorothy Medlin (Newcastle: Cambridge Scholars Publishing, 2007), pp. 168–87.

—, *The European Encyclopaedia from 1650 to the Twenty-First Century* (Cambridge: Cambridge University Press, 2019). https://doi.org/10.1017/9781108646390

Peckham, Morse, 'Dr. Lardner's *Cabinet Cyclopaedia*', *Papers of the Bibliographical Society of America*, 45 (1951), 37–58.

Rice, Adrian, 'Augustus De Morgan: Historian of Science', *History of Science*, 34 (1996), 201–40. https://doi.org/10.1177/007327539603400203

—, 'Augustus De Morgan and the Development of University Mathematics in London in the Nineteenth Century' (Ph.D. Diss., Middlesex University, 1997).

—, 'What Makes a Great Mathematics Teacher? The Case of Augustus De Morgan', *The American Mathematical Monthly*, 106 (1999), 534–552. https://doi.org/10.2307/ 2589465

—, '"Everybody Makes Errors": The Intersection of De Morgan's Logic and Probability, 1837–1847', *History and Philosophy of Logic*, 24 (2003), 289–305. https://doi.org/10.1080/01445340310001599579

Shuttleworth, Sally and Geoffrey Cantor, 'Introduction', in *Science Serialized: Representations of the Sciences in Nineteenth-Century Periodicals*, ed. by Geoffrey Cantor and Sally Shuttleworth (Cambridge, MA: MIT Press, 2004), pp. 1–16. https://doi.org/10.7551/mitpress/6080.003.0001

Stephen, Leslie, 'De Morgan, Augustus (1806–1871)', *Dictionary of National Biography*, vol. 14 (London: Smith, Elder, 1888), pp. 331–34.

Stray, Christopher, 'William Smith and his Dictionaries: A Study in Scarlet and Black', in *Classical Books: Scholarship and Publishing in Britain since 1800*, ed. by Christopher Stray (Bulletin of the Institute of Classical Studies, Supplement 101 (London: Institute of Classical Studies, University of London, 2007), pp. 35–54. https://doi.org/10.1111/ j.2041-5370.2013.tb02524.x

Turner, Mark W., 'Companions, Supplements, and the Proliferation of Print in the 1830s', *Victorian Periodicals Review*, 43 (2010), 119–32. https://doi.org/10.1353/ vpr.0.0116

Walsh, S. Padraig, *Anglo-American General Encyclopaedias, A Historical Bibliography, 1703–1967* (New York and London: Bowker, 1968).

Warwick, Lucy, 'Print and British Imperialism: The Society for the Diffusion of Useful Knowledge, 1826–46' (Ph.D. Diss., Oxford Brookes University, 2016).

Yeo, Richard, 'Reading Encyclopaedias: Science and the Organization of Knowledge in British Dictionaries of Arts and Sciences, 1730–1850', *Isis*, 82 (1991), 24–49.

—, *Encyclopedic Visions, Scientific Dictionaries and Enlightenment Culture* (Cambridge: Cambridge University Press, 2001).

Zabell, S. L., 'De Morgan and Laplace: A Tale of Two Cities', *Journal électronique d'Histoire des Probabilités et de la Statistique*, 8 (2012), 1–29, www.jehps.net.

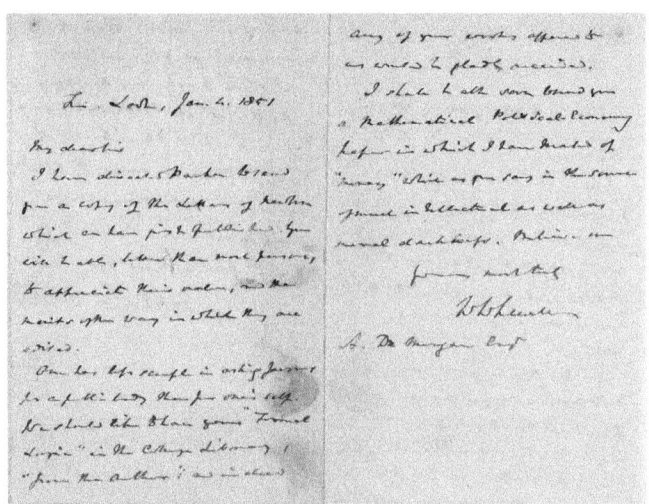

Fig. 7 In this letter from January 1851, tipped in *Correspondence of Sir Isaac Newton and Professor Cotes* (1850), William Whewell tells De Morgan that he has directed the publisher to send him a copy of this recently published edition of Newton's letters, requests a copy of De Morgan's *Formal Logic* for the Trinity College Library 'from the author', and promises to send him a new paper on the subject of money, 'which as you say is the source of much intellectual as well as moral darkness'. ([DeM] L [Newton] SSR, reproduced by permission of Senate House Library, University of London.)

5. Augustus De Morgan: Meta-Scientific Rebel

Lukas M. Verburgt

> I shall be amused if you succeed in persuading the world that Bacon had little to do with the modern progress of science.
>
> —William Whewell to Augustus De Morgan[1]

Baconianism and the British Meta-Scientific Tradition

Augustus De Morgan lived in what for science and philosophy were interesting times. During the so-called 'Second Scientific Revolution',[2] natural philosophy transformed into science which, in turn, was slowly divorced from philosophy. Looking at the world scientifically or philosophically eventually became two very different things, except for the polymath, a special kind of scholarly persona which for several decades remained an intellectual possibility, albeit an

---

1 Letter from William Whewell to Augustus De Morgan, 18 January 1859, Trinity College, Cambridge (henceforth TCC), Whewell Papers, O.15.47/25. Adapted with the permission of the University of Chicago Press from Verburgt, "Scientific Method, Induction, and Probability: The Whewell–De Morgan Debate on Baconianism, 1830s–1850s," published originally in *HOPOS: The Journal of the International Society for the History of Philosophy of Science*. © 2024 International Society for the History of Philosophy of Science. All rights reserved.

2 For more or less obvious reasons—think only of the doubts that have been raised about the meaningfulness of the term 'Scientific Revolution'!—this label has never really stuck or been much used. Still, Enrico Bellone's book on it, published under the general title of *A World on Paper*, contains a lot that is of interest. See Enrico Bellone, *A World on Paper: Studies on the Second Scientific Revolution* (Cambridge, MA: The MIT Press, 1980 [1976]).

increasingly problematic one.[3] Indeed, with the notion of science not yet a straightforward or finished matter, and philosophical reflection on science not yet separated from the actual practice of science, it is unsurprising that the most important commentaries on science in this period came from polymathic figures, such as Herschel, Brewster and Whewell in Britain, Comte, Bernard and Poincaré in France and Liebig and Helmholtz in Germany. The growing rift between science and philosophy in the first half of the nineteenth century can be seen, for example, in the fact that science was often defined by pitting its virtues against the vices of philosophy. John Herschel, in his wildly popular *Preliminary Discourse on the Study of Natural Philosophy* (1830) praised the 'experimental philosopher' by contrasting him with the 'speculative philosopher', writing that only the former's thinking is 'grounded in the realities of nature' and governed by clear 'principles'.[4] Similar examples of writers contrasting scientists and philosophers can be found in the British, German and French contexts, which arguably stand in need of comparative analysis. It makes for a fascinating chapter of a larger story about the continued entanglement of science and philosophy, even as they were being prised apart in the aftermath of natural philosophy.

Within this sweeping process, the polymathic field of 'meta-science', of which Herschel and William Whewell were the major representatives in Britain, alongside David Brewster, Baden Powell and John Stuart Mill, indeed played an interesting double role. Neither philosophy nor science, but still a little bit of both, meta-science[5] or, as Whewell sometimes called it, philosophy of knowledge, created as much as it filled

---

[3] On the transformation of natural philosophy into the sciences see the essays in David Cahan, ed., *From Natural Philosophy to the Sciences. Writing the History of Nineteenth Century Science* (Chicago and London: The University of Chicago Press, 2003). A history of the downfall of the polymath as a scholarly persona has still to be written. The following works provide useful starting points: Peter Burke, *The Polymath: A Cultural History from Leonardo da Vinci to Susan Sontag* (New Haven and London: Yale University Press, 2020) and Jeroen van Dongen and Herman Paul, eds, *Epistemic Virtues in the Sciences and the Humanities* (Cham: Springer, 2017).

[4] John F.W. Herschel, *A Preliminary Discourse on the Study of Natural Philosophy* (London: Longman, Rees, Orme, Brown & Green, and John Taylor, 1830), p. 12.

[5] On this term see Richard Yeo, *Defining Science: William Whewell, Natural Knowledge and Public Debate in Early Victorian Britain* (Cambridge: Cambridge University Press, 1993), especially Chapter 3. For a critical note on the term see Steffen Ducheyne, 'Whewell's philosophy of science', in *The Oxford Handbook of British Philosophy in the Nineteenth Century*, ed. by W.J. Mander (Oxford and New York: Oxford University Press, 2014), pp. 71–88 (pp. 84–85).

the vacuum left by natural philosophy between the old ('traditional') philosophy and the new science. In brief, it promoted science as the producer of all 'real', that is, stable and certain knowledge and made it philosophy's (partly epistemological and partly methodological) task to study the nature and conditions of its product. Rather than through a transcendental inquiry, in the good Kantian fashion, which by the early nineteenth century had just crossed the Channel, meta-scientists turned to history to explain the possibility of scientific knowledge. This is to say that philosophical study took the form of systematic reflection on progress in the physical sciences. 'We purpose to collect our doctrines concerning the nature of knowledge, and the best mode of acquiring it,' wrote Whewell, 'from a contemplation of the structure and history of those sciences ..., which are universally recognised as the clearest and surest examples of knowledge.'[6] Despite all their underlying disagreements, this is what the towering figures of Herschel and Whewell are believed to have in common: unlike their shared hero Francis Bacon, they were able to ground their philosophy of science on the actual history of the sciences, as these had successfully developed since the time of Isaac Newton. This opportunity came with the demand to pursue meta-science as a combination of philosophically-informed history of science and historically-informed philosophy of science. The programmatic ambition was to renovate Bacon's *Novum Organum* (1620) by first unearthing 'the larger features of [science's] formation', then systematizing these historical features as philosophical principles, and finally showing that these principles were 'exemplified in the history of [science's] progress'.[7] The new category of 'science' was canonised, and the 'scientist' was invented, in the work of the meta-scientists as a result of a historical-philosophical looping effect: philosophy explained what history showed through philosophical reflection on the historical record.

Among meta-scientists, the central feature of science that set it apart from other historical forms of knowledge was widely believed to be its use of a single scientific method. This was not only what made possible science's progress, but also what accounted for science's unity.

---

6   William Whewell, *History of Scientific Ideas. Volume I* (London: John W. Parker, 1858), p. 8.
7   See Herschel, *Preliminary Discourse*, Part II.

'The advances which have, during the last three centuries, been made in the physical sciences,' wrote Whewell in his 1858 *Novum Organon Renovatum*, 'these are allowed by all to be real, to be great, to be striking;

> may it not be that the steps of progress in these different cases have in them something alike? May it not be that in each advancing moment of such knowledge there is some common principle, some common process? May it not be that discoveries are made by an *Organon* ['Instrument'] which has something uniform in its working?[8]

The scientific method was seen as the very hallmark of science and, consequently, discussions on methodology stood at the heart of the meta-scientific tradition. During what C.S. Peirce once called the 'Age of Methods',[9] meta-scientists across Europe set out to philosophise scientific methodology. For the first time in the long history of philosophy, 'entire books rather than prefaces or chapters'[10] were devoted exclusively to the subject of the principles and rules of scientific inquiry. This large body of work is commonly seen to have been instrumental in the establishment of science as both a specific realm of knowledge and as a dominant way of knowing, teachable to all ('accessible'), common to all branches of science ('single'), and extrapolatable from physical science to any other field ('transferable').[11]

Within Britain in the first half of the nineteenth century, debates on scientific methodology took place against the background of new Bacon scholarship, up to the point of being indistinguishable from it. Bacon was studied almost exclusively as a theorist of method and every account of scientific method—indeed, any new scientific field—had to be at least 'ceremonially Baconian'[12] if it aspired to respectability.

---

8 William Whewell, *Novum Organon Renovatum* (London: John W. Parker, 1858), p. iv.
9 See Henry Cowles, *The Scientific Method: An Evolution of Thinking from Darwin to Dewey* (Cambridge, MA: Harvard University Press, 2020).
10 Larry Laudan, 'Theories of Scientific Method from Plato to Mach: A Bibliographic Review', *History of Science*, 7:1 (1968), 1–63 (p. 29).
11 See Richard Yeo, 'Scientific Method and the Rhetoric of Science in Britain', in *The Politics and Rhetoric of Scientific Method. Histories Studies*, ed. by John A. Schuster and Richard R. Yeo (Dordrecht & Boston: D. Reidel Publishing Company, 1986), pp. 259–97 (p. 262).
12 Jonathan Smith, *Fact and Feeling: Baconian Science and the Nineteenth-Century Literary Imagination* (Madison: The University of Wisconsin Press, 1994), p. 15.

More precisely, meta-scientific writings on method often amounted to a revision of Bacon's canons of induction, sometimes implicitly and at other times explicitly, as in the case of Whewell's *Novum Organum Renovatum* (1858). This does not merely suggest that particular views on methodology always went hand in hand with, and sometimes even coincided with, specific interpretations of Bacon. It also means that the debates in which these views were put forward were themselves shaped by tacit yet dominant Baconian assumptions about the nature of science and the aims, scope and limits of methodology. Among these assumptions was the fundamental idea that scientific knowledge is acquired through induction. Because everyone also agreed that Bacon's conception of induction was essentially flawed, one main challenge for meta-scientists was that of defining what it actually was.

The reason for the dominance of Baconianism in early nineteenth-century Britain was the all-pervasive influence of Whewell and Herschel, whose commitment to Bacon can be traced back to their student days and whose mature writings can be read as a struggle over who was Bacon's legitimate heir. Other reasons, which tellingly enough have been unearthed largely on the basis of studies of Whewell's and Herschel's life and work, all have to do with the fact that positions on methodology were part of a more wide-ranging set of debates on the nature of science.[13] As such, different takes on Baconian induction as the method of science came with different commitments on broader topics ranging from the organization and religious and social implications of science to the moral character of its practitioners. Or, *vice versa*, as illustrated for instance by Whewell's opposition to the Oxford Noetics, these wider commitments were often defended in terms of abstract methodological considerations.

Perhaps the best example of the dominance of Baconianism is that disagreements within the meta-scientific debates over methodology took place against a shared background of agreement. Put more strongly, even Baconian revisionism, however non-Baconian in appearance, was pursued in terms of a commitment to Baconian induction. Whewell,

---

13 See, for example, Laura J. Snyder, *Reforming Philosophy: A Victorian Debate on Science and Society* (Chicago: The University of Chicago Press, 2006); and Richard Yeo, *Defining Science: William Whewell, Natural Knowledge and Public Debate in Early Victorian Britain* (Cambridge: Cambridge University Press, 1993).

Herschel and Mill had very different views on what induction is, but while each was convinced of the shortcomings of Bacon's canons of induction, each saw his task as being that of renovating these canons. As always, the force of the *communis opinio* became most apparent when it was challenged. This was already the case when self-declared Baconians like Whewell and Herschel accused each other of diverting too much from Baconian tradition; Whewell when he was being 'too a priori' and Herschel when he allowed for too much speculative hypothesising. It was, of course, even more so in the case of those who self-identified as anti-Baconian.

British Anti-Baconianism

Here, De Morgan enters the picture. Like David Brewster and Charles Babbage before him and William Stanley Jevons after him, De Morgan was among the few prominent British meta-scientists who advocated anti-Baconianism, thereby occupying a somewhat anomalous or rebellious position towards the dominant British meta-scientific tradition. Much like Baconianism itself, as a 'counter-tradition' anti-Baconianism was highly heterogeneous. For example, De Morgan, who was not 'among the strongest supporters of Bacon', defended Bacon on some points, and Richard Whately, who frequently scoffed at Bacon's inductive logic, published an annotated edition of Bacon's *Essays*.[14] Perhaps the best definition that can be given of British anti-Baconianism, one that (luckily!) suffices for the aims of this chapter, is that its members opposed the philosophical core and the historical significance of Baconian induction, understood as a 'new method of arriving at truth'.[15] What followed from this rejection, and what came in its place, is much harder to determine, and differed almost

---

14 Augustus De Morgan, 'The Progress of the Doctrine of the Earth's Motion between the Times of Copernicus and Galileo, Being Notes on the Ante-Galilean Copernicans', *Companion to the Almanac for 1855*, 5–25 (p. 11); Richard Whately, *Bacon's Essays: With Annotations* (London, 1856). Interestingly, several nineteenth-century authors actually saw Whately as a contemporary Bacon. See, for example, William John Fitzpatrick, *Memoirs of Richard Whately, Archbishop of Dublin. Volume I* (London: Richard Bentley, 1864), p. 55 and pp. 325–26.

15 Augustus De Morgan, *Formal Logic: Or, The Calculus of Inference, Necessary and Probable* (London: Taylor & Walton, 1847), p. 216.

from anti-Baconian to anti-Baconian. This probably explains, yet of course does not justify, the lack of any sustained, book-length account of British anti-Baconianism in the early- and mid-nineteenth century. On the basis of primary and secondary sources scattered across time and disciplines—from nineteenth-century logical treatises to recent studies of scientific biography—it should be possible, however, to begin sketching its contours. Doing so is worthwhile for at least two reasons. First, to bring into view an important and, in hindsight, pioneering meta-scientific current in early Victorian Britain. Second, to obtain a fuller and richer understanding of the intellectual landscape in this fascinating period.

A preliminary step to this larger project will be taken in this chapter by focusing on De Morgan as a prominent advocate of anti-Baconianism, and more specifically on his anti-Baconian scientific methodology. Despite the prominence and influence of British anti-Baconians, there exist relatively few studies which engage them as meta-scientists. Menachem Fisch and Simon Schaffer's *William Whewell: A Composite Portrait*, Richard Yeo's *Defining Science*, Pietro Corsi's *Science and Religion*, Nicholas Capaldi's *John Stuart Mill: A Biography*, and Laura Snyder's *Reforming Philosophy* provide in-depth, contextualizing accounts of Whewell, Mill and Baden Powell.[16] No similar books on Whately or Brewster are available yet. The same goes for De Morgan, who stands out even among these men for rarely, if ever, being recognized as a meta-scientist or, to use modern terminology, a historian and philosopher of science. There are myriad papers and chapters on what can be taken to be aspects of De Morgan's meta-scientific outlook on methodology—several on logic, a few on probability theory and statistics, and a small handful on history of science and history of mathematics—but none in which these are brought together.[17] This is not altogether surprising. One reason concerns the current disciplinary boundaries between the

---

16   It may here be remarked that Babbage's and Herschel's work as meta-scientists, or even as scientific methodologists, is also curiously little studied. But see *The Cambridge Companion to John Herschel*, ed. by Stephen Case and Lukas M. Verburg (Cambridge: Cambridge University Press, 2024).

17   A valuable exception is Joan L. Richards, *Generations of Reason: A Family's Search for Meaning in Post-Newtonian England* (New Haven: Yale University Press, 2021), although its focus seems to be less on De Morgan's meta-scientific than on his personal outlook on science.

history of mathematics, history of logic, history of science and history of philosophy, which need to be crossed in order to bring De Morgan's anti-Baconianism into view. Another reason is that De Morgan himself never wrote a book which combined the meta-scientific elements of his thinking into an integrated outlook on science and its methodology. Within De Morgan's large oeuvre, logic and probability existed adjacent to the history of science without often explicitly intersecting. De Morgan made connections only very occasionally, and when he did, it was mostly in reviews or private correspondence.

The fact that these meta-scientific connections in De Morgan's work exist, and that it is therefore possible and fruitful to think of him as a meta-scientist, can be borne out in a number of ways. The route chosen here is to focus on De Morgan's interactions on topics related to scientific methodology with William Whewell, Master of Trinity College, Cambridge, leading meta-scientist of the early Victorian era, and author of such epoch-making works as *Astronomy and General Physics Considered with Reference to Natural Theology* (1833), *History of the Inductive Sciences* (1837) and *The Philosophy of the Inductive Sciences* (1840).[18] Drawing on their published work and largely unpublished correspondence, several major points of conflict will be identified and interpreted in terms of a friendly clash over Bacon and Baconianism, itself reflective of a larger shift within nineteenth-century debates on scientific method.

De Morgan and Whewell: Scientific Friends, Meta-Scientific Rivals

Whewell's and De Morgan's personal connection began as one of teacher and pupil at Trinity College, Cambridge, where Whewell was fellow

---

18 Another possible route would be to focus on De Morgan's views on Bacon's and, especially, Newton's personality, which could be contrasted with Whewell's views on this topic. See, in this regard, Richard Yeo, 'Genius, Method, and Morality: Images of Newton in Britain, 1760–1860', *Science in Context*, 2.2 (1988), 257–84. Maria Panteki has also provided a comparative analysis of De Morgan and Whewell, but her account focuses on their respective views on mathematics education. See Maria Panteki, 'French "Logique" and British "Logic": On the Origins of Augustus De Morgan's Early Logical Inquiries, 1805–1835', in Dov M. Gabbay and John Woods, eds, *Handbook of the History of Logic. Volume 4: British Logic in the Nineteenth Century* (Amsterdam: North-Holland, 2008), pp. 381–457.

and head tutor by the time that De Morgan entered as a student there in February 1823. Like Whewell, who had graduated Second Wrangler in 1816, De Morgan began his career conventionally as a (more or less) serious reading man, coming out Fourth Wrangler in 1827. But unlike Whewell, who climbed the ladder at his alma mater all the way from sub-sizar to Knightbridge Professor of Moral Philosophy (1838-55) and Vice-Chancellor (1842-55), De Morgan afterwards pursued an unconventional university career. When in 1827 he rejected the religious tests necessary to obtain a fellowship or a regular post, he knew there was no future for him at Cambridge. Instead, in 1828, De Morgan was appointed to the first Chair of Mathematics at the newly founded, religiously neutral London University, where he rapidly developed into a successful teacher and prolific writer.

During the 1830s–60s, when both men were at the height of their powers, Whewell and De Morgan stood on almost opposite sides on the intellectual, political and social landscape. Whewell was 'a high Tory Anglican' who made it his life's task to defend the 'elite exclusivity' of Oxbridge, whereas De Morgan was a religious radical 'committed to educating all of England's people'.[19] Perhaps the single most telling fact, in this regard, is that Whewell always remained behind the walls of Trinity College, Cambridge and De Morgan, like Babbage before him, moved to metropolitan London. This difference was reflected in the many contrasting aspects of their lives and work, whether it was the kind of mathematics they pursued—traditional British mixed mathematics versus formal Continental analysis—the type of publication venue through which they communicated their views—text-books and relatively expensive treatises versus hundreds of contributions to the cheap *Penny Cyclopaedia*—or the reasons they had for criticising the Royal Society—its inability to guard against scientific charlatans versus its failure to replace aristocratic dilettantes.[20]

---

19   Richards, *Generations of Reason*, p. 245. See also Crosbie Smith and M. Norton Wise, *Energy and Empire: A Biographical Study of Lord Kelvin* (Cambridge: Cambridge University Press, 1989), Chapter 6.

20   The number of essays relevant to these differences is enormous; particularly important are Timothy L. Alborn, 'The Business of Induction: Industry and Genius in the Language of British Scientific Reform, 1820–1840', *History of Science*, 34:1 (1996), 91–121; William J. Ashworth, 'The Calculating Eye: Baily, Herschel, Babbage and the Business of Astronomy', *The British Journal for the History of Science*, 27:4 (1994), 409–41; and Richard Yeo, 'Genius, Method, and Morality'.

Nonetheless, from the early 1830s onwards, Whewell and De Morgan were 'scientific friends' with an epistolary relationship.[21] Their correspondence, which started in 1832 and lasted until 1866, the year of Whewell's death, shows an intellectual kinship based on shared interests in, rather than doctrinaire agreement on, a wide-ranging set of topics, including Kantian philosophy, history of science, especially Newton, and Aristotelian logic. The fact that this kinship could blossom despite all their differences of opinion seems to have been due to two factors. First, De Morgan was ultimately sympathetic to certain viewpoints standing at the heart of Whewell's vision, save for the religious-conservative implications Whewell attached to them: a romantic *idealism* that held truth to be grounded in ideas, seen as products of the genius's mind, a *gradualism* that saw the human understanding of those ideas developing over time, and an advocacy of mathematics as a tool for training the mind to arrive at 'necessary truths' on the basis of clear and precise reasoning.[22] Second, Whewell seems not only to have appreciated De Morgan for his obvious talents, but also to simply have liked him for his wit and humour, which De Morgan felt comfortable enough to let flow freely in his letters to Whewell. This comes out especially strongly when De Morgan's letters are compared with Whewell's correspondence with someone like Robert Leslie Ellis, another former pupil whose vision was much more Whewellian than De Morgan's but who never achieved a similar kind of intimacy.[23] Ellis always closed his letters with a 'your humble servant'; for De Morgan, his initial and surname sufficed.

It is possible to identify some more direct and specific mutual influences between Whewell and De Morgan. However, it is important to recognise from the outset that these should neither be overstretched nor assumed to have been premised on or resulted in any sweeping

---

21   Isaac Todhunter, *William Whewell D.D., Master of Trinity College, Cambridge. An Account of His Writings. With Selections from His Literary and Scientific Correspondence*, vol. 1 (London: Macmillan, 1876), p. 60.
22   See, for example, Augustus De Morgan to William Whewell, 30 April 1844, TCC, Add.Ms.a.202/100, and Augustus De Morgan to Robert Leslie Ellis, 24 June 1854, TCC, Add.Ms.c.67/111.
23   See Lukas M. Verburgt, ed., *A Prodigy of Universal Genius: Robert Leslie Ellis, 1817–1859* (New York: Springer, 2022), Part II ('Letters').

agreement. As in the case of the 'Cambridge Network', 'Breakfast Club' or 'Analytical Society', of which Whewell and De Morgan are sometimes said to have been like-minded members, the underlying differences matter more than the apparent similarities.[24] One way to start bringing these out is to unearth their clash over (anti-)Baconianism in the history and philosophy of science. Like a nineteenth-century Aristotle and Galileo, to use a good old Kuhnian phrase, Whewell and De Morgan could look at the same thing and see something entirely different. Rather than a pendulum, in their case this became most apparent when they were looking at that thing called 'the scientific method'.

De Morgan, Whewell and Nineteenth-Century British Logic

The history of scientific method in nineteenth-century Britain begins with Richard Whately's widely popular *Elements of Logic*, first published as a book in 1826 and appearing in many reprint editions throughout the nineteenth and early twentieth centuries.[25] By the early nineteenth century, the study of formal (i.e. deductive, Aristotelian or syllogistic)[26] logic in Britain had endured, in the words of Sir William Hamilton, a century and a half of 'perversion and neglect'.[27] Its decline had been

---

24 This is one of the important takeaways from William J. Ashworth's *The Trinity Circle: Anxiety, Intelligence and Knowledge Creation in Nineteenth-Century England* (Pittsburgh: University of Pittsburgh Press, 2021). On the 'Cambridge Network', 'Breakfast Club' and 'Analytical Society' see, respectively, W.F. Cannon, 'Scientists and Broad Churchmen: An Early Victorian Intellectual Network', *Journal of British Studies*, 4:1 (1964), 65–88; Laura J. Snyder, *The Philosophical Breakfast Club: Four Remarkable Friends Who Transformed Science and Changed the World* (New York: Broadway Books, 2011); P.C. Enros, 'The Analytical Society (1812–13): Precursor of the Renewal of Cambridge Mathematics', *Historia Mathematica*, 10:1 (1983), 24–47.

25 On Whately's *Elements of Logic* see James Van Evra, 'Richard Whately and Logical Theory', in *Handbook of the History of Logic. Volume 4: British Logic in the Nineteenth Century*, ed. by Dov M. Gabbay and John Woods (Amsterdam: North-Holland, 2008), pp. 75–92, and Calvin Jongsma, 'Richard Whately's Revitalization of Syllogistic Logic', in *Aristotle's Syllogism and the Creation of Modern Logic: Between Tradition and Innovation, 1820–1930*, ed. by Lukas M. Verburgt and Matteo Cosci (London: Bloomsbury Academic, 2023).

26 For Whately, deductive reasoning meant syllogistic reasoning only. Hence, in his *Elements of Logic*, logic is synonymous with Aristotle's syllogism.

27 Sir William Hamilton, 'IV. – Logic. In Reference to the Recent English Treatises on that Science', in Sir William Hamilton, *Discussions on Philosophy and Literature,*

due to a complex combination of factors, but a key role was played by Francis Bacon's *The Great Instauration*, alongside John Locke's *An Essay on Human Understanding*.[28] The second part of Bacon's six-part programme, the *Novum Organum*, which took its title from Aristotle's work on logic, the 'Organon', argued that the cornerstone of traditional deductive logic—the Aristotelian syllogism—was useless for the pursuit of natural knowledge. Instead, the *Novum Organum* introduced a system of inductive reasoning to supersede Aristotle's, suitable for the modern age of the 'sciences of nature'. Where Aristotle's old system, based on syllogisms, derived conclusions which were logically consistent with an argument's premises, Bacon's new system investigated the fundamental premises themselves on the basis of inductive inference from the data ('natural histories') of the natural world. Following Locke and Bacon, writing in the seventeenth century, eighteenth-century Scottish Common Sense philosophers like Thomas Reid and Dugald Stewart ridiculed syllogistic logic, finding in Newton's *Principia Mathematica* an exemplar of sound inductive reasoning.[29]

Whately's *Elements of Logic* was successful in reviving the study of deductive logic not because of its positive definition of what it is, but primarily because of its negative description of what it is *not*. The clear and accessible way in which Whately drew logic's boundaries provided him with solid ground on which to argue that the seventeenth- and eighteenth-century objections all resulted from a failure to recognise logic's nature and scope. In brief, 'by representing Logic as furnishing the sole instrument for the *discovery of truth* in all subjects, and as teaching the use of the *intellectual faculties* in general', Bacon, Locke, and the Common Sense philosophers had 'raised expectations which

---

Education and University Reform. Chiefly from the Edinburgh Review; Corrected, Vindicated, Enlarged, in Notes and Appendices. 2nd edn (London: Longman, Brown, Green & Longmans, 1853), pp. 118–175 (p. 119).

28 For chapters of this history see, for instance, the essays in Marco Sgarbi and Matteo Cosci, eds *The Aftermath of Syllogism: Aristotelian Logical Argument from Avicenna to Hegel* (London: Bloomsbury Academic, 2018).

29 See, in this regard, Larry Laudan, 'Thomas Reid and the Newtonian Turn of British Methodological Thought', in Robert E. Butts and John W. Davis, *The Methodological Heritage of Newton* (Toronto: University of Toronto Press, 2016), pp. 103–131, and Richard S. Olson, *Scottish Philosophy and British Physics, 1740–1870* (Princeton: Princeton University Press, 1975), especially Chapters 9 and 10.

could not be realized'.³⁰ Consequently, not only did deductive logic come to be regarded as 'utterly futile and empty'; sight was also lost of the 'boundless field' of unexplored territory within logic's 'legitimate limits'.³¹ Rather than actually exploring it, Whately took upon himself the task of mapping this territory—that is, 'of completing and properly filling up the masterly sketch' made of it by Aristotle some two thousand years ago.³² On Whately's definition of logic as both a 'science' and an 'art', deductive logic is not just *a* method of reasoning, but *the* method of analysing the mental process involved in all correct reasoning ('science'); similarly, the syllogism is not just *an* argumentative form, but *the* form to which all correct reasoning may be reduced and which thus serves the purpose of a test to try the validity of any argument ('art'). Moreover, for Whately logic was concerned, rather narrowly, with the process of reasoning, and not with the subject matter reasoned about. This meant that the *Elements of Logic* excluded as 'extra-logical' topics like concepts and judgments, and as 'non-logical' alleged other forms of reasoning, whether it was non-syllogistic deductive or inductive reasoning.

According to Whately, induction referred to two distinct activities: the process of collecting facts so as to obtain or evaluate premises for reasoning, and the process of inferring conclusions from those facts.³³ The first activity, however useful, is not a form of reasoning at all, and thus not within the scope of logic. And as a process of inference, Whately argued contra Aristotle and Bacon, induction is simply a so-called enthymematic deduction—a syllogism with the major premise suppressed. Hence, Whately, enthusiastic as he was to defend deductive logic, went so far as to claim that deductive logic was entirely independent from induction—i.e. that all reasoning is syllogistic—and to deny that induction is a uniquely legitimate form of inference at all, let alone a logic all in itself. This controversial view was expressed famously by John Stuart Mill in his 1828 review of Whately's *Elements of Logic* in the *Westminster Review*: '[T]o *reason by induction* is a recommendation which

---

30 Richard Whately, *Elements of Logic*. 9th edn (London: J. Mawman, 1848), p. x.
31 Whately, *Elements of Logic*, p. x; Richard Whately, *Elements of Logic*. 2nd edn (London: J. Mawman, 1827), p. 7. Tellingly, the word 'boundless' appearing in the first and second edition was changed into 'extensive' in later editions.
32 Whately, *Elements of Logic*, p. 7.
33 See Whately, *Elements of Logic*, Book IV, Chapter 1.

implies as thorough a misconception of the meaning of the two words, as if the advice were, to *observe by syllogism.*'[34]

Whately's defence of deduction at the expense of induction did not merely inspire some logicians, like Hamilton and De Morgan, to advance deductive logic. It also motivated others, such as Herschel, Whewell and Mill, to show that an inductive logic *was* possible. This led to the emergence of two opposing camps within British logic—the deductive ('formal') and inductive ('scientific')—which were not on speaking terms because they rather literally spoke different languages. What John Venn wrote about logic in the 1870s also applied to the situation in the 1830s-60s:

> It would not be going too far to say that the principal difficulty in the way of a student of Logic at the present day (at any rate in England) consists not so much in the fact that the chief writers upon the subject contradict one another …, for an opportunity of contradiction implies agreement up to a certain stage, as in the fact that over a large region they really hardly get fairly within reach of one another at all.[35]

Importantly, those belonging to the inductive camp, like Whewell, all carried out their projects in terms of a renovation of Bacon's *Novum Organum*. This meant that British meta-scientific debates on methodology in the first half of the nineteenth century were conducted on the ('anti-Whatelyian') premise that induction was *the* form of scientific reasoning. As a result, the work of those belonging to the deductive camp, like De Morgan, was considered anti-Baconian not so much because it explicitly ridiculed Bacon—Baconians often did that

---

34 [John Stuart Mill], 'Review of Whately's *Elements of Logic*', in J.M. Robson, *The Collected Works of John Stuart Mill. Volume XI: Essays on Philosophy and the Classics* (Toronto: Toronto University Press, 1978), pp. 3–35 (p. 15). Mill's *System of Logic* famously turned Whately's view on its head, arguing that all deductive reasoning is grounded on induction. See, in this regard, Geoffrey Scarre, *Logic and Reality in the Philosophy of John Stuart Mill* (Dordrecht and Boston: Kluwer Academic Publishers, 1989), Chapters 1-3.

35 John Venn, 'Consistency and Real Inference', *Mind*, 1.1 (1876), 43–52 (p. 43). For a recent and more general discussion of logic in the nineteenth century see Jeremy Heis, 'Attempts to Rethink Logic', in *The Cambridge History of Philosophy in the Nineteenth Century (1790–1870)*, ed. by A.W. Wood and S.S. Hahn (Cambridge: Cambridge University Press, 2012), pp. 95–132.

too, with Whewell even sounding anti-Baconian to De Morgan's ears.[36] Rather, it was deemed anti-Baconian insofar as it was at odds with the conditions on which the search for science's methodology was carried out by Baconians. This becomes clear from the De Morgan-Whewell exchange, and arguably provides a clue as to why De Morgan's ideas on scientific methodology were largely neglected, both in his own time and by historians of Victorian science.

Whewell's Baconianism

It is well known that from his days as an undergraduate at Trinity in the 1810s to his final years as Master of that college in the 1860s, Whewell considered his project to be the reform of Bacon's inductive philosophy, which was to provide the groundwork for the reshaping of science, morality, politics and economics.[37] The task of reforming induction, which Whewell at times called the 'true faith', consisted roughly of two parts. The first was *defining* a 'true idea of induction', a philosophical task which Whewell himself took up; the second was that of *propagating* it as widely as possible through examples from specific sciences, 'to get *the people* into a right way of thinking about induction', for which Whewell solicited the help of others from his circle, such as Richard Jones for political economy and Robert Leslie Ellis for probability theory.[38]

One all-important part of this mission was to battle against those 'downwards mad'[39] who preferred a deductive approach to the sciences, that is, who held it possible to obtain natural knowledge through deductive reasoning. Aristotle himself had already been too 'fascinated & misled by the demonstrating powers of his syllogistic'.[40]

---

36 For instance, in his 1860 review in *The Athenæum* of Whewell's *On the Philosophy of Discovery*, De Morgan wrote: 'We cannot afford space to illustrate the way in which Dr. Whewell has reinforced our [negative] opinions on the history of Francis Bacon's philosophy'. – Augustus De Morgan, '*The Philosophy of Discovery, Chapters Historical and Philosophical*. By W. Whewell', *The Athenæum*, 1694, 14 April 1860, pp. 501–03 (p. 502).
37 See Snyder, *Reforming Philosophy*, chapter 1.
38 Notebook dated 28 June 1830, TCC, Whewell Papers, R.18.17/12, pp. v–ix.
39 See William Whewell to Richard Jones, 20 January 1833, TCC, Whewell Papers, Add.Ms.c.51/149, and William Whewell to Richard Jones, 22 July 1831, Whewell Papers, TCC, Add.Ms.c.51/110.
40 Richard Jones to William Whewell, 2 March 1831, TCC, Whewell Papers, Add. Ms.c.52/23.

But the most prominent of the 'deductive savages'[41] was undoubtedly Whately – who, as Whewell remarked at one point, was even worse than Aristotle because he was 'far more immersed in verbal trifling'.[42] Early in 1831, Whewell's close friend and collaborator Richard Jones wrote to Whewell after seeing the third edition of the *Elements of Logic*, ridiculing Whately's 'strange notion' that induction was a type of deductive reasoning and dismissing it.[43] Jones considered it yet 'another foolish sneer at those who think that inductive reasoning can ever be reduced to scientific form'.[44] Moreover, in following David Ricardo's theory of political economy, Whately and his fellow 'Oriel Noetics' at Oxford were the ones who were 'overrating [deduction's] pretensions', not someone like Bacon when he passed judgment on Aristotle.[45]

Given the meta-scientific context of the 1830s, Whewell and Jones saw Whately's characterisation of induction as much more than just a technical point of logic. First of all, they worried that if people accepted Whately's view, they might be led to the erroneous conclusion that the sciences—as the Oriel Noetics claimed—are essentially deductive and concerned with deducing conclusions from axioms and principles. Second, and more importantly, they were convinced that this deductive mode of thinking entailed dangerous moral and religious attitudes. Whewell's first reference in print to Whately's work appeared in his widely read Bridgewater treatise of 1833, *Astronomy and General Physics considered With Reference to Natural Theology*. Here, he influentially divided the (meta-)scientific community into two kinds of thinkers, with the deductive type (or 'mere Mathematicians')—the majority—possessing

---

41 William Whewell to Richard Jones, 19 February 1832, TCC, Whewell Papers, Add. Ms.c.51/129.
42 William Whewell to Richard Jones, 7 April 1843, TCC, Whewell Papers, Add. Ms.c.51/227.
43 Richard Jones to William Whewell, 24 February 1831, TCC, Whewell Papers, Add. Ms.c.52/20.
44 Richard Jones to William Whewell, 24 February 1831, TCC, Whewell Papers, Add. Ms.c.52/20.
45 Richard Jones to William Whewell, 2 March 1831, TCC, Whewell Papers, Add. Ms.c.52/23. On Jones's Baconian views on political economy and Whewell's and Jones's opposition to the 'Oriel Noetics' see, for example, Harro Maas, '"A Hard Battle to Fight": Natural Theology and the Dismal Science, 1820–50', *History of Political Economy*, 40:5 (2008), 143–167, and Paul Oslington, 'Natural Theology, Theodicy, and Political Economy in Nineteenth-century Britain: William Whewell's Struggle', *History of Political Economy*, 49:4 (2017), 575–606.

mental habits that 'impoverished their religious feeling' and their 'ability to appreciate moral evidence', and the inductive type (or 'Discoverers')—an elite group—displaying these virtues.[46]

Whewell illustrated this difference by using Whately's *Elements of Logic* for his own purposes, remarking that 'all which mathematics or logic can do, is to develop and extract those truths, as conclusions, which were in reality involved in the principles on which our reasoning proceeded'.[47] The implication was not just that new knowledge could only be attained on the basis of induction—or, more precisely, that there was a strong distinction between the original discovery of laws of nature by 'Discoverers' and the explication of their consequences and applications by 'mere Mathematicians'. Whewell also deemed the laborious and humbling process of ascending from observation to general principles to be simply more virtuous than the formal and dispassionate work of mathematicians. Euler, Laplace and Lagrange, in dealing with higher-level generalisations (e.g. laws of motion and gravitation), treated these as self-evident. They did not realise that, in discovering these laws, Newton had embarked on a pilgrimage and, hence, were unable to appreciate the moral and spiritual aspects of the proper pursuit of science.[48] For Whewell, the worst of the 'downwards mad' was not Whately but men like Laplace and his British followers, such as Babbage and, quite possibly, De Morgan: not only did they link mathematical deduction to scientific discovery, they also sought to

---

46 Yeo, *Defining Science*, p. 123. See William Whewell, *Astronomy and General Physics, Considered with Reference to Natural Theology* (London: William Pickering, 1833), Book III, Chapters 5–6. The phrases 'Mere Mathematicians' and 'Discoverers' appear in Hugh James Rose to William Whewell, 27 March 1833, Whewell Papers, TCC, Add.Ms.a.211 /143. On Whewell's inductive-deductive distinction see Joan L. Richards, 'The Probable and the Possible in Early Victorian England', in *Victorian Science in Context*, ed. by Bernard Lightman (Chicago: The University of Chicago Press, 1997), pp. 51–71, especially pp. 57–62.

47 Whewell, *Astronomy and General Physics*, pp. 335–36. Whewell gave the following quote from Whately's *Elements of Logic* in a footnote: 'Since all reasoning may be resolved into syllogisms, and since in a syllogism the premises do virtually assert the conclusions, it follows at once, that no new truth can be elicited by any process of reasoning.' – Whately, *Elements of Logic*, p. 215.

48 Secord has argued that Herschel's *Preliminary Discourse* was read not as a contribution to abstract philosophy but as a 'conduct manual'. See James A. Secord, *Visions of Science: Books and Readers at the Dawn of the Victorian Age* (Oxford: Oxford University Press, 2014), p. 81.

speed up this process through 'mental labor-saving techniques' which increased the 'accessibility of science' and facilitated its progress.[49]

Whewell's argument in his Bridgewater treatise was extraordinary for turning something as dry as induction and deduction into an epoch-making watershed. What it achieved was that promoting inductive or deductive reasoning, or even expressing a view on their relationship, was no longer just a theoretical matter. Instead, to work on deductive logic also meant to implicitly position oneself on much broader meta-scientific themes. Thereby, scientific method effectively became a topic reserved for those who believed in the possibility of an inductive logic. It is indeed telling, in this regard, that none of the British meta-scientists involved in debates on scientific method ever wrote on, or took an active interest in, developments in deductive logic. Despite their rejection of Whately's outlook, both Whewell and Mill were happy to concede deductive logic to Whately, who in turn conceded it pretty much to Aristotle. Rather than deduction *per se*—which for Whewell stood to induction as mathematics to scientific discovery—it was the deductive habit of mechanical formalisation that was fundamentally at odds with Whewell's project of renovating Bacon's inductive philosophy.

At the heart of this project stood Whewell's so-called antithetic epistemology. This said that all human knowledge is obtained through induction and demands the combination of ideas ('ideal') and facts ('empirical'). These ideas, which he called 'Fundamental Ideas' (Space, Time, Cause, etc.), are actively supplied by the human mind itself and not passively received from observations of the world. At the same time, these ideas make it possible to have scientific knowledge of the world outside the mind insofar as they make experience possible by allowing us to give form to our sensations. Because Whewell's Fundamental Ideas closely resembled Kant's forms of intuition and categories, as discussed in the *Critique of Pure Reason*, Whewell was criticised by his contemporaries for trying to import Kant into British philosophy.[50]

---

49   William J. Ashworth, 'Memory, Efficiency, and Symbolic Analysis. Charles Babbage, John Herschel, and the Industrial Mind', *Isis*, 87:4 (1996), 629–53 (p. 629); Michael Shortland and Richard Yeo, 'Introduction', in *Telling Lives in Science: Essays on Scientific Biography*, ed. by Michael Shortland and Richard Yeo (Cambridge: Cambridge University Press, 1996), pp. 1–44 (p. 20).

50   On the Whewell-Kant connection see, for example, Steffen Ducheyne, 'Kant and Whewell on Bridging Principles Between Metaphysics and Science', *Kant-Studien*,

De Morgan, in his 1840 review of the *Philosophy*, expressed surprise that 'the doctrines of Kant and Transcendental Philosophy are now promulgated in the university which educated Locke'.[51] But Kant was ultimately a metaphysician and Whewell a philosopher of science. Many of Whewell's Fundamental Ideas did not function as conditions of experience but as conditions for having knowledge within specific sciences; although it is possible to experience the world without having the Idea of Chemical Affinity, it is impossible to have knowledge of certain chemical processes without this Idea. Moreover, unlike Kant, Whewell believed that Fundamental Ideas (as well as the 'conceptions' included within them, such as 'force' as a modification of the Idea of Cause) emerged over the course of the development of science. 'The Ideas', he wrote, 'were in the human mind before [experience]; but by the progress of scientific thought they are unfolded into clearness and distinctness.'[52]

On the basis of this philosophical outlook, Whewell developed his inductive scientific methodology, dubbed 'Discoverers' Induction' in a letter to De Morgan from January 1859.[53] It was Baconian in a twofold sense. First, it agreed with what Bacon had said about induction, primarily that induction involved more than simple enumeration of instances, i.e. that it is something else than drawing a general proposition from particular cases. Second, it improved upon Bacon's method on the understanding that Bacon had never completed it and that if he had done so he would have paid more attention to the 'ideal' side of knowledge. At the core of Whewell's account stood the view that, in induction, 'there is a New Element added to the combination [of particular instances or cases] by the very act of thought by which they [are] combined'.[54] This 'act of thought' is a process which Whewell called 'colligation', the mental operation of bringing together a number

---

102.1 (2012), 22–45.
51 Augustus De Morgan, '*The Philosophy of the Inductive Sciences*. By W. Whewell', *The Athenæum*, 672, 12 September 1840, pp. 707–09 (p. 707). (Herschel, for one, was much harsher in his judgment about Whewell's 'a priorism'.) Whewell replied to De Morgan's anonymous review in a privately printed pamphlet, explaining the novelty of his approach as compared to Kant. See Yeo, *Defining Science*, p. 13.
52 William Whewell, *On the Philosophy of Discovery: Chapters Historical and Critical* (London: John W. Parker, 1860), p. 373.
53 William Whewell to Augustus De Morgan, 18 January 1859, TCC, 0.15.47/25.
54 Whewell, *Philosophy*, II, p. 213.

of facts by 'superinducing' upon them 'a conception of the mind ... which did not exist in any of the observed facts'.[55] Or, in more traditional logical terms:

> It has been usual to say of any general truths, established by the consideration and comparison of several facts, that they are obtained by *Induction*; but the distinctive character of this process has not been well pointed out .... The *Logic of Induction* has not yet been constructed. ... In each inductive process, there is some general idea introduced, which is given, not by the phenomena, but by the mind. The conclusion is not contained in the premises, but includes them by the introduction of a new generality.[56]

According to Whewell, this is what happened in all scientific discoveries, as the cases of Kepler and Newton showed. What made them great scientists was not their unearthing of new facts, nor their mathematical calculations; it was their explicating of new conceptions needed to colligate these facts into general laws.[57] But how did they arrive at these conceptions? Whewell offered several suggestions, each of which revolved around the decidedly non-Baconian notion of 'sagacity' or 'inventive genius':

> The necessity of a *conception* which must be furnished by the mind ... could hardly have escaped the eye of Bacon, if he had cultivated more carefully the ideal side of his own philosophy. And any attempts which he could have made to construct such conceptions *by mere rule and method*, must have ended in convincing him that *nothing but a peculiar inventive talent* could supply that which was ... contained in the facts, and yet was needed for the discovery.[58]

---

55 Whewell, *Philosophy*, II, p. 213.
56 William Whewell, 'Remarks on Mathematical Reasoning and on the Logic of Induction', in *The Mechanical Euclid*, 3rd edn (London: J. W. Parker, 1838), pp. 147–87 (pp. 177–78). This passage was reproduced verbatim in Whewell's *Philosophy*.
57 This two-step process is described in Book XI ('Of the Construction of Science') of Whewell's *Philosophy*.
58 Whewell, *Philosophy*, II, p. 402, my emphases. For the uses of 'genius' in nineteenth-century British science and philosophy see Simon Schaffer, 'Genius in Romantic Natural Philosophy', in *Romanticism and the Sciences*, ed. by Andrew Cunningham and Nicholas Jardine (Cambridge: Cambridge University Press, 1990), pp. 82–98, and Richard Yeo, 'Genius, Method and Morality'.

Whewell's point was not just that Bacon had failed to appreciate the 'inventive genius' which all scientific discovery requires; it was that Bacon had mistakenly believed that it was possible to 'supersede' genius by reducing its activities to a *'Technical Form'*.[59] At the same time, Whewell himself insisted that there is nothing 'accidental' about scientific discoveries, and he explicitly opposed David Brewster's competing view that most discoveries are the result of 'pure accident'.[60] The resulting tension, brought out by De Morgan, may be called 'Whewell's paradox': because sparks of creative genius are irreducible to methodological rules, the logic of induction is ultimately not completely logical.

There was no way to solve this paradox, and the best Whewell could offer were suggestions for dissolving it. Rather than giving rules to men of genius, rules might be given for the use they made of their genius. One mark of genius was a certain facility in generating a number of possible options for the appropriate conception. Because this process is not bound to rules, Whewell sometimes used the terms 'guessing' or 'conjecturing' to describe it. Whewell, however, was not the hypothetico-deductivist that some latter-day commentators made of him.[61] Since the selection and application of the appropriate conception often involved a series of different kinds of inferences (especially analogical reasoning), as Whewell argued, this stage of inductive discovery was not a matter of non-rational guesswork. The same obviously held for the next stage, where conceptions—in the form of hypotheses or theories—are confirmed on the basis of several tests, namely prediction, consilience and coherence. But it was undeniable that Whewell, in renovating Bacon, had stretched Baconian inductive logic to its utmost limits: it was now a matter of discoverers having '*good* metaphysics in their heads' and 'binding their metaphysics' to the facts through a process that was rule-governed only to a certain degree.[62]

---

59 Whewell, *Philosophy*, II, p. 402.
60 See David Brewster, 'On the History of the Inductive Sciences', *Edinburgh Review*, 66 (1837), 110–51 (p. 121).
61 For a critical discussion of twentieth-century readings of Whewell as a hypothetico-deductivist see, for instance, Laura J. Snyder, '"The Whole Box of Tools": William Whewell and the Logic of Induction', in *Handbook of the History of Logic. Volume 4: British Logic in the Nineteenth Century*, ed. by Dov M. Gabbay and John Woods (Amsterdam: North-Holland, 2008), pp. 163–228.
62 William Whewell, *Novum Organon Renovatum*, p. vii.

Some of the meta-scientific implications of Whewell's views on scientific method were equally at odds with Bacon's programme. Perhaps most tellingly, in placing limits on the 'formalisation' of methodology Whewell not only denied that discovery was a mechanical process, but he also undermined the idea that it should be possible at least in principle for anyone who carefully follows the scientific method to achieve scientific breakthroughs.[63] Herschel saw in this a useful corrective to the tendency of recent utilitarian reforms to promote the accessibility of science by ascribing its progress wholly to correct method.[64] It made others wonder what it was that made Whewell still identify as a Baconian. If 'the great Baconian induction' was 'a complete failure', De Morgan wondered, why try to save his programme rather than finally abandoning it for an alternative?[65]

De Morgan's Anti-Baconianism

Unlike Whewell's, De Morgan's oeuvre and career was not an unfolding of a meta-scientific plan cooked up as an undergraduate and self-consciously carried out as the years passed. Neither is it possible, at least not as strongly as in the case of Whewell, to read every single publication of De Morgan, who published even more than Whewell, as a contribution to such a plan. Nonetheless, there is arguably a common thread running throughout De Morgan's wide-ranging writings—books, encyclopedia entries, and reviews—on logic, probability theory, and history of science in regard to scientific methodology. Moreover, when contrasting his views on scientific methodology with those of Whewell it becomes possible to approach De Morgan as a meta-scientist and to see him rebelling against the Baconianism that dominated British meta-science in the first half of the nineteenth century. This has a wider significance because it suggests that, however 'excessively Baconian',

---

63 See Simon Schaffer, 'Scientific Discoveries and the End of Natural Philosophy', *Social Studies of Science*, 16.3 (1986), 387–420.
64 See [John W.F. Herschel], 'Review of the *History and Philosophy of the Inductive Sciences*', *Quarterly Review*, 135 (June 1841), 96–130.
65 Augustus De Morgan, '*The Philosophy of Discovery*', p. 503.

the 'methodological orthodoxy' in the early Victorian period did not go unchallenged.[66]

De Morgan's views on scientific methodology were anti-Baconian in a twofold sense. First, De Morgan dismissed the historical significance and philosophical correctness of Bacon's methodology, as put forward in the *Novum Organum*. Second, his views conflicted with the Baconianism of Bacon's nineteenth-century heirs. This Baconianism rested on a specific, limited interpretation of the Baconian philosophical corpus, fitted to their meta-scientific agendas. Indeed, in at least one crucial respect De Morgan remained more loyal to Bacon than a Whewell or a Herschel; he continued the search, albeit in a decidedly non-Baconian fashion, for a way to put scientific methodology into a 'Technical Form', to provide a 'machinery' for arriving at natural knowledge. At the core of his anti-Baconianism stood the conviction that Bacon and the Baconians focused too much on observation and too little on logic and mathematics as instruments of scientific discovery. Newton may have been careful at observation, having 'few superiors' in the 'inductive process', but 'it was his power of deduction which made him what he was'.[67] What De Morgan wrote about Bacon in his 1858 review of *The Works of Francis Bacon* also applied to Whewell and other Baconians:

> He averred that logic and mathematics should be the handmaids, not the mistresses, of philosophy. He meant that they should play a subordinate and subsequent part in the dressing of the vast mass of facts by which discovery was to be rendered equally accessible to Newton and to us. Bacon himself was very ignorant of all that had been done by mathematics; and, strange to say, he especially objected to astronomy being handed over to the mathematicians. Leverrier and Adams, calculating an unknown planet into visible existence by enormous heaps of algebra, furnish the last comment of note on this specimen of the goodness of Bacon's views.[68]

---

66 Charles Gillispie, *The Edge of Objectivity* (Princeton: Princeton University Press, 1960), p. 314; Richard Yeo, 'An Idol of the Market-place: Baconianism in Nineteenth Century Britain', *History of Science*, 23.3 (1985), 251–98 (p. 252).

67 Augustus De Morgan, '*History of the Inductive Sciences from the Earliest to the Present Times*. By W. Whewell', *The Athenæum*, 541, 10 March 1838, pp. 179–81 (p. 180).

68 Augustus De Morgan, '*The Works of Francis Bacon*, ed. by James Spedding, R. Leslie Ellis, and Douglas D. Heath. 5 vols.', *The Athenæum*, 1612, 18 September 1858, pp. 367–68 (p. 367).

These and other historical facts should be philosophically accounted for in scientific method. Doing so meant that Baconianism had to be abandoned, and that something else had to come in its place. Baconians like Whewell were quick to suspect a blatant case of 'downwards' thinking of the worst, Continental kind. But also, for them it was far from clear what De Morgan's vison on science exactly amounted to, let alone how it translated into an alternative scientific methodology or what its wider meta-scientific ramifications were. This is still very much an open question.[69]

De Morgan recognised the inadequacy of Bacon's inductive philosophy as well as the need for an alternative which could overcome its deficiencies. Unlike any of the Baconians, De Morgan was willing to break with British tradition and pursue this search in defiance of even ceremonial Baconianism. Instead, De Morgan thought about the history and philosophy of science in terms not of 'Bacon's rules' but 'Newton's practice'. What does this mean? First, that De Morgan denied that—historically speaking—Newton, in writing the *Principia*, had followed Bacon's inductive canons. Second, that—philosophically speaking—there are no rules for arriving at discoveries, such as that of universal gravitation, and scientific method should not aim to provide them.[70] Taken together: 'If Newton had taken Bacon for his master, not he, but somebody else, would have been Newton.'[71] The same can be put in positive terms. First, De Morgan believed that, despite his own famous 'Hypotheses *non* fingo', Newton had employed hypotheses and this convinced De Morgan that scientific knowledge progressed through

---

69 The present chapter contributes to taking a first step toward addressing this question. Among the other sources crucial in taking this step are: Laudan, 'Induction and Probability'; Maria Panteki, 'French "Logique" and British "Logic"', especially pp. 400–11 and pp. 423–41; Adrian Rice, 'Augustus De Morgan: Historian of Science', *History of Science*, 34:2 (1996), 201–40; Joan L. Richards, '"In a Rational World all Radicals would be Exterminated": Mathematics, Logic and Secular Thinking in Augustus De Morgan's England', *Science in Context*, 15:1 (2002), 137–64; John V. Strong, 'The Infinite Ballot Box of Nature: De Morgan, Boole, and Jevons on Probability and the Logic of Induction', *PSA: Proceedings of the Biennial Meeting of the Philosophy of Science Association*, 1976:1 (1976), 197–211; John Wettersten, *Whewell's Critics: Have They Prevented Him from Doing Good?* (Amsterdam and New York: Rodopi, 2005), Chapter 1 ('Immediate Rejection'); and Richard Yeo, 'Genius, Method and Morality'.

70 See also the section on De Morgan's philosophy of mathematics in Chapter 1 of this volume.

71 De Morgan, '*The Works of Francis Bacon*', p. 367.

deduction, especially mathematical reasoning. Second, De Morgan believed that a new scientific methodology should assist scientists in their practice of hypothesising. This brought him closer to Herschel than Whewell, who had accused Herschel of promoting a spirit of 'gratuitous theorising' in his *Preliminary Discourse* by not cautioning against anticipatory leaps to hypotheses.[72] Whewell and Herschel both made room for hypotheses in scientific methodology, but Herschel adopted a much more flexible stance toward hypothesising.[73] De Morgan's liberality, in this regard, went much farther even than Herschel's, however, as he shunned the principle that hypothetical speculation is only legitimate on inductive grounds.

Interestingly, De Morgan's next step was indebted to Herschel: he turned to the mathematical theory of probability to provide a criterion for choice between scientific hypotheses. De Morgan may have been the one to have imported this theory from the Continent into Britain; it was Herschel who, in a neglected passage in his *Preliminary Discourse*, introduced this 'refined and curious branch of mathematical enquiry'[74] into the British debate on scientific methodology. But Herschel only discussed it in relation to the calculation of observational errors. De Morgan took the bold and pioneering step—in the British context at least—of using probability theory to formalise and justify scientific inference, in the sense of weighing competing hypotheses offered to account for a given set of phenomena. This was anti-Baconian not just in the obvious sense of answering a philosophical question with mathematics. It also went against Baconian orthodoxy in two other, more profound and complexly related, ways—thereby unearthing what this very orthodoxy was. On the one hand, it questioned the idea of an inductive methodology that would necessarily lead to infallible

---

72 [William Whewell], 'Modern Science – Inductive Philosophy [Review of John F.W. Herschel's *A Preliminary Discourse on the Study of Natural Philosophy*]', *Quarterly Review*, 45 (July 1831), 374–407 (p. 400).

73 For a useful overview of different views on Whewell's and Herschel's views on hypotheses see Aaron D. Cobb, 'Is John F.W. Herschel an Inductivist about Hypothetical Inquiry?' *Perspectives on Science*, 20:4 (2012), 409–39; and Laura J. Snyder, 'Hypotheses in 19th Century British Philosophy of Science: Herschel, Whewell, Mill', in *The Significance of the Hypothetical in Natural Science*, ed. by Michael Heidelberger and Gregor Schiemann (Berlin: Walter de Gruyter, 2009), pp. 59–76.

74 Herschel, *Preliminary Discourse*, p. 217.

scientific knowledge. Firstly, because scientists are creative thinkers, not simply rule-followers; secondly, because induction can never prove the truth of a conclusion; and, finally, because all scientific knowledge is probable, not certain. On the other hand, rather than accepting creative genius as the ruleless core of an otherwise rule-bound methodology, it limited methodology to calculating the probability of the products (i.e. hypotheses) of someone's creativity.

Taken together, De Morgan's anti-Baconianism made scientific methodology revolve around uncertainty, both by accepting its place at the heart of science and by seeking mathematical ways to deal with it as accurately as possible. This points to a beautiful paradox of the nineteenth-century British meta-scientific debate, which may be called 'the paradox of Hume's ghost': those who were the most skeptical about induction, like De Morgan, were also the ones to recognise and confront the limits of inductive inference.

De Morgan Contra Whewell

There are many routes into De Morgan's meta-scientific outlook on methodology—for example via his technical work on formal logic, his contributions to the history of modern science, his involvement in scientific organizations, and his influence on pupils such as Jevons. Any full-blown account will have to explore each of these routes and identify the relevant intersections between them. The modest aim here is to bring out a few more specific aspects of De Morgan's views on scientific methodology by focusing on his exchanges with Whewell, who is taken as a representative of the dominant Baconian orthodoxy. Their interaction took place mostly through letters, some hundred of which have survived, four reviews in *The Athenæum*,[75] and occasional

---

75 For *The Athenæum*, De Morgan (anonymously) reviewed Whewell's *History of the Inductive Sciences* (1838), *The Philosophy of the Inductive Sciences* (1840), *Novum Organon Renovatum* (1859) and *On the Philosophy of Discovery* (1860)—the latter two being respectively the second and third part of the third edition of *The Philosophy of the Inductive Sciences*. Another review in *The Athenæum* of Whewell's work that has been attributed to De Morgan is of *The Mathematical Works of Isaac Barrow* (1860), edited by Whewell for Trinity College, Cambridge. See, in this context, Sloan Evans Despeaux and Adrian C. Rice, 'Augustus De Morgan's Anonymous Reviews for *The Athenæum*: A Mirror of a Victorian Mathematician', *Historia Mathematica*, 43:2 (2016), 148–71.

references in book chapters. Given this focus, it is unavoidable that some aspects receive more attention than others and that there are aspects which do not come into view at all, such as probability theory. Another reason for this limitation is that the interaction between De Morgan and Whewell was relatively one-directional: for example, there are about four times more letters from De Morgan to Whewell than vice versa.[76] Moreover, De Morgan reviewed Whewell's work but the reverse never occurred. This is interesting insofar as it points to disciplinary boundaries in the field of meta-science, and suggests that in the context of methodological debates De Morgan was even more polymathic than Whewell: as a mathematician, De Morgan was well-versed in history and philosophy of science, but as a mathematician-turned-philosopher, Whewell was not (and did not want to be) expert on mathematical developments in logic.

De Morgan and Whewell on Logic and Induction

One point of conflict between De Morgan and Whewell concerned the nature and scope of logic, more specifically of induction.[77] Their disagreement on this topic surfaced in 1849, when De Morgan complained in a letter that Whewell's notion of induction contained 'more than logic'.[78] It became public in De Morgan's review, written at the request of Whewell himself,[79] of the *Novum Organon Renovatum* of January 1859. Here, De Morgan wrote that: 'though we do not quarrel with any of his [i.e. Whewell's] conclusions'—for example, that every scientific discovery introduces a new conception—'we are entirely opposed to the use which he makes of the words *logic* and *induction*',

---

[76] This ratio is based on the Whewell-De Morgan correspondence held at Trinity College Library, Cambridge. For further information regarding this collection of letters, see Chapter 11 of this volume.

[77] The following analysis draws on the following accounts: Wettersten, *Whewell's Critics*, pp. 58–60, and Robert E. Butts, ed., *William Whewell's Theory of Scientific Method* (Pittsburgh: University of Pittsburgh Press, 1968), pp. 24–26.

[78] Augustus De Morgan to William Whewell, 20 April 1849, TCC, Whewell Papers, Add.Ms.a.202/114.

[79] See William Whewell to Augustus De Morgan, 18 January 1859, TCC, Whewell Papers, O.15.47/25

especially when combined into a *'logic of induction'*.[80] First, De Morgan criticized Whewell's vague, non-formal understanding of 'logic'. What De Morgan, following Whately who, in turn, followed Aristotelian tradition, meant by logic was the study of the logical form of statements and inferences. 'It has nothing to do,' he wrote in his 1839 *First Notions of Logic*, 'with the truth of the facts ... from which an inference is derived; but simply takes care that the inference shall certainly be true, if the premises be true.'[81] On the one hand, by introducing into logic the process by which premises are formed, Whewell made logic '[take] in much which the word excludes'. On the other, by failing to provide a way of showing the validity of conceptions, which bind together facts through generalisation, Whewell made logic 'exclude much which the word takes in'. De Morgan's was an appeal to tradition: Whewell had no right to claim the word 'logic' for something not concerned with logical truth and formal validity. Second, De Morgan criticised Whewell's use of the term 'induction' for taking it beyond its traditional meaning. According to De Morgan, Whewell used it too liberally as including 'the use of the whole box of tools',[82] from the 'old' to the 'new', that is, from the generalisation of particulars to the formation and testing of the general notion under which these particulars are to be brought. Again, De Morgan did not find fault with Whewell's conclusions, but insisted that Whewell had no right to redefine a canonical term to make it suit his own purposes:

> Let induction mean, as it always has done, the generalization by collection of particulars: let the act of the discoverer, by which he divines the general notion under which the particulars can be brought, receive its own proper name. ... We put it to him [Whewell], whether it would not be desirable to restrict the words *logic* and *induction* to the meanings now well agreed upon, and to find better names for the whole process, and also for the particular part which entirely depends on the acumen of the discoverer.[83]

---

80 [Augustus De Morgan], Review of William Whewell's *Novum Organum Renovatum*, *The Athenæum*, 1628 (8 January 1859), 42–44 (p. 43). De Morgan quotes ('art of discovery ...') from Whewell, *Novum Organon Renovatum*, p. v.
81 Augustus De Morgan, *First Notions of Logic (Preparatory to the Study of Geometry)*, 2nd edn (London: Taylor & Walton, 1840), p. 3.
82 Augustus De Morgan, *Formal Logic*, p. 216.
83 Augustus De Morgan, *'Novum Organum Renovatum'*, p. 44.

De Morgan's position, which distinguished logic and induction from discovery, arguably reflected a clash of underlying outlooks. The following illustrations should suffice here. For Whewell, it was not a criticism at all that his 'logic of induction' did not belong to or sit well with the 'old logic', since it was premised precisely on a Baconian break with that very tradition. As he wrote to De Morgan in a letter from January 1859:

> My object was to analyse ... the method by which scientific discoveries have really been made; and I call this method *Induction*, because all the world seemed to have agreed to call it so, and because the name is not a bad name after all. That it is not exactly the Induction of Aristotle, I know; nor is it that described by Bacon .... I am disposed to call it *Discoverers' Induction* .... I do not wonder at your denying [it] a place in Logic; and you will think me heretical and profane, if I say, *so much the worse for Logic*.[84]

Similarly, De Morgan's argument that Whewell's notion of induction was not logical would not have shocked Whewell, as Whewell disagreed with De Morgan's logical notion of induction. What De Morgan understood by induction was 'Perfect Induction', which can only be done when dealing with a limited number of observed particulars. For example, Kepler discovered that Mars moves in an ellipse, that the earth moves in an ellipse, and so on, and from this he inferred that all the planets move in ellipses. For Whewell, there was no real inference involved here, since the conclusion contained nothing that was not already asserted in the premises. Whewell's discoverers' induction also covered what De Morgan called 'Imperfect Induction', namely the mental process, or *'mysterious step'*,[85] of inferring from known to unknown cases. 'So much the worse for Logic' if it excluded this crucial element of human reasoning.

De Morgan and Whewell agreed that induction in the sense of mere summary generalisation from observed particulars played a negligible role in scientific discovery. For De Morgan, this meant that logic had nothing to do with the process of arriving at new knowledge of the world, and that discovery consisted in something else entirely—a 'third

---

84 William Whewell to Augustus De Morgan, 18 January 1859, TCC, Whewell Papers, O.15.47/25.
85 Whewell, *On the Philosophy of Discovery*, p. 284.

method', one 'not within the ken of Bacon', which revolved around the probability of hypotheses.[86] For Whewell, it meant that logic had to be broadened to include rules for both deductive (i.e. syllogistic) and inductive reasoning:

> By *Logic* has generally been meant a system which teaches us to arrange our reasonings that their truth or falsehood shall be evident in their form. In *deductive* reasonings ... the device [for this] is the *Syllogism* .... [The *Logic of Induction*] in like manner supplies the means of ascertaining the truth of our *inductive* inferences.[87]

Nevertheless, by 1860, Whewell does seem to have bitten the bullet of De Morgan's point that in scientific discovery there is more than what is traditionally called induction. '[T]he philosophy at which I aimed was not the philosophy of Induction, but the *Philosophy of Discovery*' and, as De Morgan was happy to observe in his review of *On the Philosophy of Discovery, Chapters Historical and Philosophical*, 'the title of the book is modified accordingly'.[88]

De Morgan and Whewell on Deduction and Probability

Another major point of conflict remained in place: Whewell's and De Morgan's positions vis-à-vis deductive logic. Like all Baconians, Whewell followed Whately in equating it with syllogism, which he regarded as a completed tool of very limited usefulness. Whewell did publish one ten-page article on Aristotelian logic, if only to attribute to Aristotle the misguided claim that induction *is* a syllogism.[89] De Morgan, instead, went over, under and beyond Whately, taking deductive logic far beyond the syllogism in terms of depth and scope.[90] Despite his

---

86 De Morgan, '*Novum Organum Renovatum*', p. 44. More on this topic below.
87 Whewell, *Novum Organon Renovatum*, p. 106; Augustus De Morgan, 'The Philosophy of Discovery, Chapters Historical and Critical. By W. Whewell', *The Athenæum*, 1694, 14 April 1860, pp. 501–03 (p. 503).
88 Whewell, *On the Philosophy of Discovery*, p. v.
89 See William Whewell, 'Criticism of Aristotle's Account of Induction', *Transactions of the Cambridge Philosophical Society*, 10.1 (1850), 63–72. This largely forgotten paper was later published as an Appendix to Whewell's *Philosophy of Discovery* of 1860.
90 On De Morgan as a logical innovator see Daniel D. Merrill, *Augustus De Morgan and the Logic of Relations* (Dordrecht: Kluwer Academic Publishers, 1990) and Michael E. Hobart and Joan L. Richards, 'De Morgan's Logic', in *Handbook of the*

appeal to logical tradition in criticising Whewell, De Morgan was an innovator who obviously did not believe that the 'old logic' could not be improved.[91] Indeed, he did just that in major works such as *Formal Logic* (1847) and *Syllabus of a Proposed System of Logic* (1860), tellingly opening his entry on 'Logic' for the *English Cyclopaedia* with the statement that recent innovations suggested 'that Kant's dictum about the perfection of the Aristotelian logic may possibly be false'.[92] The point of his *ad antiquitatem* was that Whewell's 'logic of induction' could not be considered a contribution to logic in the traditional sense of a formal study of deductive reasoning. Among the innovations which De Morgan did consider legitimate contributions to logic were those that sought to improve this study without thereby breaking away from Aristotle's conception of logic. One example was his own logic of relations, of which he believed the syllogism to be a special case.

A key feature of De Morgan's logical work was the use of mathematics to remove the limitations of the syllogism for deductive logic. More important than this, at least with an eye to unearthing De Morgan's views on scientific method, is his controversial use of one specific branch of mathematics, namely probability theory, in his logical work.[93] 'Many will object to this theory as extralogical', De Morgan wrote:

> But I cannot see on what definition ... the exclusion of it can be maintained. ... I cannot understand why the study of the effect which partial belief of the premises produces with respect to the conclusion, should be separated from that of the consequences of supposing the former to be absolutely true.[94]

History of Logic. Volume 4: British Logic in the Nineteenth Century, ed. by Dov M. Gabbay and John Woods (Amsterdam: North-Holland, 2008), pp. 283–330.
91 See Chapter 2 of this volume.
92 Augustus De Morgan, 'Logic (1860)', in Peter Heath, ed., *On the Syllogism and Other Logical Writings by Augustus De Morgan* (London: Routledge & Kegan Paul, 1966), pp. 247–66 (p. 247).
93 The important works, in this context, are De Morgan's book-length article in the *Encyclopedia Metropolitana* (1837), the volume *An Essay on Probabilities* (1838) and several chapters in *Formal Logic* (1847). For an in-depth discussion of De Morgan's introduction of probability into logic, see Adrian Rice, '"Everybody Makes Errors": The Intersection of De Morgan's Logic and Probability, 1837–1847', *History and Philosophy of Logic*, 24:4 (2003), 289–305.
94 De Morgan, *Formal Logic*, p. v.

On the basis of his new system of the numerically definite syllogism, where all terms are quantified, De Morgan observed that, although in the Aristotelian syllogistic no inference can be drawn from 'Some Xs are Ys' and 'Some Ys are Zs', the following inference is nonetheless valid: 'Some Xs are Ys, some Ys are Zs, therefore, there is some probability that some Xs are Zs.' It was here that De Morgan began to apply the techniques of mathematical probability theory to logic, for instance finding the probability that some Ys will be both Xs and Zs, given that the distribution of Xs and Zs among the Ys is unknown. The point of this endeavour was not to offer a full-blown theory of probable inference; instead, it was to illustrate that innovating deductive logic was not mere trifling—as Whewell believed—but could help model how people of flesh and blood could reason under conditions of uncertainty. More specifically, it suggested that it was possible to calculate what degree of rational belief someone should attach to a conclusion derived from pieces of less than certain knowledge. This points to one crucial sense in which De Morgan did not just innovate but redefined formal deductive logic: however formal, it sought to capture how rational human beings, including scientists, reason.

De Morgan's introduction of probability into logic was connected to his views on scientific methodology—i.e. his 'third method'—via his ideas on inverse probability or probability of causes. This field, which would today be called mathematical statistics,[95] dealt with the evaluation, in terms of probabilities, of competing hypotheses about the unknown causes of observed events. In De Morgan's own words: 'An event has happened, such as might have arisen from different causes: what is the probability that any one specified cause did produce the event, to the exclusion of other causes?'[96] De Morgan, approaching this situation in terms of scientific discovery, rejected the vague eliminative strategies championed by Bacon and his followers: a scientist cannot just 'lay down his this, his that, and his t'other [for example, one or two conceptions], and say, "now, one of these it must be; let us proceed to

---

95 For a discussion of De Morgan's work on 'statistical hypothesis testing', see Adrian Rice and Eugene Seneta, 'De Morgan in the Prehistory of Statistical Hypothesis Testing', *Journal of the Royal Statistical Society*, 168:3 (2005), 615–627.

96 De Morgan, *An Essay on Probabilities*, p. 53.

try which"'.[97] Rather, the best that could be done in such situations was to provide a quantitative criterion for choice between 'this, that and t'other'. Following a long line of mostly Continental mathematicians who had used probability to introduce scientific method into the realm of mathematics,[98] De Morgan believed that the probabilities of competing hypotheses could be measured and compared, not just with one another but with some standard of certainty (such as 'moral certainty'). This he did on the basis of the inverse probability techniques of Thomas Bayes and Pierre-Simon Laplace.

The core equation—letting h stand for a hypothesis and e for a body of evidence, where the conditional probability of h given e was to be interpreted as the degree of belief in the hypothesis given the evidence[99]—was used to calculate the rate at which the probability of a hypothesis increased with the number of confirming instances. However intuitive, a lot of assumptions, which would soon come to be seen as highly problematic, were needed to make this reduction of induction to deduction work. For example, perhaps most notoriously, in order to assign a value to the probability of the hypothesis before consideration of the data, namely, the prior probability $P(h)$, De Morgan and others made use of the 'Principle of Insufficient Reason'—which said that if there is no reason to favor one hypothesis over another, each should be assigned the same probability. The appeal to prior ignorance or, that is, to equally likely cases, was often confusing enough in simple cases of repeated drawings of balls from an urn with black and white balls, let alone in that of well-specified causes of complex natural events. It caused many to doubt whether a mathematical theory first developed for urn models could easily be extended, if at all, to model scientific reasoning. De Morgan, perhaps the most fervent British advocate of Continental probability, was among those—like Laplace, Condorcet and Poisson—who believed in the project of probabilising scientific method.

---

97 De Morgan, 'The Works of Francis Bacon', p. 367.
98 See, for instance, Lorraine Daston, *Classical Probability in the Enlightenment* (Princeton: Princeton University Press, 1988), Chapter 5 ('The Probability of Causes').
99 For someone like De Morgan, who treated probability as a branch of logic—and thus applicable to the relationship between propositions—this meant that propositions were assigned a definite numerical probability with respect to a body of data.

Consequently, he shared many of their assumptions and made similar mistakes, as his slightly younger peers George Boole and Robert Leslie Ellis were quick to point out.[100]

Ellis is particularly relevant, as he was one of Whewell's most dedicated protégés. Perhaps because his scientific methodology was so evidently at odds with that on which probability was constructed, Whewell showed little to no interest in probability, and when he used the term, it was often in a colloquial sense. Whewell's *Philosophy* did include discussions of such probabilistic methods as the 'method of means' and 'method of least squares', but these were brief (5 pages) and derivative.[101] It was Ellis who took up the problem of reconciling probability theory with a Whewellian philosophy of science, for which he asked Whewell's written permission.[102] Ellis's central argument was twofold. First, that what mathematicians like Laplace tried to prove mathematically, such as the regularity of nature, was true *a priori*. Second, that probability calculations rested on *a priori* truths, 'supplied by the mind itself'.[103] One implication was that probabilities cannot be said to be the 'measure of any mental state', for instance concerning the truth of an uncertain proposition. Another implication was that the theory's applicability to scientific inference was very limited, insofar as it was inadequate to the way people actually think:

> Our confidence in any inductive result varies with a variety of circumstances; *one* of these is the number of particular cases from which it is deduced. Now the measure of this confidence which the theory professes to give, depends on this number exclusively.

---

100 See, for example, George Boole, *An Investigation of the Laws of Thought on which are founded the Mathematical Theories of Logic and Probabilities* (London: Walton & Maberly, 1854), pp. 363–68, especially p. 364.

101 See William Whewell, *The Philosophy of the Inductive Sciences, Founded Upon Their History. Volume II* (London: John W. Parker, 1840), Book XIII ('Of Methods Employed in the Formation of Science'), Chapter VII ('Special Methods of Induction Applicable to Quantity'), pp. 550–56.

102 See Robert Leslie Ellis to William Whewell, TCC, Whewell Papers, Add. Ms.c.67/104. For a discussion of Ellis's work on foundations of probability theory, see Richards, 'The Probable and the Possible', pp. 64–65; and Lukas M. Verburgt, 'Robert Leslie Ellis's Work on Philosophy of Science and the Foundations of Probability Theory', *Historia Mathematica*, 40:4 (2013), 423–54.

103 Robert Leslie Ellis, 'On the Foundations of the Theory of Probabilities', *Transactions of the Cambridge Philosophical Society*, 8 (1844), 1–6 (p. 4).

> Yet no one can deny, that the force of the induction may vary, while this number remains unchanged.[104]

Ellis elaborated this point in an attack on one of De Morgan's examples in his 1837 'Theory of Probabilities', where he had calculated the probability that a vessel will have a flag on the basis of the previous ten vessels having one. But, Ellis asked, 'What degree of similarity in this new event to the previous ones, entitles it to be considered a recurrence of the same event?' The fact that this depended not only on the event, but also on the mind which contemplated it, showed that probability theory was too simplistic even to describe such an everyday situation. Likewise, regarding more complex cases based on assuming equal prior probabilities, Ellis wrote: '[M]ere ignorance is no ground for any inference whatever. *Ex nihilo nihil.*'[105] The human mind is a source of knowledge only, and precisely, insofar as it is actively involved in its creation. No doubt Whewell would have agreed.

Afterword

Most commentators have attributed the neglect of De Morgan's anti-Baconian programme either to the broader process of the downfall of classical probability or to technical mistakes. What has so far received little attention is the intellectual context in which it took shape, more specifically the fact that it was based on a meta-scientific vision that challenged the prevailing orthodoxy, represented by Whewell and his fellow Baconians.

First, De Morgan questioned not just the idea that scientific knowledge is obtained by induction alone but also the deeper conviction that it was possible to formulate a non-probabilistic method of scientific inference. Every Baconian, whether Whewell, Herschel or Mill, believed that their rules for inductive reasoning guaranteed the truth of the conclusions to which the application of these rules led. This belief, in turn, was premised on the assumption that there was no significant element of uncertainty attached to the conclusions of induction. Or, more precisely, there was

---

104 Ellis, 'Foundations', p. 4.
105 Robert Leslie Ellis, 'Remarks on an Alleged Proof of the Method of Least Squares', *Philosophical Magazine*, 37 (November 1850), 321–28 (p. 325).

such an element of uncertainty, but this pertained to the process and not the outcome of discoveries: for instance, whereas for Whewell there were no rules for a genius to arrive at conceptions, these conceptions themselves infallibly led to knowledge of necessary truths. Given such an outlook on science, it seemed an epistemic category mistake at best to even introduce probability techniques into its methodology.

Second, De Morgan went one step further by trying to reduce induction to deduction, not as Whately had done by saying that every induction is a syllogism, but by following Laplace in showing that it is based on inverse probability theory. De Morgan's alternative scientific methodology, which said that discovery is achieved by starting from a hypothesis whose probability increases as the number of confirming observations grows, was deliberately anti-Baconian in its formality. At the same time, it achieved little success—at least for a time—in large part because it failed to satisfy certain pre-formal, typically Baconian, conditions.[106] One of these conditions was that a hypothesis becomes more likely with the addition of confirming observations, but not in a linear fashion: this is because, as Whewell argued,[107] a hypothesis is made more probable by predicting surprising phenomena than by the successful prediction of unsurprising phenomena. The clash of De Morgan's 'Laplacian', quantitative probability with Whewell's 'Baconian', more qualitative view of probability was surprisingly long-standing, evidently touching on conflicting philosophical intuitions about the nature of science.[108] It continued in the 1870s–80s debates between William Stanley Jevons and John Venn, who respectively defended and attacked De Morgan, and C.D. Broad, W.E. Johnson and J.M. Keynes in the 1910s–20s. By that time, the scientific and philosophical landscape had, of course, changed considerably, and Whewell and De Morgan were names remembered only vaguely.

Quite a lot has been written recently on Whewell and his circle. Snyder has put him at the centre of a 'Breakfast Club', also consisting

---

106 See Laudan, 'Induction and Probability', pp. 193–94.
107 See, in this regard, Larry Laudan, 'William Whewell on the Consilience of Inductions', *The Monist*, 55 (1971), 368–91.
108 For the distinction between Pascalian (or Laplacian) and Baconian probability, see the work of L. Jonathan Cohen, for instance his 'Some Historical Remarks on the Baconian Conception of Probability', in L. Jonathan Cohen, *Knowledge and Language* (Cham: Springer, 1980), pp. 245–59.

of Herschel and Babbage. Ashworth, as a welcome corrective to this narrative, has zoomed in on a 'Trinity Circle', showing that from the 1820s onwards, Whewell's meta-scientific project increasingly diverged from that of Herschel and, especially, Babbage. The present chapter has attempted to add to this line of inquiry, highlighting the differences rather than commonalities between key figures in the early Victorian meta-scientific debates, by introducing De Morgan into the picture. It makes the picture more complex and, hopefully, richer. Much more work needs to be done to think through De Morgan's position vis-à-vis the Baconian tradition and the role of his anti-Baconianism in its demise. What place did he occupy on the fault-lines dividing Whewell from Babbage and Babbage from Herschel, for example? Whatever the specific answer will be, addressing such a question is likely to advance our understanding of the fascinating world of pre-Darwinian science and philosophy, as well as De Morgan's place in that world.

Bibliography

Alborn, Timothy L., 'The Business of Induction: Industry and Genius in the Language of British Scientific Reform, 1820–1840', *History of Science*, 34:1 (1996), 91–121. https://doi.org/10.1177/007327539603400104

Ashworth, William J., 'The Calculating Eye: Baily, Herschel, Babbage and the Business of Astronomy', *The British Journal for the History of Science*, 27:4 (1994), 409–41. https://doi.org/10.1017/s0007087400032428

—, 'Memory, Efficiency, and Symbolic Analysis. Charles Babbage, John Herschel, and the Industrial Mind', *Isis*, 87:4 (1996), 629–53. https://doi.org/10.1086/357650

—, *The Trinity Circle: Anxiety, Intelligence and Knowledge Creation in Nineteenth-Century England* (Pittsburgh: University of Pittsburgh Press, 2021). https://doi.org/10.2307/j.ctv1tgx06p

Bellone, Enrico, *A World on Paper: Studies on the Second Scientific Revolution* (Cambridge, MA: The MIT Press, 1980 [1976]).

Boole, George, *An Investigation of the Laws of Thought on which are Founded the Mathematical Theories of Logic and Probabilities* (London: Walton & Maberly, 1854).

Brewster, David, 'On the History of the Inductive Sciences', *Edinburgh Review*, 66 (1837), 110–51.

Burke, Peter, *The Polymath: A Cultural History from Leonardo da Vinci to Susan Sontag* (New Haven and London: Yale University Press, 2020). https://doi.org/10.12987/ 9780300252088

Butts, Robert E., ed., *William Whewell's Theory of Scientific Method* (Pittsburgh: University of Pittsburgh Press, 1968).

Cahan, David, ed., *From Natural Philosophy to the Sciences: Writing the History of Nineteenth Century Science* (Chicago and London: The University of Chicago Press, 2003).

Cannon, W.F., 'Scientists and Broad Churchmen: An Early Victorian Intellectual Network', *Journal of British Studies*, 4:1 (1964), 65–88.

Cobb, Aaron D., 'Is John F.W. Herschel an Inductivist about Hypothetical Inquiry?' *Perspectives on Science*, 20:4 (2012), 409–39. https://doi.org/10.1162/posc_a_00080

Cohen, L. Jonathan, 'Some Historical Remarks on the Baconian Conception of Probability', in L. Jonathan Cohen, *Knowledge and Language* (Cham: Springer, 1980), pp. 245–59.

Cowles, Henry, *The Scientific Method: An Evolution of Thinking from Darwin to Dewey* (Cambridge, MA: Harvard University Press, 2020). https://doi.org/10.4159/ 9780674246843

Daston, Lorraine, *Classical Probability in the Enlightenment* (Princeton: Princeton University Press, 1988).

De Morgan, Augustus. 'History of the Inductive Sciences from the Earliest to the Present Times. By W. Whewell', *The Athenæum*, 541, 10 March 1838, pp. 179–81.

—, *First Notions of Logic (Preparatory to the Study of Geometry)*, 2nd edn (London: Taylor & Walton, 1840).

—, *Formal Logic: Or, The Calculus of Inference, Necessary and Probable* (London: Taylor and Walton, 1847).

—, 'The Progress of the Doctrine of the Earth's Motion between the Times of Copernicus and Galileo, Being Notes on the Ante-Galilean Copernicans', *Companion to the Almanac for 1855*, 5–25.

—, 'The Philosophy of the Inductive Sciences. By W. Whewell', *The Athenæum*, 672, 12 September 1840, pp. 707–09.

—, 'The Works of Francis Bacon, ed. by James Spedding, R. Leslie Ellis, and Douglas D. Heath. 5 vols.', *The Athenæum*, 1612, 18 September 1858, pp. 367–68.

[—], Review of William Whewell's *Novum Organum Renovatum*, *The Athenæum*, 1628 (8 January 1859), 42–44.

[—], Review of William Whewell's *The Philosophy of Discovery*, *The Athenæum*, 1694 (14 April 1860), 501–03.

—, 'Logic (1860)', in *On the Syllogism and Other Logical Writings by Augustus De Morgan*, ed. by Peter Heath (London: Routledge & Kegan Paul), 1966, pp. 247–66.

Despeaux, Sloan Evans, and Adrian C. Rice, 'Augustus De Morgan's Anonymous Reviews for *The Athenæum*: A Mirror of a Victorian Mathematician', *Historia Mathematica*, 43:2 (2016), 148–71. https://doi.org/10.1016/j.hm.2015.09.001

Dongen, Jeroen van, and Herman Paul, eds, *Epistemic Virtues in the Sciences and the Humanities* (Cham: Springer, 2017). https://doi.org/10.1007/978-3-319-48893-6

Ducheyne, Steffen, 'Whewell's Philosophy of Science', in *The Oxford Handbook of British Philosophy in the Nineteenth Century*, ed. by W.J. Mander (Oxford and New York: Oxford University Press, 2014), pp. 71–88. https://doi.org/10.1093/oxfordhb/9780199594474.013.011

—, 'Kant and Whewell on Bridging Principles between Metaphysics and Science', *Kant-Studien*, 102:1 (2012), 22–45. https://doi.org/10.1515/kant.2011.002

Ellis, Robert Leslie, 'On the Foundations of the Theory of Probabilities', *Transactions of the Cambridge Philosophical Society*, 8 (1844), 1–6.

—, 'Remarks on an Alleged Proof of the Method of Least Squares', *Philosophical Magazine*, 37 (November 1850), 321–28.

Enros, P.C., 'The Analytical Society (1812–13): Precursor of the Renewal of Cambridge Mathematics', *Historia Mathematica*, 10:1 (1983), 24–47.

Fitzpatrick, William John, *Memoirs of Richard Whately, Archbishop of Dublin. Volume I* (London: Richard Bentley, 1864).

Gillispie, Charles, *The Edge of Objectivity* (Princeton: Princeton University Press, 1960).

Hamilton, William, 'IV. – Logic. In Reference to the Recent English Treatises on that Science', in Sir William Hamilton, *Discussions on Philosophy and Literature, Education and University Reform. Chiefly from the Edinburgh Review; Corrected, Vindicated, Enlarged, in Notes and Appendices*. 2nd edn (London: Longman, Brown, Green and Longmans, 1853), pp. 118–75.

Heis, Jeremy, 'Attempts to Rethink Logic', in *The Cambridge History of Philosophy in the Nineteenth Century (1790–1870)*, ed. by A.W. Wood and S.S. Hahn (Cambridge: Cambridge University Press, 2012), pp. 95–132. https://doi.org/10.1017/cho9780511975257.008

Herschel, John F.W., *A Preliminary Discourse on the Study of Natural Philosophy* (London: Longman, Rees, Orme, Brown & Green, and John Taylor, 1830).

—, 'Review of the *History and Philosophy of the Inductive Sciences*', *Quarterly Review*, 135 (June 1841), 96–130.

Hobart, Michael E., and Joan L. Richards, 'De Morgan's Logic', in *Handbook of the History of Logic. Volume 4: British Logic in the Nineteenth Century*, ed. by Dov M. Gabbay and John Woods (Amsterdam: North-Holland, 2008), pp. 283–330. https://doi.org/10.1016/s1874-5857(08)80010-6

Jongsma, Calvin, 'Richard Whately's Revitalization of Syllogistic Logic', in *Aristotle's Syllogism and the Creation of Modern Logic: Between Tradition and Innovation, 1820–1930*, ed. by Lukas M. Verburgt and Matteo Cosci (London: Bloomsbury Academic, 2023).

Laudan, Larry, 'Theories of Scientific Method from Plato to Mach: A Bibliographic Review', *History of Science*, 7:1 (1968), 1–63.

—, 'William Whewell on the Consilience of Inductions', *The Monist*, 55 (1971), 368–91.

—, 'Thomas Reid and the Newtonian Turn of British Methodological Thought', in *The Methodological Heritage of Newton*, ed. by Robert E. Butts and John W. Davis (Toronto: University of Toronto Press, 2016), pp. 103–31. https://doi.org/10.3138/9781442632783-007

—, 'Induction and Probability in the Nineteenth Century', in *Logic, Methodology and Philosophy of Science IV. Proceedings of the Fourth International Congress for Logic, Methodology and Philosophy of Science, Bucharest, 1971*, ed. by Patrick Suppes et al. (Amsterdam & London: North-Holland, 1973), pp. 429–38.

Maas, Harro, '"A Hard Battle to Fight": Natural Theology and the Dismal Science, 1820–50', *History of Political Economy*, 40:5 (2008), 143–67. https://doi.org/10.1215/00182702-2007-064

Merrill, Daniel D., *Augustus De Morgan and the Logic of Relations* (Dordrecht: Kluwer Academic Publishers, 1990). https://doi.org/10.1007/978-94-009-2047-7

Mill, John Stuart, 'Review of Whately's *Elements of Logic*', in J.M. Robson, *The Collected Works of John Stuart Mill. Volume XI: Essays on Philosophy and the Classics* (Toronto: Toronto University Press, 1978), pp. 3–35.

Olson, Richard S., *Scottish Philosophy and British Physics, 1740–1870* (Princeton: Princeton University Press, 1975).

Oslington, Paul, 'Natural Theology, Theodicy, and Political Economy in Nineteenth-century Britain: William Whewell's Struggle', *History of Political Economy*, 49:4 (2017), 575–606.

Panteki, Maria, 'French "Logique" and British "Logic": On the Origins of Augustus De Morgan's Early Logical Inquiries, 1805–1835', in *Handbook of the History of Logic. Volume 4: British Logic in the Nineteenth Century*, ed. by Dov M. Gabbay and John Woods (Amsterdam: North-Holland, 2008), pp. 381–457. https://doi.org/10.1016/s1874-5857(08)80012-x

Rice, Adrian, 'Augustus De Morgan: Historian of Science', *History of Science*, 34:2 (1996), 201–40. https://doi.org/10.1177/007327539603400203

—, ''Everybody Makes Errors': The Intersection of De Morgan's Logic and Probability, 1837–1847', *History and Philosophy of Logic*, 24:4 (2003), 289–305. https://doi.org/10.1080/01445340310001599579

Rice, Adrian, and Eugene Seneta, 'De Morgan in the Prehistory of Statistical Hypothesis Testing', *Journal of the Royal Statistical Society*, 168:3 (2005), 615–27. https://doi.org/10.1111/j.1467-985x.2005.00367.x

Richards, Joan L., '"In a Rational World All Radicals Would be Exterminated": Mathematics, Logic and Secular Thinking in Augustus De Morgan's England', *Science in Context*, 15:1 (2002), 137–64. https://doi.org/10.1017/s026988970200039x

—, 'The Probable and the Possible in Early Victorian England', in *Victorian Science in Context*, ed. by Bernard Lightman (Chicago: University of Chicago Press, 1997), pp. 51–71.

—, *Generations of Reason: A Family's Search for Meaning in Post-Newtonian England* (New Haven: Yale University Press, 2021). https://doi.org/10.12987/yale/9780300255492.001.0001

Scarre, Geoffrey, *Logic and Reality in the Philosophy of John Stuart Mill* (Dordrecht and Boston: Kluwer Academic Publishers, 1989).

Schaffer, Simon, 'Genius in Romantic Natural Philosophy', in *Romanticism and the Sciences*, ed. by Andrew Cunningham and Nicholas Jardine (Cambridge: Cambridge University Press, 1990), pp. 82–98.

—, 'Scientific Discoveries and the End of Natural Philosophy', *Social Studies of Science*, 16:3 (1986), 387–420.

Secord, James A., *Visions of Science: Books and Readers at the Dawn of the Victorian Age* (Oxford: Oxford University Press, 2014). https://doi.org/10.7208/chicago/9780226203317.001.0001

Sgarbi, Marco, and Matteo Cosci, eds, *The Aftermath of Syllogism: Aristotelian Logical Argument from Avicenna to Hegel* (London: Bloomsbury Academic, 2018). https://doi.org/10.5040/9781350043558

Shortland, Michael, and Richard Yeo, 'Introduction', in *Telling Lives in Science: Essays on Scientific Biography*, ed. by Michael Shortland and Richard Yeo (Cambridge: Cambridge University Press, 1996), pp. 1–44. https://doi.org/10.1017/cbo9780511525292.002

Smith, Crosbie, and M. Norton Wise, *Energy and Empire: A Biographical Study of Lord Kelvin* (Cambridge: Cambridge University Press, 1989).

Smith, Jonathan, *Fact and Feeling: Baconian Science and the Nineteenth-Century Literary Imagination* (Madison: The University of Wisconsin Press, 1994).

Snyder, Laura J., *Reforming Philosophy: A Victorian Debate on Science and Society* (Chicago: University of Chicago Press, 2006). https://doi.org/10.7208/chicago/9780226767352.001.0001

—, '"The Whole Box of Tools": William Whewell and the Logic of Induction', in *Handbook of the History of Logic. Volume 4: British Logic in the Nineteenth Century*, ed. by Dov M. Gabbay and John Woods (Amsterdam: North-Holland, 2008), pp. 163–228. https://doi.org/10.1016/s1874-5857(08)80008-8

—, 'Hypotheses in 19th Century British Philosophy of Science: Herschel, Whewell, Mill', in *The Significance of the Hypothetical in Natural Science*, ed. by Michael Heidelberger and Gregor Schiemann (Berlin: Walter de Gruyter, 2009), pp. 59–76. https://doi.org/10.1515/9783110210620.59

—, *The Philosophical Breakfast Club: Four Remarkable Friends Who Transformed Science and Changed the World* (New York: Broadway Books, 2011). https://doi.org/10.5479/ sil.1026689.39088017816588

Strong, John V., 'The Infinite Ballot Box of Nature: De Morgan, Boole, and Jevons on Probability and the Logic of Induction', *PSA: Proceedings of the Biennial Meeting of the Philosophy of Science Association*, 1976:1 (1976), 197–211.

Todhunter, Isaac, *William Whewell D.D., Master of Trinity College, Cambridge. An Account of His Writings. With Selections from His Literary and Scientific Correspondence*, vol. 1 (London: Macmillan, 1876).

Van Evra, James, 'Richard Whately and Logical Theory', in *Handbook of the History of Logic. Volume 4: British Logic in the Nineteenth Century*, ed. by Dov M. Gabbay and John Woods (Amsterdam: North-Holland, 2008), pp. 75–92. https://doi.org/10.1016/s1874-5857(08)80006-4

Venn, John, 'Consistency and Real Inference', *Mind*, 1:1 (1876), 43–52.

Verburgt, Lukas M., 'Robert Leslie Ellis's Work on Philosophy of Science and the Foundations of Probability Theory', *Historia Mathematica*, 40:4 (2013), 423–54. https://doi.org/10.1016/j.hm.2013.07.003

—, ed., *A Prodigy of Universal Genius: Robert Leslie Ellis, 1817–1859* (New York: Springer, 2022). https://doi.org/10.1007/978-3-030-85258-0

Wettersten, John, *Whewell's Critics: Have They Prevented Him from Doing Good?* (Amsterdam & New York: Rodopi, 2005). https://doi.org/10.1163/9789004359246

Whately, Richard, ed., *Bacon's Essays: With Annotations* (London, 1856).

—, *Elements of Logic*. 2nd edn (London, 1827).

—, *Elements of Logic*. 9th edn(London, 1848).

Whewell, William, *Astronomy and General Physics, Considered with Reference to Natural Theology* (London: William Pickering, 1833).

[—], 'Modern Science – Inductive Philosophy Review of John F.W. Herschel's *A Preliminary Discourse on the Study of Natural Philosophy*', *Quarterly Review*, 45 (July 1831), 374–407.

—, 'Remarks on Mathematical Reasoning and on the Logic of Induction', in *The Mechanical Euclid*. 3rd edn(London, 1838).

—, *The Philosophy of the Inductive Sciences, Founded Upon Their History*, vol. 2 (London: John W. Parker, 1840).

—, 'Criticism of Aristotle's Account of Induction', *Transactions of the Cambridge Philosophical Society*, 10.1 (1850), 63–72.

—, *History of Scientific Ideas*, vol. 1 (London: John W. Parker, 1858).

—, *Novum Organon Renovatum* (London: John W. Parker, 1858).

—, *On the Philosophy of Discovery: Chapters Historical and Critical* (London: John W. Parker, 1860).

Yeo, Richard, 'Scientific Method and the Rhetoric of Science in Britain', in *The Politics and Rhetoric of Scientific Method. Histories Studies*, ed. by John A. Schuster and Richard Yeo (Dordrecht & Boston: D. Reidel Publishing Company, 1986), pp. 259–97. https://doi.org/10.1007/978-94-009-4560-9_8

—, 'An Idol of the Market-place: Baconianism in Nineteenth Century Britain', *History of Science*, 23:3 (1985), 251–98. https://doi.org/10.1177/007327538502300302

—, 'Genius, Method, and Morality: Images of Newton in Britain, 1760–1860', *Science in Context*, 2:2 (1988), 257–84. https://doi.org/10.1017/s0269889700000594

—, *Defining Science: William Whewell, Natural Knowledge and Public Debate in Early Victorian Britain* (Cambridge: Cambridge University Press, 1993). https://doi.org/ 10.1017/cbo9780511521515

Fig. 8 Augustus De Morgan pictured in the 1860s. (Public domain, via MacTutor, https://mathshistory.st-andrews.ac.uk/Biographies/De_Morgan/pictdisplay/)

PART II

BEYOND SCIENCE

Fig. 9 A student's sketch of De Morgan teaching at University College London in 1865. (MS ADD 7, reproduced by permission of UCL Library Services, Special Collections.)

6. De Morgan and Mathematics Education

Christopher Stray

> The experience of every day makes it evident that education develops faculties which would otherwise never have manifested their existence.
>
> — Augustus De Morgan[1]

Augustus De Morgan is well known as a writer on mathematics and logic, as a historian of mathematics, and as a teacher of the subject at University College London for several decades. In this chapter, I focus on a relatively unstudied aspect of his work: his writing on the contemporary teaching and learning of mathematics.[2] In this he drew on his own education at school and, especially, at Cambridge (1823–27), but also on his own career at the University of London (1828–31), renamed University College London (1836–66), and on his knowledge of teaching at Oxford and other institutions. The basis of his assessment of educational practices was analysis and comparison in both time and space. Spatial comparison was made between London, Cambridge, Oxford and sometimes Paris; temporal comparison involved his own teaching and his own past learning. As he put it with characteristic pointedness, when he gave his introductory lecture at the University of London on 5 November 1828, 'I began to teach myself to better purpose

---

1 Augustus De Morgan, *On the Study and Difficulties of Mathematics* (London: Baldwin & Cradock, 1831), p. 3.
2 My thanks for helpful discussion to Karen Attar, Nicolas Bell, Jonathan Smith and especially Adrian Rice.

than I had been taught, as does every man who is not a fool, [when he begins to teach others,] let his former teachers be what they may'.³

Of the seven schools De Morgan attended as a child, the only one he found of any use was that run by Revd John Parsons in Redland near Bristol, which he attended from midsummer 1820 to Christmas 1822.⁴ Parsons was an Oxford graduate; though his 1805 BA was only a pass degree, he was later elected to a fellowship at Oriel College (1807–12).⁵ As in most schools in that period, Classics received much more attention than mathematics.⁶ De Morgan's account of his experiences there indicates both the dominance of Classics and his objection to the rote learning typical of the period:

> The poor ignorant Virgil and Homer scanners, and their subordinate Euclid and algebra drillers, had not the slightest idea that *a memory* is the adjunct of each faculty, that the training of one is of little or no help to another, and that the memory of words, which they over-cultivated, differs widely among young people. The allowance was forty lines a day, Latin and Greek alternately, for five days in a week, the whole to be repeated in one lot on Saturday.⁷

Why 'subordinate'? Because the education of the middle and upper classes in this period was dominated by Classics, beginning with the rote learning of composition and translation, and the mechanical learning of grammar, syntax and verse-making. This last is what is referred to by 'scanners': scanning the metre of Latin and Greek poetry, then attempting to reproduce it in 'nonsense verses' (written with no regard for sense), then 'sense verses', which made sense as well as scanning metrically.⁸ This classically-dominated scene included the public schools and

---

3   Quoted by Sophia Elizabeth De Morgan, *Memoir of Augustus De Morgan* (London: Longmans, Green, 1882), p. 29, from his 'A True and Authentic List of the Teachers of A. De Morgan', in his 'Memorandums on the Descendants of Captain John De Morgan...', UCL Special Collections, MS. ADD. 7, f. 155. Sophia omitted the phrase in parentheses from her published memoir.
4   S.E. De Morgan, *Memoir*, pp. 5–9; UCL MS Add 7, f. 154.
5   Parsons was given the living of Marden, Wiltshire in 1816, and apparently kept it till his death on 31 July 1844 (*Gentlemans Magazine*, 176 (1844), p. 327).
6   In the 1850s, De Morgan learned from Robert Leslie Ellis (Trinity 1836, Senior Wrangler 1840) that Ellis's two brothers had also been at Parsons's school (S.E. De Morgan, *Memoir*, p. 9). De Morgan and Ellis were alike in that they eventually concentrated on mathematics, while retaining a knowledge of Classics; and both men had claims to polymathy.
7   S.E. De Morgan, *Memoir*, p. 8.

Oxford; Cambridge stood out as an anomaly because of the centrality of mathematics to its curriculum, the Senate House Examination being the university's only degree examination.[9] The Cambridge colleges, however, were homes of classical teaching and learning, though some of them, notably St John's, until the 1790s the largest college in the university, were important centres of mathematical study. Boys coming up to Cambridge were first admitted to a college and then matriculated (enrolled) at the university. The annual college examinations formed part of a transition from the classical curriculum of most schools to the mathematical content of the Senate House examination, which was taken in an undergraduate's tenth term.[10]

This transition can be seen in De Morgan's progress through Trinity.[11] He was admitted to the college on 1 February 1823, and so entered in a by-term: in other words, not in the Michaelmas (autumn) term which began the academic year. In the college examination in May he was top of the second class. In her memoir, his widow Sophia De Morgan pointed out that this was only three months after his admission, but we should also take account of the fact that mathematics was a minority presence in this examination. His mother was keen for him to concentrate on Classics rather than mathematics, then to become a clergyman in the evangelical wing of the Church of England.[12] In 1824, however, when he was a 'junior sophister' (second-year undergraduate), he was in the first class in the second-year examination, which was dominated by

---

8   Two of the lower forms at Eton were named after this system, Nonsense and Sense. Cf. C.A. Stray, *Classics Transformed: Schools, Universities, and Society in England 1830–1960* (Oxford: Oxford University Press, 1998), pp. 68–71. Some schoolmasters rose above mere scanning: in 1820 De Morgan's headmaster had read out one hundred lines of Homer versified by Walter Scott, as De Morgan recalled in a letter to John Herschel of 29 April 1862 (S.E. De Morgan, *Memoir*, p. 309).

9   The examination became known as the Mathematical Tripos after the foundation of the Classical Tripos, first examined in 1824.

10 The college examinations had been instituted first at St John's in 1765, Trinity following suit in 1790. By 1830 they were in place at all the colleges. Cf. C.A. Stray, 'From Oral to Written Examination: Oxford, Cambridge and Dublin 1700-1914', *History of Universities*, 20 (2005), 76–130.

11 The best treatment of this period in De Morgan's life remains A.C. Rice, 'Augustus De Morgan and the Development of University Mathematics in London in the Nineteenth Century' (Unpublished Ph.D. Diss., Middlesex University, 1997), pp. 20–35.

12 S.E. De Morgan, *Memoir*, p. 12. He came under considerable pressure from his mother to attend the sermons of Charles Simeon, the evangelical vicar of Holy Trinity church, rather than just to 'the [college] chapel' (S.E. De Morgan, *Memoir*, p. 13).

mathematics. His tutor, John Higman, told De Morgan's mother that he was 'not only in our first class, but far, very far, the first in it'.[13]

By this time he had, according to Sophia, been captivated by George Peacock's mathematical lectures, which opened up 'new life' for him as he listened to 'Peacock's explanations'. Sophia refers to these as 'university lectures', but such lectures were in this period given only by professors, who usually charged for attendance; Peacock was not a professor until his election to the Lowndean chair of astronomy in 1837. He had been appointed an assistant tutor on being elected fellow of Trinity in 1814, so would have been giving lectures within one of the three tutorial sides (in this period undergraduates attended lectures only within their own sides). In 1823 he was appointed a tutor in succession to John Hustler, so had probably taught on Hustler's side earlier; De Morgan was perhaps given permission to attend his lectures. De Morgan's own tutor John Higman seems to have been the strongest influence on him; he was one of those who sent in testimonials for De Morgan's successful application for the London mathematics chair in 1828 (in fact he sent two), while Peacock did not send a testimonial.[14]

According to Sophia, De Morgan lived out of college for the first two years.[15] This was a common experience for Trinity undergraduates until a new court was opened in 1824, planned to cope with rising numbers of students and also to protect them from the temptations of extra-collegiate life.[16] De Morgan would normally have gone on to his third year (senior sophister) in 1824-5, but appears to have degraded, in effect taking the year off.[17] This practice became quite common in the 1820s, perhaps in response to the increasing difficulty of the Mathematical Tripos; in effect, it gave an undergraduate an additional year to prepare

---

13   S.E. De Morgan, *Memoir*, p. 12.
14   See Adrian Rice, 'Inspiration or Desperation? Augustus De Morgan's Appointment to the Chair of Mathematics at London University in 1828', *British Journal for the History of Science*, 30(3) (1997), 257–74.
15   S.E. De Morgan, *Memoir*, pp. 12–13.
16   The court was to be named King's Court or Brunswick Court, but in the end was simply called New Court. The building of the new court prompted the recording of room rents; De Morgan first appears in Easter 1826, living in Q1 Great Court, where he stayed until Lent term 1827. Sophia refers to his living 'over the gate', but Q staircase is in fact next to the Queen's Gate, on the south side of Great Court, the Great Gate being on the east side.
17   Sophia mentioned that he had been ill, perhaps as a result of overwork (*Memoir*, p. 15).

for the examination.[18] In De Morgan's case, it made sense to gain extra time after his truncated freshman year. He went on to win a college scholarship; Sophia's memoir places this in April 1825, but the college records make it clear that it was a year later.[19] He sat the Tripos in January 1827, emerging as Fourth Wrangler. Like William Whewell before him, he was defeated by men who were inferior mathematicians but superior examination candidates; this was perhaps the original source of the attacks on the domination of education by examinations that are such a feature of his later critiques of Cambridge mathematical training.

De Morgan also enjoyed music, being a talented flautist. On one occasion, musical performance combined with academic reading. As he wrote to William Rowan Hamilton in 1858:

> When I was an undergraduate, it happened to me to get very jolly in company with a party who were celebrating the new scholarship of our host. Being, as aforesaid, merry, we proceeded to sing; when it struck one of our party that we could sing as well as the choristers, a notion which came of punch and not of reason. To test the point we all got our surplices, and stood round the table, when a question arose as to what we should chant. Some one proposed $PV.VG : QV^2 :: CP^2 : DC^2$, which met with approbation. We tried to make it fit all manner of tunes; I remember 'Zitti Zitti', 'the Evening Hymn', and 'The Campbells are coming'. But we left off with a notion that Newton was not so easily set to music as we thought.[20]

---

18 This led in 1829 to a regulation that in order to check the practice, those who degraded after 30 Oct. 1830 could not obtain scholarships or sit for mathematical honours without special permission. Grace of 27 Feb. 1829, Cambridge University Archives, Degr.13.26. The term 'degrade' was glossed in a contemporary dictionary of Cambridge slang, *Gradus ad Cantabrigiam* (London: J. Hearne, 1824), pp. 43–44. The practice was disapproved of by some: in a review of J.M.F. Wright's *Alma Mater*, a memoir of his time at Cambridge 1813-19, Henry Southern referred to Wright's having *degraded* – that is, of his having descended from a struggle with his equals, to contend with the men of the year below. 'Alma Mater, or Seven Years at Cambridge', *London Magazine*, 7 (1 April 1827), 441–54 (p. 454).

19 *Memoir*, p. 15, followed by Rice, 'Augustus De Morgan', p. 32. Of the friends of about his own age mentioned in the *Memoir* (p. 16), one gained a scholarship in 1824, two in 1825.

20 De Morgan to Hamilton, 1 April 1858 (see Robert Perceval Graves, *Life of Sir William Rowan Hamilton* (Dublin: Hodges, Figgis, 1889), vol. 3, p. 546). This event can be dated to April 1824 or 1825. The proposed text was from Book I, Proposition X of the *Principia*: the theorem in which Newton derives the Inverse Square Law for elliptical orbits. In 1818 the exam subjects for the second year were changed; the new list was headed by *Principia* 1, 6-14. De Morgan's notebook on *Principia* 1.10-11 survives and is dated Sept. 1824: Senate House Library, University

'Zitti Zitti' was a famous trio from Act 2 of Rossini's *Barber of Seville* (1816); Charles Wesley's Evening Hymn was published in his *Hymns and Sacred Poems* (1740); 'The Campbells are coming' was a Scots folk tune dating from around 1715. Altogether, this was a splendid crossover event, whose bizarre nature is echoed in a letter De Morgan received in 1831 from W.H. Smyth, secretary of the Royal Astronomical Society, who remarked that 'the disputations system, being both irritable and irritating, is altogether as unsuitable for astronomers as would be the dramatising of Newton's Principia'.[21]

Most of De Morgan's writing on education dealt with university teaching in Oxford, Cambridge or London. University College was his home institution, in the sense that his whole teaching career was spent there, but Cambridge was his alma mater, and much of his writing on education consisted of a critical commentary on Cambridge from the vantage point of London. But if we compare the two universities, we need to remember that the London University, founded in 1826, took a substantial number of its staff from Cambridge, and that this complicates the task of comparison.[22] The first professors of mathematics, Latin and Greek at the London University, as it then was (De Morgan, Thomas Hewitt Key and George Long) were all alumni of Trinity College, Cambridge, as was Long's successor Henry Malden.[23] Key and Long had been founding professors at Jefferson's new University of Virginia before coming to London; in both places it was possible to avoid the religious restrictions of Anglican institutions like Oxford and Cambridge, where college fellowships were vacated on marriage and

---

of London, Add MS 775/338. This is one of 17 surviving notebooks; most are undated, but they can be assigned to the period from summer 1824 to the end of 1826 (A.C. Rice, 'Augustus De Morgan', p. 381).

21 S.E. De Morgan, *Memoir*, p. 44. Disputations, held in Latin, were a relic of the medieval oral examination system. They were used to pre-sort tripos candidates, but became clumsy and inefficient, and were abolished in 1839.
22 This is a case of what has been called 'Galton's Problem': comparing entities which interact. See Raoul Naroll, 'Galton's Problem: The Logic of Cross Cultural Research', *Social Research*, 32 (1965), 428–51.
23 The historian of University College devoted to these four a section entitled 'A Cambridge group': H.H. Bellot, *University College, London 1826-1926* (London: University of London Press, 1929), pp. 80–96. Key held the chair of mathematics in Virginia, changed to Latin when he reached London, and moved to a chair of comparative grammar when he became headmaster of the University's junior department, University College School.

renewed only on ordination. At Oxford the curriculum was dominated by Classics, mathematics being a minority subject with small numbers of both teachers and entrance scholarships; in Cambridge, the dominant subject was mathematics, its primacy not challenged by the Natural Sciences Tripos, founded in 1851, until after De Morgan's death twenty years later.[24]

De Morgan's focus on the Cambridge system was due to several factors: his own experience of it, his continuing contact with Cambridge men like Whewell and Peacock, and the fact that Cambridge was the dominant source of advanced mathematical teaching in Britain. De Morgan was uniquely placed to assess Cambridge mathematics, since his undergraduate career had been located at a time when Newtonian fluxions were giving way to the analytical/algebraic approach promoted by Peacock, Charles Babbage and John Herschel. De Morgan wrote to Whewell in 1861:

> Thank heavens I was at Cambridge at the interval between two systems, when thought about both was the order of the day even among undergraduates. There are pairs of men alive who did each other more good by discussing $\dot{x}$ versus dx, and Newton versus Laplace, than all the private tutors ever do.[25]

This transitional state is reflected in De Morgan's undergraduate notebook on the differential calculus, where he noted that there were

> Different Systems pursued – Leibnitz used differentials of descending orders. Newton used the Principle of Limits or Fluxions. Lagrange rejecting both infinitesimals and Limits has used a method purely algebraical.[26]

De Morgan outlined his negative views of the Cambridge system on numerous occasions throughout his career. Several of his reviews for

---

24 For the history of the Mathematical Tripos, see J.W.L. Glaisher, 'The Mathematical Tripos', *Proceedings of the London Mathematical Society*, 18 (1886-7), 4–38 and W.W. Rouse Ball, *A History of the Study of Mathematics at Cambridge* (Cambridge: Cambridge University Press, 1889), pp. 187-219. The outstanding modern account is Andrew Warwick, *Masters of Theory: Cambridge and the Rise of Mathematical Physics* (Chicago: University of Chicago Press, 2003).
25 De Morgan to Whewell, 20 Jan. 1861 (S.E. De Morgan, *Memoir*, p. 306).
26 Senate House Library, University of London, MS 775/332, f. 1: De Morgan notebook, [1823–24]. Cf. Rice, 'De Morgan', p. 32.

The Athenæum pursued this theme: for example, in an 1856 review of a volume containing a collection of examination papers, he opined that '[a]t Cambridge subjects are *got up* to be *written out*; by many they are *crammed*, by some they are understood'.[27]

In De Morgan's eyes, the fundamental flaw in the system of examinations employed at Cambridge and elsewhere was that they instilled the necessity for the student to 'cram' knowledge by means of hasty and unsystematic revision. Consequently, a candidate for such an examination

> employs himself in collecting, without an attempt to digest. He puts by his unfinished and half-learnt material, to await the time when the examination is close at hand. Then, in the few days or weeks which precede the trial, he makes a rush at his crude mass of ill-understood notes, and endeavours to charge his unfortunate memory with the whole of it. There is no time to think of a process, to disentangle a confusion, or to give invention a fair chance of suggesting something for future consideration. All that is wanted is, to show a mass of learning on the day of examination, to make one successful effort during a few hours.[28]

Later the same year, in an *Athenæum* review of an 1848 pamphlet proposing reforms at Oxford and Cambridge, De Morgan indicted both universities for the inadequate teaching which had prompted the rise of private tuition. This had led

> not merely to the indolent student having the dose of *cram*, to speak the slang of the seats of learning, which will just serve him for his trial, but to the candidate for honours being fed on a diet which, though there may be larger allowance of it, is often not a bit more wholesome ... There is nourishment at both Universities in plenty for those who will take it in proper quantities and allow it natural digestion; and there are many who do so take it—but they are certainly not the majority. ... Whatever of sound culture may be proposed, there are many who will neither read nor think except for the examination which is to them all in all.[29]

---

27 [De Morgan], '[Review of] The examination papers of the Society of Arts, June 1856', *The Athenæum*, 1520 (13 Dec. 1856), p. 1531.

28 De Morgan, 'On the Effects of Competitory Examinations, Employed as Instruments in Education', *The Athenæum*, 1096 (28 Oct. 1848), pp. 1076–77; repr. in *The Educational Times* (1 Dec. 1848), pp. 56–59.

29 [De Morgan], Review of C. Daubeny, *Brief Remarks on the Correlations of the Natural Sciences*, *The Athenæum*, 1070 (29 April 1848), p. 431. Charles Daubeny was

De Morgan's feelings were shared by another Trinity man, Robert Leslie Ellis (Senior Wrangler 1840). Ellis hated the obsession with competition, which could in a sense be avoided only by coming first. In 1839 he wrote of 'my ... repugnance to the wrangler making process. There is but one place for me, & that I cannot obtain.'[30]

The obsession with intense competition was not the only aspect of Cambridge mathematics of which De Morgan complained. The way in which teaching and research were organised meant that it was almost impossible for teachers and students outside Cambridge to access the publications associated with the Mathematical Tripos. This problem was identified by De Morgan in a review of George Peacock's *Algebra* published in 1835.[31] As he pointed out, Cambridge students focused on Cambridge tripos problems rather than on textbooks, and so the whole system was both intense and introverted.[32] The scale of the problem was reflected in a satirical pamphlet written by Duncan Farquharson Gregory, Fifth Wrangler of 1837, in 1838, entitled a 'Prospectus of the Society for the Translation of Cambridge Mathematical Books into Intelligible English'.[33] This opened by explaining that 'from the singularity of dialect which prevails among the Works published at this University ... these works are wholly unintelligible to people at large unconnected with Cambridge'.[34] The principal objects of the Society are then listed:

Professor of Chemistry at Oxford.

30   Cambridge, Trinity College Library, Add. MS a.82.1: R.L. Ellis, diary 8 Feb.1839. Cf. C.A. Stray, 'From Bath to Cambridge: The Early Life and Education of Robert Leslie Ellis', in *A Prodigy of Universal Genius: Robert Leslie Ellis 1817–1859*, ed. by L.M. Verburgt (Berlin: Springer Nature, 2022), pp. 3–19.

31   De Morgan, 'Peacock's Algebra', *Quarterly Journal of Education*, 9 (1835), 293–311.

32   De Morgan, 'Peacock's Algebra', p. 299. The implications of this situation were well discussed by Andrew Warwick in his *Masters of Theory*, pp. 151–54.

33   Anonymous, but ascribed in MS to Gregory and dated 21 November 1838. Copies in Cambridge University Archives, CUR 28.6.2, and in Cambridge University Library, Cam a.500.9.22; repr. in C. Whibley, *In Cap and Gown: Three Centuries of Cambridge Wit* (London: Kegan Paul, 1889), pp. 161–64. Gregory was the founding editor of the *Cambridge Mathematical Journal* (1837-), on which he worked with Archibald Smith (Senior Wrangler 1836) and Robert Leslie Ellis (Senior Wrangler 1840).

34   De Morgan had made the same point: '... the Cambridge works are found difficult by other students, except those who know the secret'. (De Morgan, 'Peacock's Algebra', p. 299.)

> I To translate the letter-press into the English language ... the Society will not consider itself restricted to the ordinary proportion of one line of explanation to seven pages of symbols.
>
> ...
>
> III Where possible, to discover and explain the author's meaning in those passages where he does not seem to have fully comprehended it himself.

The Cambridge system took some of its intensity from the fact that examiners were often chosen from very recent high-performing graduates. In 1865, after having watched and commented on Cambridge mathematics for several decades, De Morgan remarked that 'The Cambridge examination is nothing but a hard trial ... between the Senior Wrangler that is to be this present January, and the Senior Wranglers of some three or four years ago'.[35] These young examiners had a great degree of freedom in setting examination papers, and this at times led to considerable variety in the content, style and mathematical notation of questions, a variety which caused difficulty to examination candidates. In 1835, De Morgan complained:

> To have no community of system—to have the moderators of one college using one, and those of another using another—each forcing his own upon the whole University during his year of office,—to oblige the same student to read books of different notation ...—to keep him in suspense as to what notation he will be examined in ... will be no advantage to the cause of science in Cambridge.[36]

---

35 De Morgan, 'Speech of Professor De Morgan, President, At the First Meeting of the Society, January 16th, 1865', *Proceedings of the London Mathematical Society*, 1 (1865), 1-9 (pp. 3–4); cf. S.E. De Morgan, *Memoir*, p. 283. In 1869 an article on wranglers stated that 'As a general thing, the year after their MA degree finds them and other high wranglers setting Senate-house problems and riders in their turn'. (Anon., 'A Word about Wranglers', *Daily News*, 30 Jan. 1869, p. 5.)

36 De Morgan, 'Cambridge Differential Notation. On the Notation of the Differential Calculus, Adopted in Some Works Lately Published in Cambridge', *Quarterly Journal of Education*, 8 (1834), 100–10 (p. 110). This was largely a critique of Samuel Earnshaw's *On the Notation of the Differential Calculus* (Cambridge: J. and J.J. Deighton, 1832). Earnshaw had been Senior Wrangler in 1831 and was a successful coach from then till 1847. For an analysis of the problems with the Mathematical Tripos in the 1830s and 1840s, see C.A. Stray, 'The Slaughter of 1841: Mathematics and Classics in Early Victorian Cambridge', *History of Universities*, 32:2 (2022), 143–78.

This lack of central control had its advantages, as De Morgan recognised, in that a single examiner could introduce an innovation via the papers he set, as George Peacock had in 1817 when he employed analytic notation.[37] In 1848 an examining board was set up to deal with this issue of control and consistency; it issued annual reports and recommendations, and presided over an examination whose structure remained largely unchanged until 1873.[38]

In the following year, De Morgan sent a typically witty comment on Cambridge teaching to Whewell:

> The spoon will do well for inductive logic ... What is more inductive than a spoon?—moreover, spoon feeding is synthetical, which induction is—knife and fork feeding is analytical ... remember that the spoon process, as hitherto understood at Cambridge is rather ingoosative than inductive.

This letter plays on the connotation of force feeding in 'induction', while inventing 'ingoosative' to contrast 'duc(k)' with 'goose', the victim of the force-feeding that leads to the production of *pâté de foie gras*. De Morgan is also hinting at the Wooden Spoon, the title given to the undergraduate who scored the lowest marks in mathematical honours.[39] There is a lot going on in this passage, a nice example of De Morganian deep fun.[40]

While most of De Morgan's critical articles on the teaching of mathematics were devoted to Cambridge, two dealt with the Ecole Polytechnique, founded in Paris in 1794, whose institutional and intellectual style offered an interesting contrast to British universities and academies.[41] Both of these articles were published in the *Quarterly Journal of*

---

37 De Morgan, 'Wood's *Algebra*', *Quarterly Journal of Education*, 3 (1832), 276–85 (p. 276).
38 In his evidence to the 1850 Royal Commission on Cambridge, Whewell declared that 'a permanent board of examiners would be more consistent from year to year, and less affected by peculiar views and habits of the Examiners' (*Report of Her Majesty's Commissioners Appointed to Inquire into the State, Discipline, Studies, and Revenues of the University and Colleges of Cambridge: Together with the Evidence, and an Appendix* (London: HMSO, 1852), vol. 2, p. 272).
39 See C.A. Stray, 'Rank (dis)order in Cambridge 1753–1909: The Wooden Spoon', *History of Universities*, 26 (2012), 163–201.
40 Cambridge, Trinity College Library, Add. MS a.202/114: De Morgan to Whewell, 20 April 1849. The reference is perhaps to Whewell's *On the History of the Inductive Sciences* or *Philosophy of the Inductive Sciences*, new editions of which were published in 1847; Whewell's surviving letters to De Morgan do not include any from 1849.
41 In both institutions, undergraduate slang was permeated by mathematics—for example, the 'argot de l'X' of the Ecole included a formula for the curve of

Education (1831-35), edited by De Morgan's London University colleague, the professor of Greek George Long; De Morgan contributed over thirty articles to the *Journal*.[42] The first article provided a historical account of the Ecole; the second compared the level and nature of its teaching with that of Cambridge and with that of the Royal Military Academy, Woolwich.[43] The Ecole, he concluded, stood somewhere between the two: as an institution founded to train military engineers, it resembled Woolwich; as the home of high-level mathematical analysis, it rivalled Cambridge. De Morgan ended by pointing to a shared weakness:

> It is the besetting sin of *all* public places of education to become hotbeds for forcing the first order of talent to the neglect of the rest. ... It must be remembered that those who are pointed out as proofs of a good system, are generally those who would have instructed themselves under any system. The question always should be asked, How are those taught who most want teaching?[44]

This concern for the bulk of ordinary students can also be seen in one of De Morgan's earliest contributions to the *Quarterly Journal of Education*, which ended by identifying

> the fault of our schools in general. It is not recollected that they cannot expect to make learned men; but they may make good learners, and at the same time produce such a desire for knowledge as shall lead the individual to devote himself to study, where it is not matter of compulsion, as in the Universities, and still more amid the occupations of life.[45]

---

the students' uniform cap. On the Ecole and mathematics, see Ivor Grattan-Guinness, *Convolutions in French Mathematics, 1800–1840*, 3 vols. (Basel: Birkhauser Verlag, 1990).

42 His contributions are listed in Stray, 'Introduction', *The Quarterly Journal of Education*, 10 vols. (London: Routledge, 2008), vol. 1, pp. xiii–xiv. The founders of the University were also responsible for setting up the Society for the Diffusion of Useful Knowledge, to whose *Penny Cyclopaedia* (1833–43) De Morgan contributed over 700 articles: see Chapters 4 and 8 in this volume.

43 De Morgan, 'Polytechnic School of Paris', *Quarterly Journal of Education*, 1 (1831), 57–74; 'Ecole Polytechnique', *Quarterly Journal of Education*, 10 (1835), 330–40.

44 De Morgan, 'Ecole Polytechnique', pp. 339–40.

45 De Morgan, 'On Mathematical Instruction', *Quarterly Journal of Education*, 1 (1831), 264–79 (p. 279).

One of De Morgan's most eloquent attacks on the Cambridge system came in the form of a mock examination paper, a genre popular among Cambridge undergraduates:

> Q. What is knowledge?
>
> A. A thing to be examined in.
>
> Q. What is the instrument of knowledge?
>
> A. A good grinding tutor.
>
> Q. What is the end of knowledge?
>
> A. A place in the civil service, the army, the navy, & (as the case may be).
>
> Q. What must those do who would show knowledge?
>
> A. Get up subjects and write them out.
>
> Q. What is getting up a subject?
>
> A. Learning to write it out.
>
> Q. What is writing out a subject?
>
> A. Showing that you have got it up.[46]

Although much of De Morgan's criticism was directed at Cambridge, he was also unsparing in his assessment of University College London, as it became in 1836, and of the examining University of London to which it belonged from that year.[47] The denunciation of cramming quoted above

---

46 S.E. De Morgan, *Memoir*, p. 184. Sophia De Morgan called this 'an illustrative "Cambridge examination"', presumably quoting the author's own title (p. 183). A 'grinder' was a coach or private tutor (see C.A. Stray, *Grinders and Grammars: A Victorian Controversy* (Reading: Textbook Colloquium, 1995)). An earlier example of the mock-examination paper genre was dated 5 Dec. 1815 but carried the imaginary date 'Undecember 9657'. The questions included: (10) Prove all the **roots** of radical reform to be either **irrational** or **impossible**; (13) Reconcile Hoyle and Euclid, the latter of whom defines a point to be without magnitude, the former to equal five. (Cambridge University Library, Cambridge Papers, MP [unpaginated].) Edmond Hoyle's *A Short Treatise on the Game of Whist* (1742) was the standard authority on the rules of card games and was not superseded until 1864.

47 See in general Rice, 'Augustus De Morgan', pp. 167–73.

from his 1848 introductory lecture surely drew its ammunition from his own experiences in London.

Oxford teaching and examining was very different from that in Cambridge, mathematics being a minority subject, as Classics was in Cambridge.[48] In 1832 De Morgan took the opportunity afforded by reviewing a reformist pamphlet by Baden Powell, Savilian Professor of Geometry, to give a wide-ranging critique of Oxford mathematics.[49] His complaint was less with the teaching than with the dismissive attitude of most Oxford dons toward the subject. In some ways, indeed, he thought the Oxford mode of examination to be superior to those in Cambridge and London. In 1853 De Morgan was asked to specify his objections to the London examination system.[50] In his reply, De Morgan put this in a wider context:

> There are two systems in this country,—that of Oxford, in which the candidate for classical honours is examined against his subject; that of Cambridge, in which the candidate for mathematical honours is examined against his competitors. At Oxford, his class determines his qualification; at Cambridge, his place determines whether he is above or below any given competitor. At Oxford his mind may, though not without certain wholesome restraint, develop itself in reading and thought dictated by its natural bent. At Cambridge the examination realised the bed of Procrustes. The Oxford system has a tendency to develop the useful differences between the varied types of human character. The Cambridge system is an unconscious effort to destroy them.[51]

---

48   Keith C. Hannabuss, 'Mathematics', in *The History of the University of Oxford VII: Nineteenth-century Oxford, Part 2*, ed. by M.G. Brock and M.C. Curthoys (Oxford: Oxford University Press, 2000), pp. 443–55; Christopher Stray, 'Curriculum and Style in the Collegiate University: Classics in Nineteenth-Century Oxbridge', *History of Universities*, 16 (2001), 183–218; repr in Christopher Stray, *Classics in Britain: Scholarship, Education, and Publishing, 1800-2000* (Oxford: Oxford University Press, 2018), pp. 31–52.
49   De Morgan, 'State of Mathematical and Physical Sciences in Oxford', *Quarterly Journal of Education*, 4 (1832), 191–208.
50   Sophia (*Memoir*, p. 222) identified the recipient of De Morgan's reply as 'Professor Michael Foster', but this cannot be the celebrated physiologist, who was born in 1836 and became a professor only in 1869, nor can it be De Morgan's colleague G.C. Foster, whose chair was awarded in 1865, unless the letter is misdated. She may have confused Foster with his father, also Michael Foster (1810–80), who had been a student at the London University soon after De Morgan was appointed.
51   De Morgan to Foster, 15 Nov. 1853 (S.E. De Morgan, *Memoir*, pp. 225–26).

Taking a wider view, De Morgan told William Whewell in 1861 that 'There seems to be a complete acquiescence in the maxim that Oxford shall settle what the world shall think, and Cambridge shall settle who is to be Senior Wrangler. It is getting worse and worse from day to day'.[52]

This survey of De Morgan's writing on mathematics education began by invoking the remark he made in his first lecture as professor of mathematics at what was then the London University in 1828, that he was determined to teach his students better than his own teachers had taught him. The principle he then enunciated can be applied to his own career, for he surely had opportunity to improve not only on his past teachers, but on his own past self, during his long career in London.[53]

When De Morgan began teaching in London in 1828, he faced both challenges and opportunities. The challenges came from the uncertainty about his future students' mathematical knowledge, and about his own capacity, as a novice, to teach them. The opportunities arose from the same source: he was a beginner, but in charge of his course, so could make his own curriculum. He began by setting this out, with a list of topics for each of the two years, the junior (first year) and senior (second year) classes each being divided into two semesters. Having found that his students had a wide ability range, he soon split each class into two divisions; he will have remembered from his Cambridge days that college lectures suffered from being mixed-ability classes, in which progress was made at the speed of the slowest. In his first year, then, De Morgan was very much feeling his way; but in the second, he had the advantage of previous experience, and of knowing more about those who were now in his senior class. The London University attached great importance to the utility of examinations, and De Morgan spent a lot of his lecture time on questioning his classes. This oral examination was then supplemented by written tests, which became a marked feature of the University's procedure. The beginning of De Morgan's career came

---

52   De Morgan to Whewell, 20 Jan. 1861 (S.E. De Morgan, *Memoir*, pp. 305–06).
53   On his appointment, see the detailed account in Rice, 'Inspiration or Desperation?'. For more on De Morgan's teaching career, see Adrian Rice, 'What Makes a Great Mathematics Teacher? The Case of Augustus De Morgan', *American Mathematical Monthly*, 106 (1999), 534–52.

at the point when both Oxford and Cambridge moved decisively away from oral examinations to written tests using printed papers.[54]

De Morgan's return in 1836 to University College London, as it had become, prompted him to give a second inaugural lecture.[55] In this he expressed his support for some aspects of the Cambridge system of which he had been, in many respects, a critic. In particular, he spoke in favour of the Cambridge study in depth of a few subjects, rather than the wide curricular range proposed for the new University of London, to which University College belonged. On the other hand, he approved of the plan to make the new University an examining institution which left the business of teaching to its constituent colleges (University College and King's College). In this respect, it followed the practice of Cambridge.

Much more evidence is available for De Morgan's second stint as professor of mathematics than for the first. This is because 320 exercise books survive in which he made notes for his lectures.[56] He made them available for students to read, and was thus perhaps harking back to his own Cambridge experience in the 1820s, when student reading centred on the perusal of a mass of manuscript texts written both by college tutors and by private tutors (coaches).[57] In these notebooks we can see De Morgan's exposition at work, as he strove to provide material for students to study at home, one of the two kinds of study he approved of; in 1848 he recommended 'diligent study in the retirement of the closet', along with 'haunting the benches of the lecture-room, and picking up what may chance to fall'.[58] It was in the lecture-room that they were confronted by the tall, stout form and domed forehead of the man known to several

---

54 This involved shifting from the belief that fairness consisted in tailoring questioning to individual candidates' abilities, as could be done in oral questioning, to the belief we now take for granted, that fairness could only be achieved by asking all candidates the same question. For more on this, and on the London University's procedures, see Stray, 'From Oral to Written Examination'.

55 De Morgan, *Thoughts Suggested by the Establishment of the University of London: An Introductory Lecture, Delivered at the Opening of the Faculty of Arts, in University College, Oct 16, 1837* (London: Taylor & Walton, 1837). Cf. Rice, 'Augustus De Morgan', pp. 155–62.

56 Senate House Library, University of London, MS 775. The books are of about 20pp each; the contents have been very well summarised and discussed by Adrian Rice in his 'Augustus De Morgan', pp. 174–203.

57 See Wright, *Alma Mater*, vol. 2, p. 24; Warwick, *Masters of Theory*, pp. 71, 144.

58 De Morgan, 'On the Effects of Competitory Examinations', p. 14.

generations of students as 'Gussy', the nickname itself an indication of their affection.[59] His student Stanley Jevons recalled that:

> As a teacher of mathematics De Morgan was unrivalled. He gave instruction in the form of continuous lectures delivered extempore from brief notes. The most prolonged mathematical reasoning, and the most intricate formulae, were given with almost infallible accuracy from the resources of his extraordinary memory. De Morgan's writings, however excellent, give little idea of the perspicuity and elegance of his viva voce expositions, which never failed to fix the attention of all who were worthy of hearing him.[60]

Sedley Taylor, who heard De Morgan in the early 1850s, wrote in similar terms:

> De Morgan's exposition combined excellences of the most varied kinds. It was clear, vivid, and succinct—rich too with abundance of illustration always at the command of enormously wide reading and an astonishingly retentive memory. A voice of sonorous sweetness, a grand forehead, and a profile of classic beauty, intensified the impression of commanding power which an almost equally complete mastery over Mathematical truth, and over the forms of language in which he so attractively arrayed it, could not fail to make upon his auditors.[61]

Taylor's comment on De Morgan's voice is worth noting: the two men shared a love of music (and membership of the Cambridge University Musical Society). Taylor himself published on acoustics, his analysis based on the pioneering work of Hermann von Helmholtz.[62]

---

59 In his memoir of Cambridge life, J.M.F. Wright remembered that his sympathetic mathematics lecturer John Brown was affectionately known as 'Johnny', whereas for the formal and pompous classical lecturer, James Monk, 'we never used any darling diminutives' (*Alma Mater*, vol. 1, p. 123).

60 W.S. Jevons, 'De Morgan, Augustus', *Encyclopaedia Britannica*, 11th edn (Cambridge: Cambridge University Press, 1910), pp. 8–10 (p. 8).

61 Sedley Taylor, 'Augustus De Morgan', *Cambridge University Reporter*, 3 May 1871, pp. 337–38 (p. 337). Taylor was never formally a pupil of De Morgan's, but attended University College School before matriculating at Cambridge in 1855.

62 Sedley Taylor, *Sound and Music: An Elementary Treatise on the Physical Constitution of Musical Sounds and Harmony, Including the Chief Acoustical Discoveries of Professor Helmholtz* (London: Macmillan, 1873).

Conclusion

De Morgan's students' memories of his teaching resonate with the impression one gains from his letters and his published work.[63] To read De Morgan's critiques of mathematical education in Oxford, Cambridge, London and Paris is to be impressed by the sharpness of his perceptions and the down-to-earth clarity with which he expresses them. It makes one regret that he never published his very last introductory lecture at University College, delivered in 1862:

> Seldom was an address listened to within those walls with a more lively interest, or with such mirth and hearty acclamation ... The illustrations which half filled the lecture were taken from common sayings, old ballads, and nursery rhymes.[64]

De Morgan was asked to print this lecture, but never found time to do so. As so often, his subject was the method of examining at Cambridge; and as we have seen, this was the major target of his critiques for three decades. As Sophia De Morgan remarked, this was a branch of his favourite subject, education, which we might see as the topic that linked his interests in mathematics and its history: a history which he helped to make.

Bibliography

Anon., *Gradus ad Cantabrigiam* (London: J. Hearne, 1824).

Anon., Obituary of Rev. John Parsons, *Gentlemans Magazine*, 176 (1844), p. 327.

Anon., 'A Word about Wranglers', *Daily News*, 30 January 1869, p. 5.

Bellot, H. Hale, *University College, London 1826–1926* (London: University of London Press, 1929).

De Morgan, Augustus, 'Polytechnic School of Paris', *Quarterly Journal of Education*, 1 (1831), 57– 74.

—, 'On Mathematical Instruction', *Quarterly Journal of Education*, 1 (1831), 264–79.

---

63 For students' memories, see Rice, 'Augustus De Morgan', pp. 239–52.
64 *Memoir*, p. 278. Some of De Morgan's notes for this survive: UCL Special Collections, MS Add. 2.

—, *On the Study and Difficulties of Mathematics* (London: Baldwin & Cradock, 1831).

—, 'Wood's *Algebra*', *Quarterly Journal of Education*, 3 (1832), 276–85.

—, 'State of Mathematical and Physical Sciences in Oxford', *Quarterly Journal of Education* 4 (1832): 191–208.

—, 'Cambridge Differential Notation. On the Notation of the Differential Calculus, Adopted in Some Works Lately Published in Cambridge', *Quarterly Journal of Education*, 8 (1834), 100–10.

—, 'Peacock's Algebra', *Quarterly Journal of Education*, 9 (1835), 293–311.

—, 'Ecole Polytechnique', *Quarterly Journal of Education*, 10 (1835), 330–40.

—, *Thoughts Suggested by the Establishment of the University of London: An Introductory Lecture, delivered at the opening of the Faculty of Arts, in university College, Oct 16, 1837* (London: Taylor & Walton, 1837).

—, Review of C. Daubeny, *Brief Remarks on the Correlations of the Natural Sciences*, The Athenæum, 1070, 29 April 1848, p. 431.

—, 'On the Effects of Competitory Examinations, Employed as Instruments in Education', *The Athenæum*, 1096, 28 October 1848, pp. 1076–77; reprinted in *The Educational Times*, 1 December 1848, pp. 56–59.

—, 'The examination papers of the Society of Arts, June 1856', *The Athenæum*, 1520, 13 December 1856, p. 1531.

—, 'Speech of Professor De Morgan, President, At the First Meeting of the Society, January 16th, 1865', *Proceedings of the London Mathematical Society*, 1 (1865), 1–9.

De Morgan, Sophia Elizabeth, *Memoir of Augustus De Morgan, With Selections from his Letters* (London: Longmans, Green, 1882).

Earnshaw, Samuel, *On the Notation of the Differential Calculus* (Cambridge: J. and J.J. Deighton, 1832).

Glaisher, James W. L., 'The Mathematical Tripos', *Proceedings of the London Mathematical Society*, 18 (1886–87), 4–38.

Grattan-Guinness, Ivor, *Convolutions in French Mathematics, 1800–1840*, 3 vols. (Basel: Birkhauser Verlag, 1990).

Graves, Robert Perceval, *Life of Sir William Rowan Hamilton*, 3 vols (Dublin: Hodges, Figgis, 1882–89).

Hannabuss, Keith C., 'Mathematics', in M.G. Brock and M.C. Curthoys, eds, *The History of the University of Oxford VII: Nineteenth-century Oxford, Part 2* (Oxford: Oxford University Press, 2000), pp. 443–55.

Jevons, William Stanley, 'De Morgan, Augustus', *Encyclopaedia Britannica*, 11th edn (Cambridge: Cambridge University Press, 1910), pp. 8–10.

Naroll, Raoul, 'Galton's Problem: The Logic of Cross cultural Research', *Social Research*, 32 (1965), 428–51.

Report of Her Majesty's Commissioners Appointed to Inquire into the State, Discipline, Studies, and Revenues of the University and Colleges of Cambridge: Together with the Evidence, and an Appendix (London: HMSO, 1852).

Rice, Adrian C., 'Augustus De Morgan and the Development of University Mathematics in London in the Nineteenth Century' (Ph.D. Diss., Middlesex University, London, 1997.)

—, 'Inspiration or Desperation? Augustus De Morgan's Appointment to the Chair of Mathematics at London University in 1828', *The British Journal for the History of Science*, 30:3 (1997), 257–74. https://doi.org/10.1017/s0007087497003075

—, 'What Makes a Great Mathematics Teacher? The Case of Augustus De Morgan', *The American Mathematical Monthly*, 106 (1999), 534–52. https://doi.org/10.2307/2589465

Rouse Ball, Walter William, *A History of the Study of Mathematics at Cambridge* (Cambridge: Cambridge University Press, 1889).

[Southern, Henry], 'Alma Mater, or Seven Years at Cambridge', *London Magazine*, 7, 1 April 1827, pp. 441–54.

Stray, Christopher A., *Grinders and Grammars: A Victorian Controversy* (Reading: The Textbook Colloquium, 1995).

—, *Classics Transformed: Schools, Universities, and Society in England 1830–1960* (Oxford: Oxford University Press, 1998).

—, 'Curriculum and Style in the Collegiate University: Classics in Nineteenth-century Oxbridge', *History of Universities*, 16 (2001), 183–218, reprinted in *Classics in Britain: Scholarship, Education, and Publishing, 1800–2000* (Oxford University Press, 2018), pp. 31–52. https://doi.org/10.1093/oso/9780199248421.003.0006

—, 'From Oral to Written Examination: Oxford, Cambridge and Dublin 1700–1914', *History of Universities*, XX (2005), 76–130. https://doi.org/10.1093/oso/9780199289288.003.0004

—, 'Introduction', *The Quarterly Journal of Education*, 10 vols. (London: Routledge, 2008), vol. 1, pp. v–xvii.

—, 'Rank (dis)order in Cambridge 1753-1909: The Wooden Spoon', *History of Universities*, 26 (2012), 163–201. https://doi.org/10.1093/acprof:osobl/9780199652068.003.0003

—, 'From Bath to Cambridge: The Early Life and Education of Robert Leslie Ellis', in *A Prodigy of Universal Genius: Robert Leslie Ellis 1817–1859*, ed. by Lukas M. Verburgt (Berlin: Springer Nature, 2021), pp. 3–19. https://doi.org/10.1007/978-3-030-85258-0_1

—, 'The Slaughter of 1841: Mathematics and Classics in Early Victorian Cambridge', *History of Universities*, 35:2 (2022), 143–78. https://doi.org/10.1093/oso/ 9780192884220.003.0005

Taylor, Sedley, 'Augustus De Morgan', *Cambridge University Reporter*, 3 May 1871, pp. 337– 38.

Taylor, Sedley, *Sound and Music: An Elementary Treatise on the Physical Constitution of Musical Sounds and Harmony, Including the Chief Acoustical Discoveries of Professor Helmholtz* (London: Macmillan, 1873).

Warwick, Andrew, *Masters of Theory: Cambridge and the Rise of Mathematical Physics* (Chicago: University of Chicago Press, 2003). https://doi.org/10.7208/chicago/ 9780226873763.001.0001

Whewell, William, *On the History of the Inductive Sciences*, new edn (London: J.W. Parker, 1847).

—, *Philosophy of the Inductive Sciences*, new edn (London: J.W. Parker, 1847).

Whibley, Charles, *In Cap and Gown: Three Centuries of Cambridge Wit* (London: Kegan Paul, 1889).

Wright, John M. F., *Alma Mater, or Seven Years at the University of Cambridge, by a Trinity-Man*, 2 vols (London: Black, Young and Young, 1827).

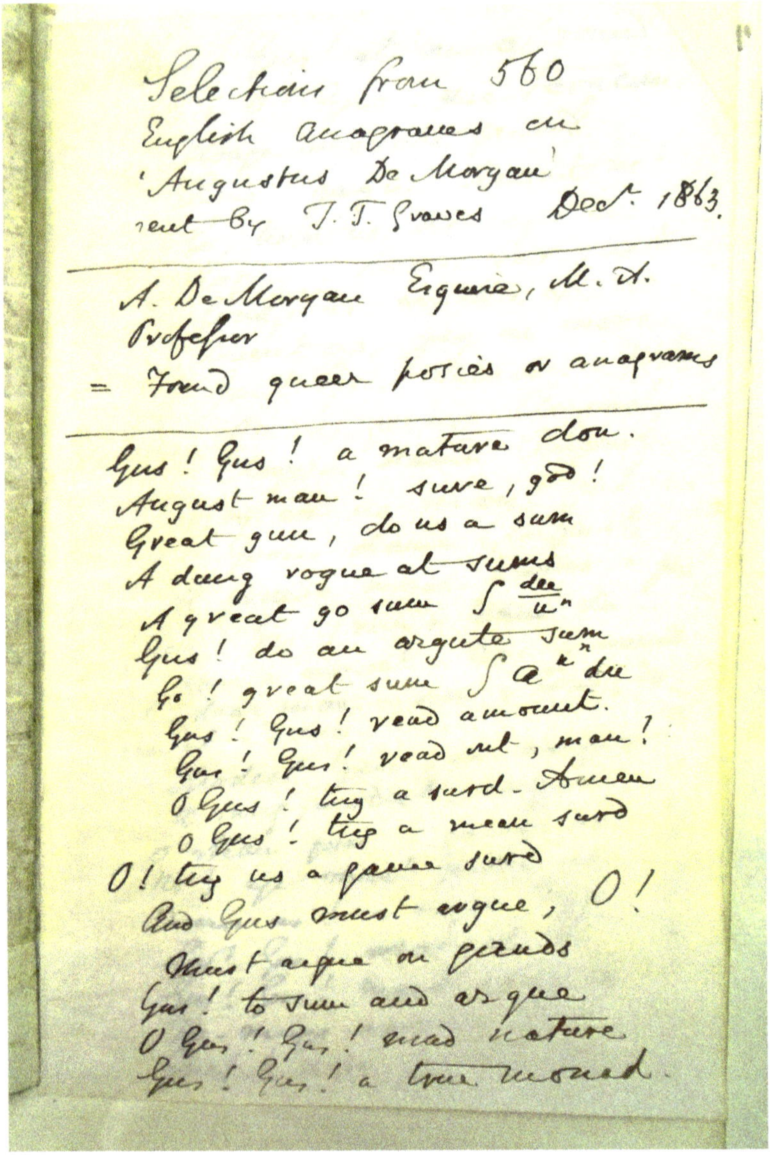

Selection from 560
English anagrams on
'Augustus De Morgan'
sent by J. T. Graves Decr. 1863.

A. De Morgan Esquire, M. A.
Professor
= Found queer poesies on anagrams

Gus! Gus! a mature don.
August man! sure, go!
Great gun, do us a sum
A dang rogue at sums
A great go sum $\int \frac{du}{u^n}$
Gus! do an argute sum
Go! great sum $\int a^n du$
Gus! Gus! read amount.
Gus! Gus! read out, man!
O Gus! try a surd - Amen
O Gus! try a mean surd
O! try us a game surd
And Gus must argue, O!
Must argue on grounds
Gus! to sum and argue
O Gus! Gus! read nature
Gus! Gus! a true mound.

Fig. 10 Part of a list of 560 English anagrams of De Morgan's name compiled by his friend, fellow mathematician and bibliophile John Thomas Graves in December 1863. Graves created over two thousand permutations in several languages, of which around twenty are featured in *A Budget of Paradoxes*. (MS ADD 7, reproduced by permission of UCL Library Services, Special Collections.)

7. De Morgan's *A Budget of Paradoxes*

Adrian Rice

> Great gun, do us a sum!
> — One of several anagrams of
> 'Augustus De Morgan' in *A Budget of Paradoxes*

Introduction

Even the briefest study of Augustus De Morgan's work shows him to have been an intriguing character whose enormous intellect was matched by a sharp wit and keen sense of humour. Although he is best known today for his eponymous laws, for a generation or so after his death one of his main claims to fame was work of a very different character. This was a 500-page book entitled *A Budget of Paradoxes*, published posthumously in 1872, one year after his death, which remains his most accessible and often-quoted work.[1] *A Budget of Paradoxes* arose from De Morgan's journalistic contributions to the weekly London-based literary magazine, *The Athenæum* (an antecedent of today's *New Statesman*), which supplied Victorian Britain with news of the latest developments in the arts, science, and politics.

Over a period of thirty years between 1840 and 1869, De Morgan wrote about one thousand book reviews for *The Athenæum* on a wide variety of topics, from mathematics to history, music and literature.[2] His

---

1 The book was originally published as: Augustus De Morgan, *A Budget of Paradoxes* (London: Longmans, Green, 1872). A two-volume second edition from Open Court Publishing, edited and annotated by David Eugene Smith, appeared in 1915. Unless stated otherwise, all citations from the *Budget* will be from the original 1872 edition.

2 Sloan Evans Despeaux and Adrian C. Rice, 'Augustus De Morgan's Anonymous Reviews for *The Athenæum*: A Mirror of a Victorian Mathematician', *Historia*

wide-ranging areas of expertise went hand-in-hand with his standing as one of nineteenth-century Britain's most knowledgeable bibliophiles.[3] By the end of his career, De Morgan had accumulated a library that stood at nearly four thousand items, many of which were acquired as complimentary copies of works he had reviewed for *The Athenæum*.

In August 1863, in a letter on the subject of *The Athenæum* to Henry Brougham, a former Lord Chancellor and one of the founders of University College London, De Morgan wrote:

> I am on the point of giving, in that paper, a series headed "A Budget of Paradoxes" giving a list, with comments, of all the circle-squarers, universe-builders, &c who are in my library. I think I shall have about 200, including all the rational *paradoxers*, as I call them, who are not much known ... They are a rare lot.[4]

By 'budget', De Morgan meant simply a collection or assortment, whereas the word 'paradox' was used in its archaic sense of 'something which is apart from general opinion, either in subject-matter, method, or conclusion'.[5] So, for example, theories such as heliocentricism or Darwinism would originally have been classed as paradoxes, because they did not then belong to the mainstream of accepted scientific thought. But although works by mainstream authors do appear in the *Budget*, the focus of De Morgan's attention throughout is the published output of mathematical cranks, frauds and pseudoscientists.[6] He wrote concerning selection:

> To this my answer is, that no selection at all has been made. The books are, without exception, those which I have in my own

---

Mathematica, 43 (2016), 148–71.

3   See Chapter 10 in this volume.
4   University College London (UCL), Brougham Correspondence, no. 10,307: Letter from Augustus De Morgan to Lord Brougham, 13 Aug. 1863.
5   Augustus De Morgan, *A Budget of Paradoxes* (London: Longmans, Green, 1872), p. 2.
6   While all of the books reviewed in the *Budget* were unorthodox, a few were significant enough to have merited learned discussion in De Morgan's more serious publications. For example, his 1848 paper 'An Account of the Speculations of Thomas Wright of Durham' (*Philosophical Magazine*, 3rd ser., 32 (1848), 241–52) concerned Wright's *Original Theory or New Hypothesis of the Universe* (1750), which had proposed novel ideas concerning nebulae and the Milky Way, later adopted by William Herschel and subsequently by the mainstream astronomical community. De Morgan's entry on Wright's book in the *Budget* was far more cursory (*Budget*, p. 90).

library; and I have taken *all*—I mean all of the kind: Heaven forbid that I should be supposed to have no other books!⁷

De Morgan's *Budget of Paradoxes* column first appeared in *The Athenæum* on Saturday 10 October 1863, running for about three and a half years until 30 March 1867. De Morgan clearly relished his new career as a columnist, since it provided the perfect forum for his whimsical, humorous, eccentric, but above all, entertaining style of writing. As his widow, Sophia, informs us in her preface to the eventual published book: 'the Budget was in some degree a receptacle for the author's thoughts on any literary, scientific, or social question'.⁸ It also proved provocative, since several of its readers were compelled to write to *The Athenæum*, either to protest at their inclusion or to demand insertion in a subsequent issue.

After the publication of the concluding column in 1867, De Morgan continued to collect material and write further additions to the *Budget*, which he inserted piecemeal into what grew to be a lengthy and complex manuscript. It was this manuscript which served as the basis of the book that appeared in 1872 'reprinted, with the author's additions, from the "Athenæum".' Although noting apologetically that De Morgan's work on the book's preparation was incomplete at the time of his death, its editor, Sophia De Morgan, expressed the hope that 'it will be welcomed as an old friend returning under great disadvantages, but bringing a pleasant remembrance of the amusement which its weekly appearance in the *Athenæum* gave to both writer and reader'.⁹

That hope was certainly achieved. Subsequent reviews were glowing in their praise and the fondness with which the book was regarded was near universal. The mathematician William Clifford described the book as 'by very far the most individual book of the age', while P. G. Tait called it 'absolutely unique'.¹⁰ Indeed the Scottish mathematician Alexander Macfarlane later recalled that when he was a student at the University of Edinburgh, Tait once said to him: 'If you wish to read something

---
7   *Budget*, pp. 5–6.
8   Sophia Elizabeth De Morgan, 'Editor's Preface', in *Budget*, pp. v–vii (p. vi).
9   *Budget*, p. vii.
10 [William Kingdon Clifford], Review of *A Budget of Paradoxes*, *The Academy* (15 Aug. 1873), 306–07 (p. 306) [repr. in *Mathematical Papers* (London: Macmillan, 1882), pp. 559–61 (p. 559)]; Peter Guthrie Tait, 'De Morgan's *Budget of Paradoxes*', *Nature* (30 Jan. 1873), 239–40 (p. 239).

entertaining, get De Morgan's *Budget of Paradoxes* out of the library.'[11] What then was the attraction of *A Budget of Paradoxes*, and what made it so unusual? To answer these questions, let us take a brief look at what David Eugene Smith described as 'one of the most delicious bits of satire of the nineteenth century'.[12]

A Brief Survey of the *Budget*

De Morgan opens his *Budget of Paradoxes* with the following idiosyncratic words:

> If I had before me a fly and an elephant, having never seen more than one such magnitude of either kind; and if the fly were to endeavour to persuade me that he was larger than the elephant, I might by possibility be placed in a difficulty. The apparently little creature might use such arguments about the effect of distance, and might appeal to such laws of sight and hearing as I, if unlearned in those things, might be unable wholly to reject. But if there were a thousand flies, all buzzing, to appearance, about the great creature; and, to a fly, declaring, each one for himself, that he was bigger than the quadruped; and all giving different and frequently contradictory reasons; and each one despising and opposing the reasons of the others—I should feel quite at my ease. I should certainly say, My little friends, the case of each one of you is destroyed by the rest. I intend to show flies in the swarm, with a few larger animals, for reasons to be given.[13]

The book that follows is essentially an ordered list of reviews of 270 publications presented roughly in chronological order, with the earliest item being an incunabulum from 1489 and the latest a work dating from 1866. The reviews vary in length from one or two lines to several pages and are interspersed here and there with tangential remarks, excerpts from articles, witty verse, puzzles, puns, amusing anecdotes, and very many digressions. These include discussions on probability and the

---

11 Alexander Macfarlane, *Lectures on Ten British Mathematicians of the Nineteenth Century* (New York: Wiley, 1916), p. 25.
12 David Eugene Smith, 'Preface to the Second Edition', in Augustus De Morgan, *A Budget of Paradoxes*, 2nd edn (Chicago: Open Court Publishing, 1915), pp. vi–viii (p. viii).
13 *Budget*, p. 1.

law of large numbers, the Buffon needle problem, Wilson's theorem in number theory, the controversy over the discovery of Neptune, the merits of a proposed system of decimal coinage, the veracity of the story of Newton and the apple, the etymology of various words, and several metaphysical considerations. For example, in a review of a philosophical work, *The Mystery of Being; Or are Ultimate Atoms Inhabited Worlds?*, De Morgan parodies a verse by Jonathan Swift:

> Great fleas have little fleas, upon their backs to bite 'em,
> And little fleas have lesser fleas, and so *ad infinitum*.[14]

A multitude of rogue theories are presented from a host of academic disciplines including mathematics, physics, astronomy, medicine, economics, logic, philosophy, and theology. Such theories include attempts to trisect the angle, to construct a perpetual motion machine, to prove that the earth is flat, to disprove the law of universal gravitation and, of course, to square the circle—and all are beautifully sent up by De Morgan in his wonderfully satirical style. *Comets Considered as Volcanoes, and the Cause of their Velocity and Other Phenomena Thereby Explained* receives the review: 'The title explains the book better than the book explains the title';[15] while a reference to an alleged eight-volume work on parallel lines produces the remark: 'Surely this is a misprint; *eight* volumes on the theory of parallels? If there be such a work, I trust I and it may never meet, though ever so far produced.'[16] Perhaps the most poignant remarks concern a pamphlet by a Mr. James Sabben, *A Method to Trisect a Series of Angles Having Relation to Each Other; Also Another to Trisect Any Given Angle* (1848), described by its author as 'The consequence of years of intense thought'. De Morgan's review, in its entirety, reads: 'Very likely, and very sad.'[17]

One of the most famous works reviewed is Robert Chambers' *Vestiges of the Natural History of Creation*. Published anonymously in

---

14   *Budget*, p. 377. Swift's original verse in *On Poetry: A Rapsody* (1733) was:
    So, Nat'ralists observe, a Flea
    Hath smaller Fleas that on him prey,
    And these have smaller Fleas to bite 'em,
    And so proceed *ad infinitum*.
15   *Budget*, p. 303.
16   *Budget*, p. 137.
17   *Budget*, p. 255.

1844, this controversial book began the popularisation of the idea of the transmutation of species and helped pave the way, in the public eye at least, for the subsequent acceptance of Darwin's theories of evolution by natural selection. Unsurprisingly, speculation about the book's authorship was rife for many years, with the truth only revealed in 1884, thirteen years after the deaths of both Chambers and De Morgan.[18] In his review, De Morgan confessed:

> I never hear a man of note talk fluently about it without a curious glance at his proportions, to see whether there may be ground to conjecture that he may have more of 'mortal coil' than others ... [w]ith a hole behind which his tail peeped through.[19]

His only criticism of the book concerned an inaccuracy in its very first sentence, which read: 'It is familiar knowledge that the earth which we inhabit is a globe of somewhat less than 8,000 miles in diameter, being one of a series of eleven which revolve at different distances around the sun.'[20] 'The *eleven*!' De Morgan exclaimed, 'Not to mention the Iscariot which Le Verrier and Adams calculated into existence, there is more than a septuagint of *new* planetoids.'[21]

He expressed similar sentiments in a review of *The Decimal System as a Whole* (1856), a book written in support of the adoption of the decimal system, but which perhaps took the idea a little too far:

---

18 James A. Secord, *Victorian Sensation: The Extraordinary Publication, Reception, and Secret Authorship of Vestiges of the Natural History of Creation* (Chicago: University of Chicago Press, 2003).
19 *Budget*, p. 211.
20 [Robert Chambers], *Vestiges of the Natural History of Creation* (London: John Churchill, 1844), p. 1.
21 *Budget*, p. 211. There is a lot going on in this sentence. The eleven celestial bodies referred to by the author of *Vestiges* are seven planets (Mercury, Venus, Earth, Mars, Jupiter, Saturn and Uranus) and the four then-known asteroids (Ceres, Pallas, Juno and Vesta). The planet Neptune, discovered in 1846 via the mathematical calculations of Urbain Le Verrier and John Couch Adams, constituted the twelfth known body to be orbiting the sun, hence De Morgan's reference to the twelfth disciple, Judas Iscariot. A further biblical allusion occurs in his use of the word 'septuagint'. Although usually understood, when capitalised, to mean the Greek translation of the Old Testament, the word literally means seventy. By the time De Morgan was writing, in the mid-1860s, astronomers had discovered around 100 asteroids in the solar system. He would thus have been justified in viewing *Vestiges'* enumeration of eleven planetary bodies as somewhat out of date.

> The proposition is to make everything decimal. The day, now 24 hours, is to be made 10 hours. The year is to have ten months, Unusber, Duober, &c. Fortunately there are ten commandments, so there will be neither addition to, nor deduction from, the moral law. But the twelve apostles! Even rejecting Judas, there is a whole apostle of difficulty. These points the author does not touch.[22]

Several paradoxes focus on attempts to apply mathematics, not always successfully, to areas hitherto untouched by the subject. One such area was psychology, on which the German philosopher and psychologist Johann Friedrich Herbart published *De Attentionis Mensura Causisque Primariis* in 1822. In it Herbart laboured to derive mathematical formulae for various attributes of concentration. His success was marred only by their unintelligibility, as De Morgan writes:

> As a specimen of his formula, let t be the time elapsed since the consideration began, β the whole perceptive intensity of the individual, ϕ the whole of his mental force, and z the force given to a notion by attention during the time t. Then,
>
> $$z = \phi(1 - e^{-\beta t})$$
>
> Now for a test. There is a *jactura*, v, the meaning of which I do not comprehend.[23] If there be anything in it, my mathematical readers ought to interpret it from the formula
>
> $$v = \frac{\pi \phi \beta}{1 - \beta} e^{-\beta t} + C e^{-t}$$
>
> and to this task I leave them, wishing them better luck than mine.[24]

Theology was another subject which appeared ripe for mathematicising, as evinced by such works as John Craig's *Theologiae Christianiae Principia Mathematica* (1699) and Richard Jack's *Mathematical Principles of Theology* (1747). Not only were these titles in direct emulation of Newton's *Philosophiae Naturalis Principia Mathematica*, but their content was clearly influenced by the then novel idea that concepts such as force were not only measurable but subject to mathematical laws of quantifiable

---

22 *Budget*, p. 301.
23 As a scholar well versed in Classics, De Morgan would have understood *jactura* to be a Latin word meaning loss or expense; his confusion presumably arose from the question of how this term related to the subject of concentration.
24 *Budget*, p. 150.

variation. In his book, Craig, himself an able mathematician, took as an axiom that the rate at which suspicions against historical evidence increase is proportional to the square of the time elapsed. On this hypothesis, he attempted to calculate how long it would take the evidence of Christianity to die out, coming up with a date of AD 3150, which he also gave as the year of the second coming. 'It is a pity that Craig's theory was not adopted,' De Morgan commented drily, 'it would have spared a hundred treatises on the end of the world'.[25]

But while some authors wished to emulate the work of Isaac Newton, there were plenty in the eighteenth and nineteenth centuries who wished to prove him wrong. The Italian Caelestino Cominale published his *Antinewtonianismus* in two hefty volumes in 1754 and 1756, in which he sought to refute the Newtonian theories of light, inertia, vacua and gravitation. Although they occupied a place on De Morgan's shelves, he had to admit: 'I never attempted these big Latin volumes, numbering 450 closely-printed quarto pages. The man who slays Newton in a pamphlet is the man for me.'[26] Such a man was one Captain Walter Forman, a retired Royal Naval officer from Shepton Mallet, who published an anti-Newtonian pamphlet in 1833, *A Letter to the Secretary of the Royal Astronomical Society, in Refutation of Some Absurd & Mistaken Notions, upon Philosophical Subjects which are Held in Common by that Society and by all the Newtonian philosophers*.

De Morgan clearly found this easier reading than its Latin counterpart, since he tells us that he was 'happy to state that there is no truth in the rumour of the laws of gravitation being about to be repealed. We have traced this report, and find it originated with a gentleman living near Bath (Captain Forman, R.N.), whose name we forbear to mention.'[27] A quicker read still was a flyer handed out on the street sometime in 1847 which read:

> Important discovery in astronomy, communicated to the Astronomer Royal, December 21st, 1846. That the Sun revolve round the Planets in $25748\frac{2}{5}$ years, in consequence of the combined Attraction of the Planets and their Satellites, and that the Earth revolve round the Moon in 18 years and 228 days.[28]

---

25 *Budget*, p. 77.
26 *Budget*, p. 96.
27 *Budget*, p. 185.
28 *Budget*, p. 253.

The Morbus Cyclometricus

If subverters of Newtonian physics were a sizeable constituent of De Morgan's paradoxers, by far the most numerous were those amateur mathematicians who claimed to have squared the circle or, to put it in modern terminology, to have found a rational value of π. The *Budget* dealt with more than fifty such works, including two books by J. P. de Fauré, an obscure Swiss mathematician, published in 1747 and 1749. Both claimed to have proved that the true value of π is 256/81. But De Morgan observed, much to his surprise, that the second of the two volumes carried printed endorsements from two far more eminent Swiss mathematicians, namely Johann Bernoulli the younger and Samuel König. However, on closer inspection, he noticed that these endorsements were very cunningly worded. Bernoulli's testimonial reads: 'Under the assumptions framed in this memoir, it [the conclusion] is so obvious that ... it needs neither evidence nor authority to be recognized by everyone',[29] while König's is even more evasive: 'I concur with the judgment of Mr. Bernoulli, in consequence of these assumptions.'[30] This verbal dexterity prompted De Morgan to write: 'It should seem that it is easier to square the circle than to get round a mathematician.'[31]

But by far the most notorious circle-squarer, as well as the most voluminous contributor to the *Budget of Paradoxes*, other than De Morgan himself, was a certain James Smith. One of the most bizarre, irrepressible and pugnacious figures in the history of mathematics, Smith was in De Morgan's words a man who 'more visibly than almost any other known to history, reasoned in a circle by way of reasoning on a circle'.[32] De Morgan was clearly wholly unprepared for the magnitude of Smith's mathematical inabilities:

> I had not anything like an adequate idea of Mr. James Smith's superiority to the rest of the world in the points in which he is superior. He is beyond doubt the ablest head at unreasoning, and the greatest hand at writing it, of all who have tried in our day

---

29 'Suivant les suppositions posées dans ce Mémoire, il est si évident que ... cela n'a besoin ni de preuve ni d'autorité pout être reconnu par tout le monde.'
30 'Je souscris au jugement de M. Bernoulli, en conséquence de ces suppositions.'
31 *Budget*, p. 90.
32 *Budget*, p. 331.

to attach their names to an error. Common cyclometers sink into puny orthodoxy by his side.[33]

Smith had been a successful Liverpool merchant who chaired the local marine board for several years as well as being a member of the Mersey Docks and Harbour Board.[34] He possessed an avocational interest in mathematics and, following his retirement in 1855, became interested in the problem of squaring the circle. Believing that he had discovered the true value of π to be precisely $3\frac{1}{8}$, in 1859 Smith privately published *The Problem of Squaring the Circle Solved*, which De Morgan reviewed in *The Athenæum* on 5 March of that year. Not surprisingly, his review was hardly a ringing endorsement of Smith's work, although it did state that

> we by no means desire to prevent any one who is not deep in mathematics from squaring the circle. It is a mode of meddling with unknown things which cannot do any harm, except to the speculator himself. If Mr. Smith will study geometry, he will find out his own fallacy fast enough.[35]

Smith did not find his own fallacy and appears to have entered into a private correspondence with De Morgan in which the mathematician attempted (valiantly, though unsuccessfully) to persuade Smith of the errors in his reasoning.[36] It quickly became evident that Smith's 'proof' rested on an initial assumption that $\pi = 3\frac{1}{8}$, followed by the deduction of a few consequences consistent with that hypothesis. To a mathematician and logician of De Morgan's calibre, this was too much. He later wrote:

> Euclid assumes what he wants to *disprove*, and shows that his *assumption* leads to absurdity, and so *upsets itself*. Mr. Smith assumes what he wants to *prove*, and shows that *his* assumption makes *other propositions* lead to absurdity. This is enough for all who can reason. Mr. James Smith cannot be argued with ...[37]

---

33 *Budget*, p. 317.
34 E. I. Carlyle, 'Smith, James (1805–1872)', rev. by Adrian Rice, *Oxford Dictionary of National Biography* (Oxford: Oxford University Press, 2004), https://doi.org/10.1093/ref:odnb/25824.
35 [Augustus De Morgan], Review of *The Problem of Squaring the Circle Solved*, *The Athenæum* (5 Mar. 1859), 319.
36 No letters in this correspondence appear to be extant from either party. However, see contemporaneous letters from Smith to William Hepworth Dixon, editor of *The Athenæum*, arguing against De Morgan's position on the question of squaring the circle (University College London, MS ADD 118).
37 *Budget*, p. 327.

De Morgan cannot, therefore, have been surprised that Smith remained unconvinced by his arguments; but he was taken aback two years later when Smith published De Morgan's private letters to him in *The Quadrature of the Circle: Correspondence between an Eminent Mathematician and J. Smith, Esq.*[38] De Morgan responded with an extended review in *The Athenæum*, later reprinted in the *Budget*. Charging Smith with violating 'the decencies of private life' by the unauthorised publication of private correspondence, De Morgan declared that he 'deserves the severest castigation; and he will get it'.[39] Yet he could resort to little but ridicule. While mockingly noting that Smith was clearly suffering from a condition he denoted as 'the *morbus cyclometricus*,' (the circle-measuring disease), De Morgan reassured readers of the *Budget* that Smith 'is not mad. Madmen reason rightly upon wrong premises: Mr. Smith reasons wrongly upon no premises at all.'[40]

Smith replied with letters and advertisements published in various journals and newspapers, including *The Athenæum*, where he outlined his argument and claimed 'for your readers the right and the opportunity of judging for themselves, whether they would desire to class me in the category of either fools or madmen'.[41] De Morgan responded by including in the *Budget* a simple proof sent to him by William Rowan Hamilton that the circumference is greater than $3\frac{1}{8}$ diameters, based only on the first four books of Euclid, saying: 'We give it in brief as an exercise for our juvenile readers ...' He continued:

> It reminds us of the old days when real geometers used to think it worth while seriously to demolish pretenders. Mr. Smith's fame is now assured: Sir W. R. Hamilton's brief and easy exposure will procure him notice in connexion with this celebrated problem, to the historians of which we now hand him over.[42]

---

38 James Smith, *The Quadrature of the Circle: Correspondence between an Eminent Mathematician and J. Smith, Esq* (London: Simpkin, Marshall, 1861). De Morgan's copy encloses press clippings, letters from William Rowan Hamilton to Augustus De Morgan, 24 Apr.–13 May 1861) and copies of letters from Hamilton to the editor of the *Athenæum* (1-6 May 1861). See Senate House Library, University of London, [DeM] L.6 [Quadrature] SSR.
39 *Budget*, p. 319.
40 *Budget*, pp. 318, 319.
41 James Smith, [Advertisement] 'The Quadrature of the Circle. To the Editor of the Athenæum', *The Athenæum* (25 May 1861), 679.
42 [Augustus De Morgan], 'The Circle', *The Athenæum* (8 June 1861), 764.

Other mathematical colleagues rallied to support De Morgan when Smith submitted a paper on the subject for presentation at the 1861 meeting of the British Association for the Advancement of Science in Manchester. After realising that Smith's legitimate-sounding 'The Relations of a Circle Inscribed in a Square' contained nothing but nonsensical ramblings about π, the Association rejected it, conscious of the serious harm such a paper could do to the reputation of mathematics at such a prestigious meeting. Indeed, in his presidential speech to the mathematical and physical section of the Association, mathematician and astronomer George Airy made a point of insisting 'that such communications should not be made to [this] Section, as they were a mere loss of time'.[43]

Despite this curt dismissal, Smith continued to publish books, pamphlets and letters, in which extracts from correspondence with De Morgan, Hamilton and others were interspersed with further fruitless attempts to argue his case. As time passed, occasional bizarre outbursts of frustration punctured these mathematical arguments, such as Smith's peculiar depiction of De Morgan as a 'Mathematical elephant ... pumping your brains ... behind the scenes'.[44] De Morgan retorted: 'an odd thing for an elephant to do, and an odd place to do it'.[45]

De Morgan continued to wrestle good-naturedly with him for several more years via both the *Budget* and the columns of his *Athenæum* reviews, although towards the end of his life, he did wonder whether Smith's incessant correspondence would cease only on the occasion of his (De Morgan's) death:

> And this time may not be far off: for I was X years old in A.D. X^2; not 4 in A.D. 16, nor 5 in A.D. 25, but still in one case under that law. And now I have made my own age a problem of quadrature, Mr. J. Smith may solve it. But I protest against his method of assuming a result, and making itself prove itself: he might in this way, as sure as eggs is eggs (a corruption of X is X), make me 1,864 years old, which is a great deal too much.[46]

---

43 George Biddell Airy, 'Address by G.B. Airy, Astronomer Royal, President of the [Mathematics and Physics] Section', in *Report of the Thirty-First Meeting of the British Association for the Advancement of Science* (London: John Murray, 1862), pp. 1–2 (p. 2).
44 James Smith, *The Quadrature of the Circle* (Liverpool: Edward Howell, 1865), p. 55.
45 [Augustus De Morgan], Review of *The Quadrature of the Circle*, *The Athenæum* (27 May 1865), 717.
46 *Budget*, p. 332.

Astronomers and Alcohol

De Morgan used the *Budget* not only as a forum for the debunking of pseudoscientific nonsense, but as a repository for some of his own discursions on the history of science and mathematics. Thus, among many other things, we find a lengthy discussion on the history and calculation of the date of Easter, a description of Napier's theological work on the book of Revelations, and a charming account of the now-famous Spitalfields Mathematical Society, which existed from 1717 to 1845. Established as a club for the improvement of the studious artisan, especially the silk weavers of East London, the Society's great rule was: 'If any member be asked a question in the Mathematics by another, he shall instruct him in the plainest and easiest method he can, or forfeit one penny'.[47] Of their weekly meetings in Crispin Street, East London, De Morgan noted 'that each man had his pipe, his pot, and his problem'.[48]

The Society's members included the optician John Dollond, the mathematician Thomas Simpson, and De Morgan's own father-in-law, William Frend. Its final president was Benjamin Gompertz, an actuarial mathematician and friend of De Morgan, who provided some fascinating information on the background to a particular song, apparently sung at one of the Society dinners around 1800, and reproduced in full in *A Budget of Paradoxes*.[49] Entitled 'The Astronomer's Drinking Song' it was presumably sung to the tune of 'The Vicar of Bray':[50]

> Whoe'er would search the starry sky,
> Its secrets to divine, sir,
> Should take his glass—I mean, should try
> A glass or two of wine, sir!
> True virtue lies in golden mean,
> And man must wet his clay, sir;
> Join these two maxims, and 'tis seen
> He should drink his bottle a day, sir!

---

47 J. W. S. Cassels, 'The Spitalfields Mathematical Society', *Bulletin of the London Mathematical Society*, 11 (1979), 241–58 (p. 244). See also Larry Stewart and Paul Weindling, 'Philosophical Threads: Natural Philosophy and Public Experiment among the Weavers of Spitalfields', *British Journal for the History of Science*, 28 (1995), 37–62 (pp. 41–42).
48 *Budget*, p. 232.
49 *Budget*, pp. 234–36.
50 We give an edited version: the full song has eleven verses.

When Ptolemy, now long ago,
Believed the earth stood still, sir,
He never would have blundered so,
Had he but drunk his fill, sir:
He'd then have felt it circulate,
And would have learnt to say, sir,
The true way to investigate
Is to drink your bottle a day, sir!

Copernicus, that learned wight,
The glory of his nation,
With draughts of wine refreshed his sight,
And saw the earth's rotation;
Each planet then its orb described,
The moon got under way, sir;
These truths from nature he imbibed
For he drank his bottle a day, sir!

Poor Galileo, forced to rat
Before the Inquisition,
E pur si muove was the pat
He gave them in addition:
He meant, whate'er you think you prove,
The earth must go its way, sirs;
Spite of your teeth I'll make it move,
For I'll drink my bottle a day, sirs!

Great Newton, who was never beat
Whatever fools may think, sir;
Though sometimes he forgot to eat,
He never forgot to drink, sir:
Descartes took nought but lemonade,
To conquer him was play, sir;
The first advance that Newton made
Was to drink his bottle a day, sir!

D'Alembert, Euler, and Clairaut,
Though they increased our store, sir,
Much further had been seen to go
Had they tippled a little more, sir!
Lagrange gets mellow with Laplace,
And both are wont to say, sir,
The *philosophe* who's not an ass
Will drink his bottle a day, sir!

The fact that smoking and drinking were permitted at meetings of the Spitalfields Society contrasted sharply with the more sober gatherings of its eventual successor, the London Mathematical Society. Writing in 1866, De Morgan stated proudly that

> There is a new Mathematical Society, and I am, at this present writing, its first President. We are very high in the newest developments, and bid fair to take a place among the scientific establishments. ... But not a drop of liquor is seen at our meetings, except a decanter of water: all our heavy is a fermentation of symbols; and we do not draw it mild.[51] There is no penny fine for reticence or occult science; and as to a song! not the ghost of a chance.'[52]

Conclusion

As a literary work, *A Budget of Paradoxes* is something of a paradox (in the modern sense) itself: a learned volume yet written in a popular style, serious but funny, respectful but subversive. And although time has inevitably dated its content in terms of allusions to the science, literature, and politics of the day, some parts are still extremely amusing. De Morgan clearly had not only a wicked sense of humour, but a very engaging style as well. He obviously enjoyed telling a good story and one feels very much when reading the book as if he was writing exactly as he would have spoken.

Writing in *Nature* in January 1873, P. G. Tait was fulsome in his praise of the *Budget*. 'Nothing,' he opined, 'in the slightest degree approaching it in its wonderful combinations has ever, to our knowledge, been produced.'[53] He observed that De Morgan's keen intellect, jocular writing style and even-tempered disposition made him the ideal author of such a book: 'And every page of it shows that he thoroughly enjoyed his task.'[54] In his mode of presentation, Tait likened De Morgan to a puppeteer, playing 'with his puppets, showing off their peculiarities,

---

51   In this pun-loaded sentence, 'heavy' is a Scottish term for a type of medium-strength beer known in England as 'bitter', as opposed to 'mild' ales, which are generally lower in alcohol content.
52   *Budget*, p. 236.
53   Tait, 'De Morgan's *Budget of Paradoxes*', p. 239.
54   Tait, p. 239.

posing them, helping them when diffident, restraining them when noisy, and even occasionally presenting himself as one of their number'.[55] He went on to note the uniformly good-humoured nature displayed in De Morgan's writing, 'so that the only incongruities we are sensible of are the sometimes savage remarks which several of his pet bears make about their dancing master'.[56]

In a review in *The Academy*, published seven months later, William Clifford praised the book for 'helping us to extend the habits of right thinking which we have got by practice in one subject over the whole range of our knowledge', and said it should be read 'by those who care to be led into right thinking and warned from wrong'.[57] Yet he noted a logical error in De Morgan's reasoning at one point in the book. One work reviewed in the *Budget*, *From Matter to Spirit*, concerned alleged manifestations of spiritual beings.[58] Its anonymous author was Sophia De Morgan ('C. D.'), with a preface by her husband ('A. B.') in which, although he displayed a cautious scepticism, he maintained that some of the phenomena could only have been caused by either an 'unseen intelligence' or something as yet unconceived by man.[59] But as Clifford pointed out: 'This apparently suspended judgment involves and hides the assumption that the said phenomena cannot possibly be referred to certain well known and commonly conceived things—the art of the conjuror, and the delusion of contagious excitement.'[60]

In 1940 Dirk Struik noticed a further error, this time historical.[61] In two separate sections of the *Budget*, De Morgan regales the reader with a humorous anecdote regarding an alleged encounter between the French philosopher Denis Diderot and the Swiss mathematician Leonhard Euler:

> The following story is told by Thiébault, in his *Souvenirs de vingt ans de séjour à Berlin*, published in his old age, about 1804.[62] ... Diderot paid a visit to the Russian Court at the invitation of

---

55 Tait, p. 239.
56 Tait, p. 239.
57 [Clifford], Review of *Budget*, p. 307.
58 For more on this book, see Chapter 9 of this volume.
59 [Sophia Elizabeth De Morgan], *From Matter to Spirit* (London: Longman, Green, Longman, Roberts, & Green, 1863), pp. v–xlv (p.xxvii).
60 Clifford, p. 307.
61 Dirk J. Struik, 'A Story Concerning Euler and Diderot', *Isis*, 31 (1940), 431–32.
62 Dieudonné Thiébault (1733–1807), a French man of letters who spent many years at the court of Frederick the Great.

the Empress. He conversed very freely, and gave the younger members of the Court circle a good deal of lively atheism. The Empress was much amused, but some of her councillors suggested that it might be desirable to check these expositions of doctrine. The Empress did not like to put a direct muzzle on her guest's tongue, so the following plot was contrived. Diderot was informed that a learned mathematician was in possession of an algebraical demonstration of the existence of God, and would give it him before all the Court, if he desired to hear it. Diderot gladly consented: though the name of the mathematician is not given, it was Euler. He advanced towards Diderot, and said gravely, and in a tone of perfect conviction: *Monsieur,*

$$\frac{a + b^n}{n} = x,$$

donc Dieu existe; répondez! Diderot, to whom algebra was Hebrew, was embarrassed and disconcerted; while peals of laughter rose on all sides. He asked permission to return to France at once, which was granted.[63]

This story was later repeated faithfully in print in the first half of the twentieth century by several historians of mathematics, including Florian Cajori, David Eugene Smith and E. T. Bell, as well as by the mathematical populariser Lancelot Hogben in his *Mathematics for the Million*.[64] But it does not make sense. Euler would have known that Diderot was one of the most intelligent people on the continent and that he was familiar, not just with algebra, but with many areas of mathematics, having written among other things on involutes and probability. For his part, Diderot would not have been swayed for one moment by such an unconvincing trick as Euler's equation. But more than this, at no point in Thiébault's original account is Euler actually mentioned at all—his insertion in the story is purely an invention by De Morgan.[65] Surprising for a historian who was usually scrupulous about checking his sources, it serves as a rare example of De Morgan not letting the facts get in the way of a good story.

---

63 *Budget*, pp. 250–51. The story is told again on p. 474.
64 Florian Cajori, *A History of Mathematics*, 2nd ed. (New York: Macmillan, 1919), p. 233; David Eugene Smith, *History of Mathematics* (Boston: Ginn, 1923), vol. 1, pp. 522–23; E. T. Bell, *Men of Mathematics* (New York: Simon & Schuster, 1937), pp. 146–47; Lancelot Hogben, *Mathematics for the Million* (London: Allen & Unwin, 1936), pp. 13–14.
65 Dieudonné Thiébault, *Mes Souvenirs de vingt ans de séjour à Berlin*, 5 vols. (Paris: Buisson, 1804–05), vol. 3, pp. 140–43.

By Struik's time, of course, knowledge of many of the allusions current at the time of the *Budget*'s writing were quickly fading from living memory. For this reason the *Budget* did not age well, and half a century after its first publication in *The Athenæum*, David Eugene Smith was writing: 'Many books that were then current have now passed out of memory, and much that agitated England in De Morgan's prime seems now like ancient history.'[66] So much of the content is framed in the context of political or cultural references that were soon largely forgotten. How many today would know, for example, that when De Morgan refers to 'the lady in Cadogan Place' he is writing about Mrs. Wititterly from *Nicholas Nickleby*?[67] Or that 'Miss Pickle, in the novel of that name' is actually Mrs. Grizzle from *The Adventures of Peregrine Pickle*, a picaresque tale by Tobias Smollett, first published in 1751?[68] For reasons such as these, Smith produced, in 1915, a new edition of the *Budget*, in two volumes with numerous additional footnotes containing valuable contextual information and references.

Reviewing this new edition for the American Mathematical Society in 1916, Louis Karpinski wryly observed: 'A modern De Morgan would, in two volumes like these, have room only for titles of published nonsense.'[69] Continuing the theme in the journal *Science*, Karpinski lamented the torrent of 'paradoxical nonsense, foisted upon the press by authors ignorant of what has been done by others in the fields in which these authors would instruct the public'.[70] Among the offenders, he listed 'philosophers ignorant of the work of Georg Cantor and Dedekind who wish to instruct mathematicians about the nature of the number idea and the psychology of number, school superintendents who are profoundly ignorant of the fundamental ideas of arithmetic who wish to write text-books on arithmetic, old maids living in a two-room flat

---

66 Smith, 'Preface to the Second Edition', p. vii.
67 *Budget*, p. 4. The fact that De Morgan added a footnote in the 1860s to clarify this phrase suggests that he considered the allusion too obscure for contemporary readers to figure out.
68 *Budget*, p. 89. De Morgan provided no footnote or explanatory passage for this reference, suggesting that the book was considered well enough known for the allusion to stand unexplained.
69 [Louis C. Karpinski], Review of *A Budget of Paradoxes*, *Bulletin of the American Mathematical Society*, 22 (1916), 468–71 (p. 471).
70 [Louis C. Karpinski], Review of *A Budget of Paradoxes*, *Science*, 42 (19 Nov. 1915), 729–31 (p. 730).

on the fifteenth floor of a New York apartment who wish to instruct the parents of the United States on the art of bringing up a large family of children ...' and the scores of delusional 'mathematicians' who claimed to have proved Fermat's Last Theorem.[71]

Today, the situation is little different. With his long view of history, De Morgan might say "Twas ever thus.' And with a multitude of contemporary unorthodox theories like climate-change denial, creationism and scientology, it is clear that the phenomenon of the paradoxer has not died down with time. Indeed, with real science often indistinguishable in the public mind from pseudoscience, it is arguable that the need for a modern-day De Morgan to set the record straight has never been greater.

Bibliography

Airy, George Biddell, 'Address by G.B. Airy, Astronomer Royal, President of the [Mathematics and Physics] Section', in *Report of the Thirty-First Meeting of the British Association for the Advancement of Science* (London: John Murray, 1862), pp. 1–2.

Bell, Eric Temple, *Men of Mathematics* (New York: Simon & Schuster, 1937).

Cajori, Florian, *A History of Mathematics*, 2nd edn (New York: Macmillan, 1919).

Carlyle, E. I., 'Smith, James (1805–1872)', rev. by Adrian Rice, *Oxford Dictionary of National Biography* (Oxford: Oxford University Press, 2004), vol. 51, p. 189. https://doi.org/10.1093/ref:odnb/25824

Cassels, J. W. S., 'The Spitalfields Mathematical Society', *Bulletin of the London Mathematical Society*, 11 (1979), 241–58.

[Chambers, Robert], *Vestiges of the Natural History of Creation* (London: John Churchill, 1844).

Clifford, William Kingdon, Review of *A Budget of Paradoxes*, *The Academy* (15 Aug. 1873), 306–07 [repr. in *Mathematical Papers* (London: Macmillan, 1882), pp. 559–61].

De Morgan, Augustus, *A Budget of Paradoxes* (London: Longmans, Green, 1872). 2nd edn, ed. by David Eugene Smith, 2 vols. (Chicago: Open Court Publishing, 1915).

—, Review of *The Problem of Squaring the Circle solved*, *The Athenæum* (5 Mar. 1859), 319.

---

71 [Karpinski], Review, *Science*, p. 730.

—, 'The Circle', *The Athenæum* (8 June 1861), 764.

—, Review of *The Quadrature of the Circle*, *The Athenæum* (27 May 1865), 717.

[De Morgan, Sophia Elizabeth], *From Matter to Spirit. The Result of Ten Years' Experience in Spirit Manifestations. Intended as a Guide to Enquirers* (London: Longman, Green, Longman, Roberts & Green, 1863).

Despeaux, Sloan Evans, and Adrian C. Rice, 'Augustus De Morgan's Anonymous Reviews for *The Athenæum*: A Mirror of a Victorian Mathematician', *Historia Mathematica*, 43 (2016), 148–71. https://doi.org/10.1016/j.hm.2015.09.001

Hogben, Lancelot, *Mathematics for the Million* (London: George Allen & Unwin, 1936).

Karpinski, Louis C., Review of *A Budget of Paradoxes*, *Science*, 42 (19 Nov. 1915), 729–31.

—, Review of *A Budget of Paradoxes*, *Bulletin of the American Mathematical Society*, 22 (1916), 468–71.

Macfarlane, Alexander, *Lectures on Ten British Mathematicians of the Nineteenth Century* (New York: Wiley, 1916).

Secord, James A., *Victorian Sensation: The Extraordinary Publication, Reception, and Secret Authorship of Vestiges of the Natural History of Creation* (Chicago: University of Chicago Press, 2003). https://doi.org/10.7208/chicago/9780226158259.001.0001

Smith, David Eugene, *History of Mathematics*, 2 vols (Boston: Ginn, 1923).

Smith, James, [Advertisement] 'The Quadrature of the Circle. To the Editor of the Athenæum', *The Athenæum* (25 May 1861), 679.

—, *The Quadrature of the Circle* (Liverpool: Edward Howell, 1865).

Stewart, Larry, and Paul Weindling, 'Philosophical Threads: Natural Philosophy and Public Experiment Among the Weavers of Spitalfields', *British Journal for the History of Science*, 28 (1995), 37–62. https://doi.org/10.1017/s0007087400032684

Struik, Dirk J., 'A Story Concerning Euler and Diderot', *Isis*, 31 (1940), 431–32.

Tait, Peter Guthrie, 'De Morgan's *Budget of Paradoxes*', *Nature* (30 Jan. 1873), 239–40.

Thiébault, Dieudonné, *Mes souvenirs de vingt ans de séjour à Berlin*, 5 vols. (Paris: Buisson, 1804–05).

Fig. 11 No. 57–58 Russell Square has served as the headquarters of the London Mathematical Society since 1998, when it was re-named De Morgan House, after the Society's founding President. With its distinctive late Georgian architecture and imposing terraced houses, Russell Square is both an archetypal Bloomsbury location and an area De Morgan would have known well. (By Nicholas Jackson - Own work, CC BY-SA 3.0, via Wikimedia Commons, https://commons.wikimedia.org/w/index.php?curid=26549887)

8. Augustus De Morgan and the Bloomsbury Milieu

Rosemary Ashton

> Our house was so near the college that my husband could come home in the intervals between his morning and afternoon lectures, instead of remaining away from 8 A.M. till 5 P.M., as he was obliged to do afterwards when we lived at a greater distance from Gower Street.
>
> — Sophia Elizabeth De Morgan[1]

In 1906 E. V. Lucas, in his book *A Wanderer in London*, described Bloomsbury as follows:

> It is a stronghold of middle-class respectability and learning. The British Museum is at its heart; its lungs are Bedford Square and Russell Square, Gordon Square and Woburn Square: and its aorta is Gower Street, which goes on for ever. Lawyers and law students live here, to be near the Inns of Court; bookish men live here, to be near the Museum; and Jews live here, to be near the University College School, which is non-sectarian. Bloomsbury is discreet and handy; it is near everything, and although not fashionable, any one, I understand, may live there without losing caste.[2]

This is an accurate sketch of the geographical, social, and professional character of Bloomsbury at the beginning of the twentieth century. Only two years before Lucas published his book, the sisters Vanessa and Virginia Stephen and their brothers moved from Kensington after

---

1   Sophia Elizabeth De Morgan, *Memoir of Augustus De Morgan* (London: Longmans, Green, 1882), p. 88.
2   E. V. Lucas, *A Wanderer in London*, rev. edn (London: Methuen, 1913), pp. 189–90.

the death of their father Leslie Stephen, choosing Bloomsbury to start their new independent lives. A society of *avant garde* writers and artists soon joined them, forming what is widely known as the 'Bloomsbury Group'. But Lucas gives us a glimpse of the area in the *preceding* century, when it gained its early character, becoming established as the home of several important progressive educational and cultural institutions and individuals. One progressive individual who made his home in Bloomsbury in the 1820s as a very young man was Augustus De Morgan.

In the early years of the nineteenth century Bloomsbury expanded as the large Bedford Estate in the western part of the area was developed into streets and garden squares, while in the eastern half the Foundling Hospital Estate was also turned into residential streets. In 1800 the land to the north of the old British Museum in Montagu House on Great Russell Street consisted of open fields leading as far north as the villages of Hampstead and Highgate. At that time only two large institutions existed in Bloomsbury (the name is believed to derive from the manor house—'bury'—of William Blemond, who acquired land around what is now Bloomsbury Square in 1201).[3] These were the Foundling Hospital, established by Thomas Coram in 1745, and the British Museum, founded in 1753. From the 1820s to the 1840s the new, much extended British Museum was built in neoclassical style by Robert Smirke on the site of Montagu House, which was demolished to make way for the new building.

Francis Russell, the fifth Duke of Bedford, had demolished his London house in neighbouring Bloomsbury Square in the year 1800, and had begun laying out the area of his land to the north, which was developed over the next forty or more years by successive Dukes of Bedford to include new residential streets and squares such as Tavistock Square, Gordon Square, Bedford Square, and the imposing Russell Square. From the 1820s onwards, new institutions appeared in Bloomsbury; these were many and varied, but almost all of them had

---

3   See Eliza Jeffries Davis, *The University Site, Bloomsbury* (Cambridge: Cambridge University Press, 1936), p. 30; Richard Tames, *Bloomsbury Past: A Visual History* (London: Historical Publications, 1993), pp. 8–9.

pioneering founders with radical and progressive principles in the fields of education or medicine.⁴

London as a whole doubled its population in the first half of the nineteenth century, from approximately one million in 1800 to over two million in 1850. Some hitherto rural areas, including Pimlico, the Portland Estate around the new Regent's Park, and Bloomsbury, grew with particular rapidity during this period.⁵ A new method of land development emerged; speculative builders leased large quantities of land from landowners in order to build houses which were then let to their new inhabitants on long leases. Whole swathes of streets and squares were developed according to this system. Two such builders—James Burton, and slightly later, Thomas Cubitt—were responsible for much of the building of Bloomsbury.⁶ In October 1826 the *Morning Chronicle* carried an article, 'Increase of London, from the Rage for Building', which marvelled at the building boom going on in the capital. The author singles out Bloomsbury for particular mention:

> Upon whatever side we turn ourselves towards the suburbs, we find not only houses, but whole streets, squares, villages, and we might almost say towns, raised as if the architects had become possessed of the lamp of Aladdin. Taking the Strand as a centre, and looking north upon that space bounded by the New Road [renamed Euston Road in 1857] and Tottenham and Gray's Inn Roads [respectively the western and eastern boundaries of Bloomsbury], we are struck with astonishment to see the ground which, thirty years ago, formed the garden and meadows of Montague [sic] House, now covered with spacious and even magnificent houses, and laid out in squares and streets not to be surpassed, if they are equalled by any portion of the metropolis.⁷

---

4   For a detailed history of these institutions as Bloomsbury evolved in the nineteenth century, see Rosemary Ashton, *Victorian Bloomsbury* (New Haven and London: Yale University Press, 2012), and the website of the UCL Bloomsbury Project, led by Rosemary Ashton: www.ucl.ac.uk/bloomsbury-project.
5   See Donald J. Olsen, *Town Planning in London: The Eighteenth and Nineteenth Centuries*, 2nd edn (New Haven and London: Yale University Press, 1982).
6   For the work of Burton (1761–1837) in Bloomsbury, see Dana Arnold, *Rural Urbanism: London Landscapes in the Early Nineteenth Century* (Manchester: Manchester University Press, 2005); for Cubitt (1788–1855), see Hermione Hobhouse, *Thomas Cubitt: Master Builder* (London: Macmillan, 1971).
7   *Morning Chronicle*, 25 Oct. 1826, p. 3.

As Lucas pointed out eighty years later, these new Bloomsbury houses were intended for middle-class professionals, not the aristocracy, which had its grand squares further west and closer to the centre of power in Westminster. Geographical position largely determined the demographic. The expanding new British Museum on Great Russell Street was attractive to visitors, readers who wished to use its unrivalled library, and hopeful employees, many of whom took houses or parts of houses as they were built in the new streets surrounding the Museum. The already well-established Inns of Court to the south-east employed many legal men who were happy to move with their families to a pleasant street or square close to their workplace. In *The Pickwick Papers* (1836-7), Dickens observed the local life of lawyers, and in particular of their clerks, rather lower down in the social order, but still part of the increasing and diversifying middle class now colonising the area:

> There are several grades of lawyers' clerks. There is the articled clerk, who has paid a premium, and is an attorney in perspective, who runs a tailor's bill, receives invitations to parties, knows a family in Gower Street, and another in Tavistock Square; who goes out of town every long vacation to see his father, who keeps live horses innumerable; and who is, in short, the very aristocrat of clerks.[8]

Dickens knew of what he wrote. In 1827, not long after his father's brief spell as a debtor in Marshalsea Prison in 1824, Dickens had left school and been taken on as a junior clerk by a firm of solicitors based in Gray's Inn Road. He had then become a parliamentary reporter on the *Morning Chronicle*, living in bachelor lodgings in Furnival's Inn until his marriage to Catherine Hogarth in April 1836; in April 1837 they took a house at 48 Doughty Street in the south-east of Bloomsbury, adjoining the Foundling Estate. This was not Dickens's first experience of living in Bloomsbury. In December 1823, when he was eleven, the family, hoping to avert the Marshalsea fate, had moved to 4 Gower Street North, where Mrs Dickens briefly and unsuccessfully tried to save the family finances

---

8   Charles Dickens, *The Posthumous Papers of the Pickwick Club* (London: Chapman & Hall, 1837), ch. 31.

by opening a school.⁹ Later, from 1851 to 1858, Dickens and his family lived in Tavistock House on the north-eastern edge of Tavistock Square.

As Bloomsbury grew physically in the 1820s, plans were underway for a remarkable institution to be sited near the top of the everlasting Gower Street—at that time divided into Gower Street North at the top end where the street met the New Road, a middle stretch named Upper Gower Street, and the lower portion, plain Gower Street, which ran south to Bedford Square. The new institution would thus be located just five minutes' walk to the north of the British Museum. On 6 June 1825 *The Times* printed an article entitled 'The London College'. It described a meeting held two days earlier at the Crown and Anchor Tavern in the Strand of 'about 120 of the gentlemen who have taken a principal interest in the formation of the London College, or University'. In the chair was Henry Brougham, the radical Whig lawyer and politician who had acquired a formidable reputation on account of his bravura performance in 1820 as the defence lawyer for Queen Caroline, consort of George IV. After the death of George III, the new king had demanded that Caroline be put on trial for adultery in order to prevent her from attending his coronation. The process, which took the form of a 'Bill of Pains and Penalties' brought in the House of Lords, was debated from August to November 1820 but abandoned when the Tory government led by Lord Liverpool realised that the bill, which passed by a very slim majority in the Lords, would fail in the House of Commons. Partly because of George IV's unpopularity in the country, but also in large measure thanks to Brougham's witty cross-examining of witnesses, Caroline was taken to the people's hearts as she was daily cheered and the king jeered at by large crowds of Londoners.¹⁰

During the 1820s Brougham was a prominent member of several reforming educational movements, including Dr George Birkbeck's new Infant School Society and the London Mechanics' Institution, which Birkbeck opened in 1823 to offer instruction to working men. Brougham himself was the leading spirit in the foundation in 1826 of the Society for the Diffusion of Useful Knowledge (SDUK), which used new printing

---

9   See Michael Slater, *Charles Dickens* (New Haven and London: Yale University Press, 2009), pp. 19–22, 27.
10 See R.A. Melikan, 'Pains and Penalties Procedure: How the House of Lords "Tried" Queen Caroline', *Parliamentary History*, 20:3 (2001), 311–32.

techniques to print cheap educational pamphlets for working people on a variety of subjects including hydrostatics, hydraulics, pneumatics, electricity, galvanism, and the workings of the latest industrial invention, the steam engine.[11] Brougham was doing all this at the same time as organising preparations for the Gower Street college, drumming up publicity and financial support, appointing professors, and also finding time to make influential speeches in the House of Commons in favour of radical legal reforms. The plan for London's new university was devised by Brougham along with the Scottish poet Thomas Campbell. It was supported by members of the radical Whig wing in parliament, including Lord John Russell, son of the sixth Duke of Bedford. Other supporters were the fiercely anti-establishment MP Joseph Hume, George Birkbeck, Zachary Macaulay, the Benthamite James Mill, and reform-minded lawyers like Stephen Lushington, who had worked alongside Brougham as one of Queen Caroline's counsellors in 1820.

All of those present at the meeting in the Crown and Anchor, according to *The Times*, were agreed on 'the necessity of establishing for the great population of this metropolis a college, which would comprehend all the leading advantages of the two great universities', while allowing students to live at home with their parents, so avoiding the expense of an education at Oxford or Cambridge. Taking its cue from the Scottish universities, at which Brougham and Campbell had both been educated, and from German universities which Campbell had visited and admired, the University of London would expand the traditional syllabus to include new subjects under the headings 'science, literature, and the arts'. Most boldly and radically of all, no theology would be taught, and the new university would have no chapel. There would be no religious tests for entry or graduation, such as those operating at the two existing English universities, where students were obliged to sign the Thirty-Nine Articles of the Church of England before being allowed to take their degree and teaching fellows were likewise obliged to sign. In Gower Street there was to be, as *The Times* reported, quoting Brougham at the meeting, 'no barrier to the education of any sect among His Majesty's subjects'.[12] Roman Catholics, non-conformists

---

11 See Ashton, *Victorian Bloomsbury*, pp. 58–81 (Chapter 2: 'Steam Intellect: Diffusing Useful Knowledge').
12 *The Times*, 6 June 1825, p. 4.

including Methodists, Baptists, Unitarians, as well as Jews, Muslims, Hindus and those of no faith, would be welcomed.

Brougham announced that the institution would be financed by raising transferable shares of one hundred pounds each. A committee was appointed to take the plan further, and letters of support were read out from the Dukes of Bedford and Norfolk, the former not only a local landowner, but also a reforming Whig, and the latter a Roman Catholic who would naturally take an interest in the opening of higher education to his co-religionists.[13] The editor of *The Times*, Thomas Barnes, was a friend of Brougham's and of the fledgling institution. He reported encouragingly on its progress from 1825 until October 1828, when the doors opened at the new building in Gower Street, designed in neo-classical style by William Wilkins, who went on a few years later to create the National Gallery building in Trafalgar Square.

The founders of the university had four clear aims. The first was to offer higher education in the largest and most advanced city in the world, it being a shameful anomaly that London had no university, unlike Paris, Prague, Florence, and other great European cities, not to mention the four ancient universities in Scotland. Secondly, the intention was to educate the sons of the expanding middle class, including the manufacturing class, who might be priced out of Oxbridge or who belonged to dissenting religious groups. The most radical principle was the non-sectarian one later noted by Lucas. (This forward-looking aim was almost immediately the cause of problems when it came to choosing professors, as the well-meaning founders differed between those who thought there should be no discrimination on the grounds of religious belief, so that even an Anglican clergyman—several of whom unexpectedly applied for professorships—would not be barred, and others who wished for religion of every denomination to be a barrier to appointments.)[14] Finally, another innovation, and one which was soon emulated by the many new universities which sprang up in the later nineteenth century, was the enlargement of the curriculum beyond the traditional classical, mathematical and theological education offered by

---

13 *The Times*, 6 June 1825, p. 4.
14 See H. Hale Bellot, *University College London 1826-1926* (London: University of London Press, 1929), the most comprehensive account of the early years – and struggles – of the university.

Oxford and Cambridge. The new institution planned to have classes in geography, architecture, geology, modern history, various branches of science and medicine, and modern languages and literatures, including English, French, German, Italian, Spanish and Hebrew. Brougham and his friends were not quite so radical as to promote the higher education of women, but it was at University College London (the name adopted in 1836 in place of the University of London) that women did first register for degrees in 1878.[15]

Site-clearing began in 1826, and by April 1827 enough had been achieved to allow a ceremony marking the laying of the foundation stone of the building. Cheered on by a long account by the Brougham-friendly *Times*, the event took place on 30 April 1827 in front of a reported crowd of 'upwards of 2,000'. The Duke of Sussex, the only one of George III's sons to associate with liberal causes such as anti-slavery and Roman Catholic rights, laid the first stone with a frank speech praising the 'present undertaking' as being likely to 'excite the old Universities to fresh exertions, and force them to reform abuses'. At the dinner following the ceremony, held in the Freemasons' Hall on Great Queen Street, just south of Bloomsbury, the Duke of Sussex, Grand Master of the Freemasons, took the chair, and over four hundred supporters and subscribers attended. Radical and in a sense anti-establishment though the new venture was, its leaders were careful to propose toasts to the King and the Royal Family, and when Brougham came to give his speech, he was keen to balance the Duke of Sussex's attack on Oxbridge with an assurance that, though the aim was to 'spread the light of knowledge over the world' and overcome the sneers and jibes of the 'enemies of human improvement, light, and liberty', he was by no means 'inimical to the two great English Universities', which he hoped would 'flourish as heretofore'.[16]

The dinner broke up after further toasts, including one to Birkbeck's Mechanics' Institution and another to Brougham's Society for the Diffusion of Useful Knowledge. It was clear to friends and foes alike that the new university was intimately connected with the widening

---

15 Hale Bellot, *University College London*; see also Negley Harte and John North, *The World of UCL 1828–1990*, rev edn (London: University College London, 1991), which brings the history of University College London nearer to the present day.

16 *The Times*, 1 May 1827.

of education for the non-academic classes as well as for the aspiring student sons of the London middle class. Brougham chaired both organisations, and many of the same liberal politicians and lawyers were on the committees of both. The SDUK produced from March 1827 a series of sixpenny treatises, mainly on scientific subjects. Brougham wrote the first, *A Discourse of the Objects, Advantages, and Pleasures of Science*, which sold over 40,000 copies within a few years.[17] With characteristic confidence Brougham offered an up-to-date survey of science, from mathematics to natural philosophy, the solar system, electricity, and the workings of the steam engine. The tracts were written with the honourable intention of 'conveying knowledge to uneducated persons, or persons imperfectly educated', and they were written by experts in their fields. De Morgan was soon one of them, serving on the committee for many years and writing hundreds of essays and articles on mathematics; he saw the SDUK as part of what he called approvingly 'the social pot-boiling', and preserved his own copies of the papers of the Society.[18]

It was not long before sceptical observers of both the SDUK and the university went into print with parodies and caricatures of the 'March of Intellect' or 'March of Mind' mantras of Brougham and his fellow members of the 'Steam Intellect Society'. The cleverest, most famous, and most humorous of these was *Crotchet Castle*, the comic novel published in 1831 by Thomas Love Peacock. The SDUK and Brougham himself are targets for Peacock's not entirely unjust satire. Chapter 2 begins with the Reverend Doctor Folliott bursting out in indignation:

> I am out of all patience with this march of mind. Here has my house been nearly burned down, my cook taking it into her head to study hydrostatics, in a sixpenny tract, published by the Steam Intellect Society, and written by a learned friend who is for doing all the world's business as well as his own, and is equally well

---

17 See Ashton, *Victorian Bloomsbury*, p. 61. The fullest account of the SDUK is to be found in an unpublished M.A. thesis held by the University of London Library at Senate House: Monica C. Grobel, 'The Society for the Diffusion of Useful Knowledge 1826–1846 and its Relation to Adult Education in the First Half of the Nineteenth Century, alias "The Sixpenny Science Company" alias "The Steam Intellect Society"', 1933.

18 De Morgan's collection of SDUK papers is in the UCL Special Collections. His remark about 'social pot-boiling' is quoted in his wife Sophia's *Memoir of Augustus De Morgan* (London: Longmans, Green, 1882), p. 51.

qualified to handle every branch of human knowledge. I have a great abomination of this learned friend... My cook must read his rubbish in bed; and as might naturally be expected, she dropped suddenly fast asleep, overturned the candle, and set the curtains in a blaze.[19]

As for the building going up on Gower Street, there was more at stake for hostile observers, of whom there were many, than merely sneering at the idea of cooks neglecting their tasks or workmen downing tools to read improving books. Many believed that religion was in danger. They liked to call Wilkins's building 'pagan', partly in allusion to its architectural style, but more because they saw the principles on which the new university was founded as a threat to Church and State. The reason for raising finance through selling shares was that the new institution could expect neither blessing nor funding from the political and religious establishment. George IV could hardly have been expected to welcome an enterprise managed by his nemesis Brougham. Moreover, the Tory government and the leaders of the Church of England were suspicious of the group of radical and liberal lawyers and politicians who were setting up this new seat of learning at a time when the government was attempting to resist introducing political and social reforms in parliament, including the extension of the franchise and the emancipation of Roman Catholics.

The Tory press soon got to work attacking the Gower Street project and rejoicing in the opportunity to conflate the two educational projects—university and SDUK—and to write fancifully about labourers in muddy boots attending lectures. As early as February 1825 the ultra-Tory satirical newspaper *John Bull* was sneering at 'this magnificent national establishment' and its 'liberal committee' proposing to 'instruct butchers in geometry, tallow-chandlers in Hebrew'.[20] The selling of shares was looked down upon; Robert Cruikshank produced a cartoon, 'The Political Toy-Man', which showed Brougham in his lawyer's wig and gown hawking shares like a street vendor in the hallowed courts of Lincoln's Inn.[21] *John Bull* thought up nicknames for the as yet unbuilt

---

19 Thomas Love Peacock, *Crotchet Castle* (London: T. Hookham, 1831), Ch. 2, 'The March of Mind'.
20 *John Bull*, 14 Feb. 1825; quoted in Ashton, pp. 28–29.
21 The cartoon is reproduced in Harte and North, *The World of UCL*, p. 15.

university; it was 'a new Cockney College' intending to teach dustmen to speak Latin and Greek (10 July 1825), and it was to be named 'Stinkomalee' in honour of the marshy, stagnant plot of land bordering the Bedford Estate which was soon purchased at the top of Gower Street (26 December 1825). More humorously, a poem by Winthrop Mackworth Praed purportedly addressed to the fellows and professors of Oxford and Cambridge, appeared in the *Morning Chronicle* in July 1825:

> Ye Dons and ye doctors, ye Provosts and Proctors,
> Who are paid to monopolize knowledge,
> Come make opposition by voice and petition
> To the radical infidel College.[22]

At the same time the leading members of the Duke of Wellington's Tory government, with the blessing of George IV, came together with the leaders of the Church of England to set up a rival institution, to be named King's College London. Intent on spoiling the planned opening of the Gower Street institution in October 1828, an impressive group of luminaries met together on Saturday 21 June 1828 in the Freemasons' Tavern next door to Freemasons' Hall. Their purpose, as *The Times* reported on the following Monday, was to establish

> a seminary for educating the youth of the metropolis and imparting religious instruction as taught by the established church, to be entitled 'The King's College, London'. At half-past 12 o'clock his Grace the Duke of Wellington entered the hall, accompanied by the Archbishops of Canterbury and York, the Bishops of London, Chester, Winchester ..., the Lord Mayor, and several other persons of rank and distinction.[23]

Wellington explained that the new college would teach (cannily taking its lead from its godless equivalent in Bloomsbury) 'the various branches of literature and science', but also (like the ancient universities) 'the doctrine and discipline of Christianity as inculcated by the united Church of England and Ireland'. The meeting ended with the pledging

---

22 Winthrop Mackworth Praed, 'A Discourse delivered by a College Tutor at a Supper-Party', *Morning Chronicle*, 19 July 1825, p. 3.
23 'New College on the Principles of the Church of England', *The Times*, 23 June 1828, p. 5.

of subscriptions from two dozen individuals and institutions, from Wellington's £300 to £1000 each from the Archbishops of Canterbury and York.[24]

The Anglican university which opened its doors on the Strand in October 1831, only three years after the upstart 'godless' college in Bloomsbury, did not, as is clear, struggle for funds. Its governors included, *ex officio*, the Archbishops of Canterbury and York, the Home Secretary, the Speaker of the House of Commons, the Dean of St Paul's, the Dean of Westminster and the Lord Mayor of London.[25] Unlike the Gower Street establishment, King's College London was instantly given a charter. A few years later, by a piece of good will and good management on both sides, the two institutions agreed to co-operate. In 1836 both became constituent parts of a new examining and degree-awarding body, called the University of London; in order to be permitted to distribute degrees, the Gower Street institution agreed to drop its original name and become University College London instead.[26]

Throughout the spring and summer of 1828, amid all the satire and protest and rivalry, Brougham and his colleagues on the university committee determinedly set about finding professors for the various subjects. At the Annual General Meeting of Proprietors held on 27 February 1828 there was a report on the progress being made on the building itself, and on the recent success in appointing professors to the chairs of, among other subjects, Greek Language, Literature and Antiquities (George Long, 'late Fellow of Trinity College, Cambridge'); Roman Language, Literature and Antiquities (The Rev. John Williams, 'late of Balliol College, Oxford, Rector of Lampeter, Cardiganshire'); and Jurisprudence (John Austin 'of Lincoln's Inn, Barrister at Law'). Also filled were the chairs in modern languages, though not, of course, by graduates of Oxford and Cambridge, where such subjects were not taught. Interestingly, radicals of a different kind from Brougham and his fellows stepped up to fill these language posts. The chair of Italian Language and Literature went to Antonio Panizzi, a political exile from Italy who had arrived in England in 1823 penniless and with a death

---

24 *The Times*, 23 June 1828, p. 5.
25 See *The Times*, 31 May, 26 June, 15 Sept. 1828, and 31 Aug. 1829.
26 See the UCL Bloomsbury Project, www.ucl.ac.uk/bloomsbury-project/institutions/ucl.htm.

sentence from Modena for belonging to a secret revolutionary society. Against all odds, Panizzi would move on in 1831 to become an assistant librarian at the British Museum. Six years later he achieved the astonishing feat of being appointed Keeper of the Printed Books at the Museum, an Italian thus taking charge of the greatest English copyright library.[27] A second political exile, Don Antonio Alcalà Galiano, became professor of Spanish, and a third, Ludwig von Mühlenfels, who had spent nearly two years in prison in Germany a few years earlier for alleged revolutionary activities, became professor of German. Another German, the brilliant philological scholar Friedrich Rosen, was appointed to teach Oriental Literature. He was 22 years old. The chair of Hebrew went to Hyman Hurwitz, 'late Master of the Jewish Academy at Highgate' and friend of the ageing 'Sage of Highgate', the poet Samuel Taylor Coleridge, who sent the committee a testimonial for Hurwitz.[28]

Among this group of first appointments was another very young and brilliant man. The person who became the first professor of mathematics, a subject which had always been taught with distinction at Oxford and Cambridge and which attracted one of the strongest fields of candidates, was 21-year-old Augustus De Morgan, 'of Trinity College, Cambridge'.[29] De Morgan, who had recently arrived in London, having graduated in 1827 as fourth Wrangler, was chosen out of 31 candidates.[30] He was later described by a former student as 'towering up intellectually above all his fellows', known familiarly as 'Gussy', and presenting an imposing figure, 'stout and tall', with 'a superb dome-like forehead' and 'very short-sighted eyes peering forth though gold-rimmed spectacles'.[31] The 'Second Statement of Council' of the new university, published in November 1828, as the first lessons got underway, described De Morgan's duties, which were to teach three classes of students for a total of sixteen hours a week.[32] In common with all the other professors, De

---

27 See Ashton, *Victorian Bloomsbury*, pp. 46–49.
28 Ashton, *Victorian Bloomsbury*, pp. 45–46, 50–51.
29 London, UCL Special Collections, 'University of London Annual General Meeting of Proprietors, held on Wednesday, the 27th of February, 1828', pp. 4–5.
30 See Adrian Rice, 'Inspiration or Desperation? Augustus De Morgan's Appointment to the Chair of Mathematics at London University in 1828', *The British Journal for the History of Science*, 30:3 (1997), 257–74, p.261.
31 Bellot, p. 80, quoting Thomas Hodgkin's account of 1901.
32 London, UCL Special Collections, 'Second Statement of the Council of the University of London, explanatory of the Plan of Instruction', p. 10.

Morgan was to be paid in part out of the fees paid by his particular students. Though Brougham and his friends had confidently predicted a total student body of 2,000 from the start, the first session, 1828--29, disappointingly attracted only 641; the hoped-for figure of 2,000 was only achieved over eighty years later.[33] De Morgan was luckier than some of his colleagues, particularly those teaching foreign languages; his classes attracted nearly 100 students, according to his wife Sophia, writing after his death.[34] In 1832 the highest-paid professors, those teaching medicine, earned up to £700, while the professors of English, philosophy, and German were paid £30, £21 and £11.10s respectively. De Morgan did better, but never earned more than £500 in a year.[35]

Though some of the new appointees belonged to the Church of England—the 'godless' college had decided to accept teachers as well as students who professed all faiths and none—many were heterodox or privately agnostic. De Morgan was such a man. He was a theist and looked warmly on Unitarianism, but never adopted any particular creed, calling himself a 'Christian unattached'. He was principled enough to reject the idea of taking his MA and becoming a fellow at Cambridge, so moved to London after graduating in 1827 and entered Lincoln's Inn, intending to study law.[36] At this time he resided with his mother at 25 Hatton Garden, near the Inns of Court; from 1828 to 1832 he lived, still with his mother, in Bloomsbury, first at 90 Guilford Street, on Foundling Estate land, then from 1832 to 1837 at 5 Upper Gower Street, a few doors south of the university. Here their neighbours were the family of William Frend, a mathematician who, like De Morgan, had graduated with distinction from Cambridge—in his case as second Wrangler in 1780. Frend had subsequently taught mathematics and philosophy at Cambridge while also officiating as an Anglican priest in two nearby parishes. On questioning his orthodox faith and becoming a Unitarian in 1787, he resigned his parish livings and was dismissed from his tutorship—though not his fellowship—at Jesus College. Frend published an anti-war pamphlet in 1793, just as Britain and France went

---

33 See Harte and North, *The World of UCL*, pp. 42, 45.
34 Sophia De Morgan, *Memoir*, p. 30.
35 Bellot, p. 179.
36 See Leslie Stephen, 'De Morgan, Augustus (1806–1871)', rev. I. Grattan-Guinness, *Oxford Dictionary of National Biography* (Oxford: Oxford University Press, 2004), version 25 May 2006.

to war. For this he was tried by the university authorities and in 1794 was dismissed. After moving to London he wrote books on maths and other subjects and joined radical associations in their protests against William Pitt's repressive government in the 1790s.[37] In 1837 the likeminded Augustus De Morgan married Frend's oldest daughter, Sophia, and they started their married life at 69 Gower Street, further down the long Gower Street, near the British Museum.[38]

What exactly were these Bloomsbury streets like when De Morgan came to live in the area? The building boom, which had begun in the 1790s with the development by James Burton and others of the Duke of Bedford's land in the west and the open land on either side of the Foundling Hospital to the east, was far advanced in the eastern part of Bloomsbury by the late 1820s.[39] Two large residential squares, Brunswick Square and Mecklenburgh Square, were planned to flank the hospital building. The first, Brunswick Square, was developed on the west side by James Burton between 1795 and 1802. Its spacious houses and central shared garden accommodated respectable families, including those who held senior offices at the Foundling Hospital itself. The other square, Mecklenburgh, on the east side of the hospital, was built a decade later, between 1810 and 1825.[40]

Guilford Street, where De Morgan and his mother settled at No. 90 in 1828, lies just south of the Foundling Hospital and runs from west to east from the edge of Russell Square (the largest square on the neighbouring Bedford Estate) to Gray's Inn Road, the eastern boundary of both the Foundling Estate and Bloomsbury itself. The houses here, also built by Burton, were completed in 1797. When De Morgan lived there in the 1820s and 1830s, Guilford Street was both respectable and interestingly modern in its mixture of inhabitants. It was home to many professional men and their families. There were artists and architects like the engraver George Shepheard, who lived here from 1821 to 1842,

---

37 See Nicholas Roe, 'William Frend (1757–1841)', *Oxford Dictionary of National Biography* (Oxford: Oxford University Press, 2004).

38 See Chapter 9 of this volume for a study of De Morgan's wife and family. For a list of De Morgan's addresses with dates, see London, UCL Special Collections, De Morgan MS ADD 7, f. 156.

39 For the history of the Foundling Hospital, see Gillian Pugh, *London's Forgotten Children: Thomas Coram and the Foundling Hospital* (Stroud: Tempus, 2007).

40 For a detailed account of Bloomsbury's streets, squares, and buildings, see the UCL Bloomsbury Project website.

and the architect Charles Reeves; a large number of lawyers—Guilford Street being a short walk from the Inns of Court—and many doctors, including for a time Thomas Wakley, the founder and first editor of *The Lancet* in 1823, as well as physicians who worked either at University College Hospital from its opening in 1834 or at one of the many specialist hospitals in nearby Queen Square.[41]

As for Gower Street, it was inhabited by artists, architects, antiquaries, scientists, many lawyers and even more doctors. Among the early professors at the university in the late 1820s and early 1830s, both De Morgan and Panizzi lived in Gower Street (Panizzi at No. 2, Gower Street North from 1828 until he moved into lodgings inside the British Museum in 1837). Andrew Amos, professor of English Law, lived in Burton Crescent (named after its architect-builder), and Galiano lived at 19 Marchmont Street on the Foundling Estate.[42] Some of the supporters of the university also lived in the area. One of them, Brougham's friend the lawyer James Loch, had a house in Bloomsbury Square. (Brougham himself lived in Mayfair, near Berkeley Square.) By mid-century the Post Office Directories record mainly professional men and their families in Gower Street, including several keepers of departments in the British Museum; the dentist James Robinson, who conducted the first operation under anaesthetic (ether) in Britain in 1846; the parents of the painter John Millais; and from 1839 to 1842 Charles Darwin and his wife Emma.[43]

The plan from the beginning was for the university to have its own teaching hospital. This was achieved in 1834; twelve years later, in December 1846, two days after James Robinson had used ether to extract a tooth, the first surgical operation using anaesthetics in Europe was performed at the university by the combative professor of Clinical Surgery, Robert Liston, who asked his invited audience to time the operation, and who proceeded to amputate his patient's leg in twenty-five seconds.[44] The hospital was unusual in England, being a proper teaching hospital attached to a university. In Oxford and Cambridge, anatomy was taught, but intending doctors had to spend time in one of

---

41 UCL Bloomsbury Project.
42 Information on these addresses comes from the Book of Admissions to the Reading Room in the British Museum Central Archive.
43 See the UCL Bloomsbury Project website: Gower Street.
44 See Bellot, pp. 164–66.

a number of London's private schools, where surgeons and physicians associated with the old-established hospitals, St Thomas's, Guy's and St Bartholomew's, taught for large fees. Once again, Scotland offered an example which the Gower Street university and hospital were keen to follow. Edinburgh University had established its own medical school in 1736, and between them the four Scottish universities had produced nine-tenths of all medical graduates in Britain between 1750 and 1800.[45] It was hardly surprising that most of the medical professors appointed by Brougham's committee in 1828 came from Edinburgh.[46]

Needless to say, the English 'establishment', in this case the celebrated surgeons and physicians who earned a second salary in the private medical schools of London, objected to the idea of a teaching hospital which would take away some of their students. *John Bull* rose up again in 1829, taking advantage of the recent scandal in Edinburgh, where the notorious pair of body-snatchers, Burke and Hare, had murdered tramps in order to sell their bodies to Professor Robert Knox for the purposes of teaching dissection. An Edinburgh skipping song soon did the rounds:

> Up the close and down the stair,
> But and ben with Burke and Hare.
> Burke's the butcher, Hare the thief,
> Knox the boy who buys the beef.[47]

John Bull joined in with an article in January 1829 suggesting that 'Stinkomalee' might be responsible for the disappearance of prostitutes from the area 'for the purposes of dissection ... here as well as in Scotland'.[48] De Morgan, known by family and students for his light-hearted verses and comic caricatures, including self-caricatures, composed a witty response of his own, set to the tune of the Scottish song 'Comin' through the rye':

---

45 R. A. Houston and W. W. J. Knox, *The New Penguin History of Scotland from the Earliest Times to the Present Day* (London: Allen Lane in association with National Museums of Scotland, 2001), p. xlvi.
46 See James Fernandez Clarke, *Autobiographical Recollections of the Medical Profession* (London: J. & A. Churchill, 1874), pp. 299, 314–15.
47 See Ashton, *Victorian Bloomsbury*, p. 108. 'But and ben' refer to the inner and outer rooms in a simple two-roomed dwelling.
48 Ashton, *Victorian Bloomsbury*, p. 109.

> Should a body want a body
> Anatomy to teach,
> Should a body snatch a body,
> Need a body peach?[49]

In these early years the university struggled to make ends meet; donations were drying up, student numbers remained low and professors were justifiably dissatisfied with their low salaries. The answer was to open a school on the university's grounds in Gower Street; the same principles of freedom of belief as those of the parent institution would apply, and there was every reason to suppose that boys taught at the new school would in many cases automatically continue to the university itself. The school would—and indeed did—save the university from failure. Brougham was undoubtedly behind an editorial in *The Times* on 29 September 1829 which painted a disingenuously rosy picture of the state of affairs on Gower Street. It began cleverly:

> The London University has been so successful in the ends which it proposed, and has so triumphantly answered in practice the objections made to its foundation, that its distinguishing principle is not likely to be long confined to one kind of academical establishment. That characteristic being the union of public education with private residence or domestic superintendence, appears equally well adapted to a great day-school for the education of the better classes as to a College or University.

The editorial was careful not to denigrate the 'great classical day schools' already existing in many British cities, while explaining that the principles of the new school attached to the University of London would mirror those of that institution, in particular the fact that the school would be attended entirely by day pupils, not boarders.[50] The reason for stressing this fact was made clear in the prospectus for the school. It was the question of religion and religious teaching, which was already causing problems in the university and would be liable to become even more difficult when boys rather than young men were at issue. As before, Christians and non-Christians alike were eligible to apply, and it was vital to make it clear that, 'as the School is to be strictly a Day School,

---

49 See A.M.W. Stirling, *William De Morgan and his Wife* (London: Butterworth, 1922), p. 33.
50 'Editorial', *The Times*, 29 Sept. 1829, p. 2.

parents or guardians will have the opportunity of superintending the religious education of the boys as they may think proper. In this point the Teachers of the school are bound not to interfere'.[51]

As with the university, the school syllabus was wider than that of existing schools; the boys, aged 8–15, were to be taught English, arithmetic, Latin, and writing in their first two years, with Greek, French, German, history, geography and drawing added from the third.[52] The school opened on 1 November 1830 with fifty-eight pupils. It was located in a house rented by the university at 16 Gower Street. In less than two years space was found on the university premises, where the boys also had a playground which was the subject of a fine engraving by George Scharf in 1833. Scharf lived in Francis Street, close to Gower Street, and his two sons, George junior—later to become the first secretary of the National Portrait Gallery—and Henry, were among the first pupils to enrol at the University of London School (later renamed University College School).[53] Other boys were the sons of proprietors and supporters like William Wilkins, Isaac Lyon Goldsmid and George Birkbeck, and of course those of the first professors, including Andrew Amos and the flamboyant Dionysius Lardner, professor of natural philosophy and astronomy, who lectured to large audiences on the steam engine, among other subjects.[54] In due course De Morgan sent his sons William (born in 1839) and George (born in 1841) to the school. William had his father's artistic talent, becoming a celebrated designer and potter, while George, who inherited his father's mathematical genius, became co-founder of the London Mathematical Society in 1865.[55]

During his time at University College, which lasted—though not continuously—from the beginning in 1828 until 1866, De Morgan was a striking teacher. He was loved for his wit and jokes and feared for his impatience with late arrivals, two of whom got up a petition in 1838 to complain about his habit of locking the doors of the lecture room five

---

51 See H. J. K. Usher, C. D. Black-Hawkins, and G. J. Carrick, *An Angel Without Wings: The History of University College School 1830–1980* (London: University College School, 1981), p. 12.
52 Usher et al., *An Angel Without Wings*, pp. 13–14.
53 See Ashton, *Victorian Bloomsbury*, pp. 97, 103. The original Scharf engraving is in the UCL Art Museum.
54 Ashton, *Victorian Bloomsbury*, pp. 84, 102–03.
55 For more on De Morgan's children, see Chapter 9 in this volume.

minutes after the start. They pointed out that they had paid in advance for entry to the lectures and that it was difficult for students who lived some distance away always to be punctual.[56] The same uprightness and independence which had prevented him from staying on at Cambridge after graduating because of his unorthodox religious beliefs caused him to resign twice from the university. The first time was early on, in 1831, when the professor of anatomy, Granville Sharp Pattison, was dismissed for incompetence. De Morgan and the impecunious but honourable Friedrich Rosen resigned in protest at the university's highhanded treatment of Pattison.[57] Having returned, with characteristic generosity, when his successor in the chair of mathematics drowned in 1836, De Morgan taught vigorously until 1866, when he resigned for a second time over the controversy which attended the efforts of the leading Unitarian minister James Martineau to acquire the chair of philosophy.[58]

As might be expected, De Morgan was a supporter of the movement for the higher education of women, which naturally had its origins in Bloomsbury. In 1849 the wealthy dissenter Elizabeth Reid took a house in Bedford Square and opened her Ladies' College (later renamed Bedford College and integrated as a constituent college of the University of London). She could not offer full degrees, nor could she staff her college with women, of course, as none had yet been educated to higher educational standard, so she asked some of the professors at University College to walk down Gower Street and teach her girls and young women, which a good number agreed to do. Among them was De Morgan, though he left in 1850, claiming pressure of work. He had been keen to advise Mrs Reid and her committee, telling them 'never [to] begin by drawing up constitutions. They are sure to prove clogs on the wheel. Let the work begin in good earnest, and with no needless machinery.'[59] No doubt he had in mind the problems and arguments that had bedevilled the university in its first years. It was he who drew up a draft prospectus for the Ladies' College, in which it was firmly stated that 'no question whatsoever is to be asked as to the religious opinions of a pupil, nor is

---

56 London, UCL Special Collections, College Correspondence, Petition dated 22 March 1838; quoted in College Correspondence, p. 318.
57 See Ashton, *Victorian Bloomsbury*, pp. 50, 66.
58 Ashton, *Victorian Bloomsbury*, pp. 52, 66.
59 See S.E. De Morgan, *Memoir*, pp. 26–27.

any pupil to be required to attend any theological lectures which may be given'.[60]

After a forty-year career in academia and scholarly publishing, De Morgan died in 1871, aged 64. Though he had moved his family in 1844 to a larger house in Camden, he continued to be a prominent Bloomsbury figure all his life. He had been one of the youngest and brightest of all the pioneering individuals who found their way in the late 1820s to the new university, and he was one of the longest serving. Bloomsbury, as E. V. Lucas wrote, contained the British Museum at its 'heart', Gower Street as its 'aorta', and was noted as a place where 'bookish men' and people of all faiths and none could live, work and study. It had no greater representative in the nineteenth century than the brilliant, independent-minded Augustus De Morgan.

Bibliography

Arnold, Dana, *Rural Urbanism: London Landscapes in the Early Nineteenth Century* (Manchester: Manchester University Press, 2005).

Ashton, Rosemary, *Victorian Bloomsbury* (New Haven and London: Yale University Press, 2012). https://doi.org/10.12987/yale/9780300154474.001.0001

Bellot, H. Hale, *University College London 1826–1926* (London: University of London Press, 1929).

Clarke, James Fernandez, *Autobiographical Recollections of the Medical Profession* (London: J. & A. Churchill, 1874).

Davis, Eliza Jeffries, *The University Site, Bloomsbury* (Cambridge: Cambridge University Press, 1936).

De Morgan, Sophia Elizabeth, *Memoir of Augustus De Morgan* (London: Longmans, Green, 1882).

Dickens, Charles, *The Posthumous Papers of the Pickwick Club* (London: Chapman & Hall, 1837).

Harte, Negley, and John North, *The World of UCL 1828–1990*, rev. edn (London: University College London, 1991).

Hobhouse, Hermione, *Thomas Cubitt: Master Builder* (London: Macmillan, 1971).

---

60 See Ashton, *Victorian Bloomsbury*, p. 224.

Houston, R. A. and W. W. J. Knox, *The New Penguin History of Scotland from the Earliest Times to the Present Day* (London: Allen Lane in association with National Museums of Scotland, 2001).

Lucas, E. V., *A Wanderer in London*, rev. edn (London: Methuen, 1913).

Melikan, R.A., 'Pains and Penalties Procedure: How the House of Lords "Tried" Queen Caroline', *Parliamentary History*, 20:3 (2001), 311–32. https://doi.org/10.1111/j.1750-0206.2001.tb00380.x

Olsen, Donald J., *Town Planning in London: The Eighteenth and Nineteenth Centuries*, 2nd edn (New Haven and London: Yale University Press, 1982).

Peacock, Thomas Love, *Crotchet Castle* (London: T. Hookham, 1831).

Pugh, Gillian, *London's Forgotten Children: Thomas Coram and the Foundling Hospital* (Stroud: Tempus, 2007).

Rice, Adrian, 'Inspiration or Desperation? Augustus De Morgan's Appointment to the Chair of Mathematics at London University in 1828', *The British Journal for the History of Science*, 30:3 (1997), 257–74. https://doi.org/10.1017/s0007087497003075

Roe, Nicholas, 'William Frend (1757–1841)', *Oxford Dictionary of National Biography* (Oxford: Oxford University Press, 2004). https://doi.org/10.1093/ref:odnb/10169

Slater, Michael, *Charles Dickens* (New Haven and London: Yale University Press, 2009). https://doi.org/10.12987/9780300165524

Stephen, Leslie, 'De Morgan, Augustus (1806–1871)', rev. by I. Grattan-Guinness, *Oxford Dictionary of National Biography* (Oxford: Oxford University Press, 2004). https://doi.org/10.1093/ref:odnb/7470

Stirling, A. M. W., *William De Morgan and his Wife* (London: Butterworth, 1922).

Tames, Richard, *Bloomsbury Past: A Visual History* (London: Historical Publications, 1993).

University College London, UCL Bloomsbury Project (2011), www.ucl.ac.uk/bloomsbury-project.

Usher, H. J. K., C. D. Black-Hawkins, and G. J. Carrick, *An Angel Without Wings: The History of University College School 1830–1980* (London: University College School, 1981).

Fig. 12 Sophia Elizabeth De Morgan, from a photograph taken in 1886. (Public domain, from *Threescore Years and Ten. Reminiscences of the late Sophia Elizabeth De Morgan* (London: Richard Bentley, 1895).)

9. De Morgan's Family: Sophia and the Children

Joan L. Richards

> The marriage was a most happy one,
> and surrounded by a family of seven children, ...
> De Morgan sought his happiness ... in his home ...
>
> — *Nature*[1]

When Augustus De Morgan moved to London after his Cambridge education, his reputation as a creative mathematical thinker preceded him. Within months of his arrival, William Frend, an aging actuary and prominent political radical, enfolded the 'rising young man'[2] into his intellectual circle. The elderly activist and the budding mathematician were bound together by their Cambridge education. This rested on the conviction that mathematics constituted the purest form of reason: a message that inspired both Frend and De Morgan throughout their lives and committed them to bringing the people around them to the full exercise of the reason that defined their humanity.[3] Frend's political liberalism was rooted in the conviction that despite their differences, all humans were alike in being reasoning beings, and he devoted his life to breaking down the barriers that cut Jews, dissenters and Catholics off from English political life.

---

1   Robert Tucker, 'Augustus De Morgan', *Nature*, 4 Jan. 1883, pp. 217–20 (p. 220).
2   Sophia Elizabeth De Morgan, *Memoir of Augustus De Morgan* (London: Longmans, Green, 1882), p. 20. For William Frend, see Frida Knight, *University Rebel* (London: Gollancz, 1971); Joan L. Richards, *Generations of Reason: A Family's Search for Meaning in Post-Newtonian England* (New Haven: Yale University Press, 2021).
3   Richards, *Generations of Reason*.

Yet for all its liberality, the view of reason Frend and De Morgan learned at Cambridge was essentially gendered. Even as Frend fought for the rights of Jews and Muslims, it never occurred to him that women should be given the vote. De Morgan has long been lauded for teaching mathematics to Ada Lovelace, but he saw her as an exceptional being who was ultimately broken by her determination to study mathematics.[4] The reason that both Frend and De Morgan furthered throughout their lives was essentially masculine. Frend was nonetheless completely committed to raising all of his seven children to be reasoning beings. He sent all of his sons to Cambridge, and did everything he could to educate his daughters at home. Frend taught his oldest daughter, Sophia, a great deal of mathematics and astronomy, but she agreed with her husband that women could be broken by pursuing academic subjects too assiduously.[5] She was always much more interested in the reason that tied together the many different people she encountered in her life than she was in the abstractions of mathematics.

Sophia had plenty of opportunities to explore the practice of reason in her childhood home. She came of age listening to her father's conversations with the rag-tag group that flocked to the Frend house 'like martins in the summertime'[6] and learned a great deal from the subset that was at once more respectable and comprehensible. She remembered this group as 'peculiar people', all of whom 'had leading thought or special study': Greek scholar Thomas Taylor; the engraver interested in Babylonian antiquities, John Landseer; self-taught Hebrew scholar John Bellamy; mythologist Godfrey Higgins.[7] All these men believed that the truth they were seeking had been known in the past before the vagaries of human history had obscured it, and that the way

---

4   For De Morgan's views of Ada's health see Alison Winter, 'A Calculus of Suffering: Ada Lovelace and the Bodily Constraints on Women's Knowledge in Early Victorian England', in *Science Incarnate: Historical Embodiments of Natural Knowledge*, ed. by Christopher Lawrence and Steven Shapin (Chicago: University of Chicago Press, 1998), pp. 202–39.
5   S.E. De Morgan, *Memoir*, p. 176.
6   Sophia Elizabeth De Morgan, *Threescore Years and Ten: Reminiscences of the Late Sophia Elizabeth De Morgan to Which Are Added Letters to and from Her Husband the Late Augustus De Morgan, and Others*, ed. Mary De Morgan (London: Bentley, 1895), p. 97.
7   S.E. De Morgan, *Reminiscences*, p. 61.

to uncover that truth was through some form of etymology.[8] They also struggled to make sense of the wide variety of objects and artefacts left by ancient peoples. Several were also astronomers. All were completely committed to reading the past through its objects, its languages and its peoples.

For Sophia, these men's visits constituted an ongoing archaeological, philological and ethnological seminar. From them she learned to use her reason to decipher and explore the deep truths that lay hidden in the world around her. She had access to the larger world through books, but was always at least equally interested in learning from the everyday experiences of all of the people around her, be they friends, neighbours, servants or children. All could be conduits into the deep truths of reason. The lessons she learned from her father's visitors laid the groundwork for what was to be a lifetime of trying to reason her way to an understanding of the deepest truths of human existence.

Sophia was just 19 years old when her father first invited 21-year-old Augustus De Morgan to their home. The young man immediately introduced a new critical perspective into the ongoing seminar that constituted the Frend household. Augustus and Sophia were not married until ten years after they first met. The foundation of their life together was laid in the decade they spent more as siblings than as lovers in the benevolent reasoning world of William Frend.

After their marriage in 1837, the De Morgans' life together was divided along clear gender lines: Sophia managed the household, while Augustus spent his time either teaching mathematics at University College London or writing in his book-filled study. In the first thirteen years of their marriage Sophia gave birth to seven children: Alice, William, George, Edward, Anna Isabella, Christiana and Mary. The division of labour in the household meant that she was in charge of the centrally important task of raising all of these children to be reasoning human beings. Her oldest son William remembered the result as 'a curious admixture of freedom of thought and outlook far in advance' of the times, 'combined with notions of conduct which even then were

---

8   Joyce Godwin, *The Theosophical Enlightenment* (Albany: State University of New York Press, 1994), p. 76.

held to be unduly strict and old-fashioned'.⁹ A more fine-grained picture of both sides of this dynamic may be seen in the two nursery journals that Sophia kept when her first children were young. Sophia began the first of these on 1 January 1840, when Alice was a year and a half and William nearly two months old, and the second in 1842.¹⁰ Sophia was completely devoted to all of her children from the moment of their birth, but the focus of her journals was on the development of their reason. She delighted in them as infants, but was even more fascinated by the thinking they revealed as they began to talk.¹¹

Sophia's 'unduly strict and old-fashioned' approach to her children's conduct fairly leaps from the pages of her first nursery journal. Alice 'has not yet a distinct idea of obedience', she wrote as her toddler was approaching her second birthday: 'this she must learn before she learns anything else'.¹² When it came to questions of obedience, Sophia was caught between contradictory positions. On the one hand stood her husband and his mother, both of whom insisted that 'obedience must be *instantaneous*'.¹³ On the other stood her mother and sisters, who, like the childhood experts Richard and Maria Edgeworth, questioned whether obedience was truly 'the virtue of childhood'.¹⁴ They agreed that children had to learn obedience, but recommended that Sophia do all she could to avoid the issue by creating a child-centred environment in which regularity reigned.

Sophia tried, but the ideal of a child-centred ambience was difficult to achieve within the confines of the De Morgan household. The house at 69 Gower Street was not large, and its spaces needed always to be divided between Sophia's efforts to be a mother and homemaker and De Morgan's labour as a bread-winning educator and writer. Except for his nine o'clock and three o'clock lectures, he was either meeting private

---

9   A.M.W. Stirling, *William De Morgan and His Wife* (London: Butterworth, 1922), p. 48.
10   The first of Sophia's Nursery Journals [henceforth NJ] is at Barnsley, De Morgan Collection, DMF_MS_0024. The second survives only as quoted in Chapter 2 of Stirling, *William De Morgan*, 'A Nursery Journal,' pp. 38–50.
11   For the genre of parents recording child talk, see W.F. Leopold, 'The Study of Child Language and Infant Bilingualism', *Word*, 4 (1948), 1–17.
12   NJ, 30 April [1840].
13   NJ, 9 Apr [1841].
14   Maria Edgeworth and Richard Lovell Edgeworth, *Practical Education*, 2 vols. (London: J. Johnson, 1798), p. 173.

pupils or writing in his home library. Every evening he had dinner with Sophia and spent a little time with his children, but otherwise the little society comprised of Sophia, Alice, William and nursemaid Jane had always to fit itself around his needs for peace and quiet.

Alice represented an enormous problem in this dynamic. Sophia began her journal with great hope, but her narrative quickly devolved into a whole series of efforts to control her little girl's behaviour—by ignoring her, closing her into a different room, holding her hands to restrain them, or tying her to her chair. Each seemed to work for a while, but Alice kept raising the stakes. On 20 January 1840 she became so angry that 'she screamed & fought Jane [the maid],' yanked William's legs, and hit her mother.[15] This was just the opening volley of a week that devolved into 'an almost incessant scene of crying, disobeying, holding hands & forgiving'. Finally, Alice was given 'a grey powder & on Friday she was gentle & good with very few exceptions'.[16] The 'grey powder' to which Sophia resorted was undoubtedly one of the opiates that Victorians imbibed in startling amounts. Although it seems to have been effective in calming the little girl, it had the side-effect of inducing vivid dreams. Sophia did not connect the medicine with her child's nightmares, but on some level it seems that Alice did, and she added 'grey powder' to the list of things she fought against. Sophia devoted pages to trying to work out how to respond to baby Alice's tantrums.

Sophia's entire programme of obedience was essentially a way to clear a space in which to encourage the growth of her children's reason. When she started her first journal, William was not yet two months old, but Alice was beginning to talk and her mother was entranced. Much of her attention focused on recording, interpreting and revelling in the many facets of her daughter's speech: 'She used to say "oh", instead of "Yes"—She has now learnt to say "*Ye*" and I heard her today say "Oh dear!" & correct herself to "Ye! Dear."'[17] A year later, Alice could still be difficult to understand, but Sophia found it well 'worth the trouble of puzzling it out'.[18] She delighted in unexpected connections; when 'Jane

---

15 NJ, Monday [20 Jan, 1840]. Jane was Jane Day, a 30-year-old servant in the De Morgan household.
16 NJ, Tuesday Wednesday & Thursday [21-24 Jan. 1840].
17 NJ, Saturday, [July 1840].
18 NJ, 4 July 1841.

said something about *a jacket*,' Alice began singing Jack and Jill.[19] As she grew older, Alice's connections became ever more intriguing. Alice 'calls the feathery white cloud "the juice of the sky" because I told her they were wet,' Sophia proudly recorded, and was delighted when her daughter called seeds 'the eggs of the flowers'.[20] 'Alice could frame a language' she glowed when her 3-year-old said 'open a light' instead of 'light a match'.[21] Throughout the four years that she kept her journals, Sophia was fascinated by Alice's use of words.

As William became verbal, he was equally interesting, albeit in somewhat different ways. He was always more amenable than Alice, which Sophia saw as a reflection of inborn character, but which might equally be attributed to his being a boy, who was given wheelbarrows with which to play while Alice was having to sit still to have her hair combed or to hem handkerchiefs with neat little stitches. As William began to talk, Sophia noticed he was particularly interested in rhymes like 'Billy sees Clown/A-tumbling down',[22] and was amazed by his ability to remember pieces of poetry.[23] At the age of 2, she reported, he spent hours studying a book of birds and enjoyed assigning their names to those around him: 'You're a silky starling!' he told his mother; when she asked him who he was, he responded with 'a three-toed Woodpecker'.[24] Whereas in Sophia's journals Alice displays a quick-silver verbal intelligence, William emerges as an acute observer.

Her children's imaginations provided Sophia with another revealing entry into their minds. From the age of about two and a half, Alice had an imaginary companion, named Marmee, whom she would often let stand in for herself as in 'Mama, My Marmee will yore [roar] an wake up hi[s] little brother dat *tiny* boy'.[25] As her daughter grew older, Sophia began introducing other characters designed to carry messages about good behaviour. When Alice resisted getting out of her warm bed on chilly mornings, Sophia told her a 'very interesting story' in which an imaginary Louisa had cured herself of the same behaviour 'by her own

---

19   NJ, Sunday, 15 Feb. [1840].
20   Stirling, *William De Morgan*, p. 42.
21   NJ, 10 Dec. 1840.
22   Stirling, *William De Morgan*, p. 41.
23   Stirling, *William De Morgan*, p. 42.
24   Stirling, *William De Morgan*, p. 42.
25   NJ, 2 Dec. 1840.

determination'.²⁶ Sophia's other stories could be more fun. She once 'induced Willie to walk instead of being carried, by pretending that they were people travelling through a strange country in which we met all kinds of wild animals, cats were panthers, horses—lions, and dogs—tigers, etc.'²⁷ Alice and William embroidered on this suggestion with such enthusiasm that passers-by stopped to check what was happening, and even they themselves had to be calmed when fears of various 'preten' [pretend] beasts became overwhelming.

Even as Sophia was thrilling at her children's imaginations and telling stories to help them interpret the world around them, she was aware of a drawback. Following the twists and turns of her children's imaginations could be fun, but telling stories could shade into lying, and lying could blossom into an even more serious problem than disobedience. Concern about dishonesty was a persistent theme in the early Victorian world. Writers across the spectrum, from the Anglican William Whewell to the Unitarian Harriet Martineau, essentially agreed that lying was a temptation to be avoided at all costs. Sophia agreed completely, and as her children became ever more articulate and imaginative, she remained vigilant.²⁸ Weighing the wonder of her children's imaginations against the danger of their lying was always a delicate balancing act.

Sophia's concern about lies was rooted in the Lockean program of reason in which maintaining the clear connection between words and their proper meanings was absolutely essential to the pursuit of reason's truth. At work, her husband was constantly being reminded of the difficulty of maintaining those connections as his students combined symbols in meaningless ways or lost themselves in arguments and proofs.²⁹ The problem Sophia faced with her children was in many ways more complicated. Even as she delighted in the poetry of their speech, she had always to be equally alert that they did not ever completely lose sight of the connection between her words and their meanings.

---

26 NJ, 18 [Dec. 1840].
27 Stirling, *William De Morgan*, p. 41.
28 For more on Victorian views of lying in children see 'Lies and Imagination', in Sally Shuttleworth, *The Mind of the Child: Child Development in Literature, Science and Medicine, 1840–1900* (Oxford: Oxford University Press, 2010), pp. 60–74.
29 For more on Augustus De Morgan's pedagogically-inspired desire for accurate expression, as well as his fascination with language and symbolic notation, see the section on his philosophy of mathematics in Chapter 1 of this volume.

Although Sophia was the primary figure who negotiated the ups and downs of their children's lives, Augustus was also fascinated by watching their developing minds. He composed picture letters to engage the interpretative skills of children who were learning to read.[30] His letters to his friend, the Irishman William Rowan Hamilton, are filled with the kinds of challenges he liked to explore with them. 'Take a child and say "now we are going to draw a house",' De Morgan directed, 'but then draw one in which the chimney is hugely out of proportion. As soon as the child says "that chimney is *too big*",' he exulted, 'the remark was dictated by the presence and action of the notion of relative magnitude.'[31] In another, he constructed an elaborate story about a boy rolling a hoop across parish lines as a way to define the meaning of the total area of a curve that intersects itself any number of times.[32] He undoubtedly tested the self-evidence of what he saw as the 'four-colour axiom' on them.[33] Sophia watched all of these exercises with interest. Although she stopped keeping nursery journals, she remained as deeply invested in cultivating her children's reasoning powers as her husband.

De Morgan had 'in the earlier part of his life held man-like and masterful views of women's powers and privileges',[34] and he always liked pointing out that 'when we overcome a difficulty we say we *master* it, but if we fail we say we *miss* it',[35] but living with Alice had its effect. As he watched his daughter grow into a reasoning being, he agreed with Sophia that she needed a school that would give her all the opportunities to develop her reason to its fullest that her brothers would have at the University College School for boys. They threw themselves behind Elizabeth Reid to help create a secular school in which women

---

30   Senate House Library, University of London, MS 913/A/3.
31   Augustus De Morgan to William Rowan Hamilton, 31 Dec. 1863. Robert Perceval Graves, *Life of Sir William Rowan Hamilton* (Dublin: Hodges, Figgis and London: Longmans, Green, 1882–1889), vol. 3 (1889), p. 603.
32   Augustus De Morgan to William Rowan Hamilton, 26 Sept. 1849. Graves, vol. 3, pp. 278, 282. The question was posed in the *Cambridge and Dublin Mathematical Journal*.
33   For what De Morgan called the four-colour axiom, but is now known as the four-colour theorem, see: Robin Wilson, *Four Colors Suffice: How the Map Problem Was Solved* (Princeton: Princeton University Press, 2002); Rudolf Fritsch and Gerda Fritsch, *The Four-Color Theorem: History, Topological Foundations, and Idea of Proof*, trans. by Julie Peschke (New York: Springer, 1998).
34   S.E. De Morgan, *Memoir*, p. 94.
35   Stirling, *William De Morgan*, p. 32.

would hold the power. In 1849, Alice became one of the first pupils in the Ladies' College at 47 Bedford Square.[36] Long after the De Morgan girls were grown, the family connection remained strong enough for Joan Antrobus, Augustus and Sophia's great-granddaughter, to travel from South Africa to attend the College's successor tertiary institute, Bedford College, for the year 1925–26.

Establishing the Ladies' College in Bedford Square was one cause that drew in both of the De Morgans; abolition was another. Moved by the explosion on the English scene of Harriet Beecher Stowe's 1852 novel *Uncle Tom's Cabin*, which sold more than a million and a half copies there in its first year, Sophia vowed to do everything in her power to bring an end to the institution of slavery.[37] In autumn 1852 she drafted a letter to be signed by the people of England urging the people of America to give up their slaves. She acknowledged that the English shared the blame for slavery, having established the system at a time when 'Americans were not under their own laws and legislature'. Now, however, 'uninfluenced by those personal interests which involve and obscure the question on its own soil', the English had a clearer view of the pernicious effects of slavery than did those who were caught up in it. Sophia then laid out what she saw as the horrors of slavery, before closing with the hope that God 'will bring to your hearts a conviction of its enormity, & give you strength to abjure it'.[38] In her attempt to address the problem through a combination of rational argument and theistic conviction, Sophia showed herself to be her father's daughter. But her goal was political change, and that could not be achieved merely through writing. It was necessary to bring the letter to a larger audience.

Sophia shared her idea with Rachel Chadwick, who shared it with her husband, the social reformer Edwin Chadwick, and he told the great reformer, Lord Shaftesbury, about the plan. Shaftesbury wrote to *The Times* on Wednesday, 9 November. Subsequently, under the leadership

---

36 For the history of this institution, see Margaret J. Tuke, *A History of Bedford College for Women* (London: Oxford University Press, 1939). For the earlier history, see Chapter 8 of this volume and, for more detail, Rosemary Ashton, *Victorian Bloomsbury* (New Haven and London: Yale University Press, 2012), pp. 215–38.

37 See Audrey Fisch, 'Uncle Tom and Harriet Beecher Stowe in England', in *The Cambridge Companion to Harriet Beecher Stowe*, ed. by Cindy Weinstein (Cambridge: Cambridge University Press, 2004), pp. 96–112.

38 Senate House Library, University of London, MS913B/2/3, Draft proposal on slavery in Sophia's hand.

of the Duchess of Sutherland, the plea from the *women* of England to the *women* of America (changed from Sophia's 'people' of England to those of America) gained enough signatures from women across all classes in the United Kingdom, Australia, Canada, New Zealand and Palestine, to fill twenty-six folio volumes. In spring 1853 Stowe visited England in order to receive them.[39] Sophia met Stowe several times at Mrs Reid's house.

On 23 December 1853 Alice De Morgan died of phthisis, or tuberculosis. The De Morgans were shattered. Sophia had apparently been trying to combat the teenager's 'weakness and delicacy' for some time, but Augustus 'did not realise the degree of illness till the end was near, and the blow fell heavily upon him'.[40] Twenty-five years later, Sophia was still unable to write of these events, and Augustus never tried. Thus *The Old Man's Youth,* the semi-autobiographical novel Alice's brother William wrote at the end of his life, is the clearest description of the family's experience. Even seventy years later, his memories of helplessness remained so vivid that 'I *am* that boy, the growing panic of that moment is on me still, and the gloom'.[41]

When Augustus returned to his office after Alice's burial, he felt as if he 'had been suddenly carried off, all round the world, and set down again at his desk'.[42] The only work that penetrated his grief was Whewell's anonymously published *Of the Plurality of Worlds*, which speculated about whether there was life anywhere else in the universe.[43] By the time Whewell sent him the second edition of *Plurality*, De Morgan was becoming convinced that there were 'inhabitants', possibly including Alice, of other planets, with '*uses* independent of us' which he strongly suspected were 'also *trusts*, and therefore I suppose *responsibilities*'.[44]

---

39   Joan D. Hedrick, *Harriet Beecher Stowe: A Life* (New York: Oxford University Press, 1994), p. 232.
40   S.E. De Morgan, *Memoir*, p. 190.
41   William De Morgan, *The Old Man's Youth and the Young Man's Old Age* (London: Heinemann, 1921), p. 101.
42   Augustus De Morgan to William Rowan Hamilton, Jan 10, 1854. Graves, vol. 3, p. 470. De Morgan wrote this in the first person, so was 'set down again at my desk'.
43   William Whewell, *Of the Plurality of Worlds: An Essay*, ed. by Michael Ruse (Chicago: University of Chicago Press, 2001), p. 253 [295]. De Morgan later reviewed this work in Augustus De Morgan, *A Budget of Paradoxes* (Chicago: Open Court, 1915), p. 63.
44   Augustus De Morgan to William Whewell, May 21, 1854. S.E. De Morgan, *Memoir*, p. 230.

In answer to Whewell's expanses of emptiness, De Morgan offered a universe teeming with intelligences in which his daughter's life was not wasted.

Sophia agreed with her husband that the universe was filled with spirits, but her sense of its inhabitants was more domestic and immediate. In letters to Lady Byron, written after her sister Harriet died in 1836, she spent months developing a theory of life after death. In the immediate aftermath of her father's death in 1841, she found 3-year-old Alice's response affirming. In her nursery journal she described her efforts to give her little girl 'as *true* an idea' of Frend's death as was possible with a 3-year-old. She told Alice that the doctor was trying to cure her grandfather, but that he would probably fail, and that when that happened, he 'will go away to a nice place where he will be made quite well'.[45] This attempt to construct a child's-eye view of the afterlife seems to have made sense to Alice, who spent several days exploring the idea of this 'nice place', and asking 'whether the birds sang & the trees were pretty & had buds and fruit'. Ever truthful, Sophia admitted that she did not know because she had not been there, but she repeated that it was a very nice place that was filled with good people. When Alice gave her opinion that '*he gathered the fruit from one of those trees, & eat it, & dat made him quite well*', Sophia was thrilled. 'What an extraordinary idea to enter a baby's head!' she exclaimed. Even though no one had told Alice the story of Adam and Eve, the little girl was talking of 'eating the fruit of the tree of life'.[46] All of Sophia's efforts to ensure that her little girl spoke the truth were rewarded by this glimpse into what William Wordsworth described as the 'heaven that lies about us in our infancy',[47] which Sophia saw as a description of the world of the afterlife.

Beyond this incident with Alice, there is little evidence of Sophia's interest in life after death in the 1840s, but as she was living at home with her ever-increasing brood, she was, like several other spiritualists, actively experimenting with the invisible forces of mesmerism. The attraction of this approach is suggested by an experience described in the Nursery Journal, in which one-year-old William was all but killed by the standard medical treatments of leeching, blistering, lancing and

---

45 NJ, 26 March, [1841].
46 NJ, 26 March, [1841].
47 William Wordsworth, 'Ode: Intimations of Mortality'.

starving prescribed to combat a fever. Making passes over a feverish child's body was a considerably more attractive approach. Sophia was aware that mesmerism did not guarantee a cure—success required the action of invisible forces that were poorly understood—but the outcomes of many medical treatments were not predictable either.

Sophia never claimed particular mesmeric prowess. She did say that 'many patients have spoken of *light* which they said they saw streaming from my fingers' when she made passes over their bodies,[48] but even as she offered this credential, she included herself among 'those who had no power of vision', and therefore saw nothing.[49] In one instance she did claim to see the effects of her efforts. When a neighbour brought her a ten-week-old baby whose legs seemed poorly aligned, Sophia decided there was no harm in trying mesmerism before turning to the bandages the doctor had recommended. After about six passes 'from the knees to the end of the little feet', the legs began to move into their natural position, and 'the muscles gained a power which they never had before'.[50] This success carried Sophia through years of experimentation.

Over time, she began to see that the powers she had first observed in a medical context might extend to larger phenomena. In 1849, when she had induced a mesmeric trance in an effort to treat 'fits' in a 'young and ignorant girl', she found herself a startled witness to 'the state of clairvoyance'. While under the mesmeric influence, the girl talked Sophia through the streets of London to a house where she observed with minute detail the room in which Augustus was visiting one of his friends. Although she never left her chair, the girl's descriptions of the house, the room and the conversation within it were so complete, detailed and accurate that Sophia and Augustus were both convinced that she had made an actual 'mental' journey to the place she described.[51] That this unschooled girl had the power to see reality while travelling in thought was powerful support for the existence of a transcendent world

---

48 Sophia Elizabeth De Morgan, *From Matter to Spirit: The Result of Ten Years' Experience in Spirit Manifestations. Intended as a Guide to Enquirers* (London: Longman, Green, Longman, Roberts & Green, 1863), p. 45.
49 S.E. De Morgan, *From Matter to Spirit*, p. 46.
50 S.E. De Morgan, *From Matter to Spirit*, pp. 43–44.
51 Augustus's version of this experience is in a letter to the Rev. William Heald [1849] in S.E. De Morgan, *Memoir*, pp. 206-08. Sophia's version is in S.E. De Morgan, *From Matter to Spirit*, pp. 47–49.

of mind that lay behind the material one. Sophia's interest was part of a far wider contemporary fascination with spiritualism.[52]

In 1853, an American spiritualist, Mrs. Hayden, burst onto the London scene. This formidable woman was interested in moving past impersonal forces to a world of spirits who could use those forces to communicate with people. Having for decades been convinced that Harriet and her father still lived in some other-worldly realm, Sophia found the project very attractive. She wanted to believe herself sceptical, but her defences began to weaken within the first hour when Mrs. Hayden delivered the message that Harriet was 'happy'. When, on her second visit, her father tapped out *'Why do you doubt the holy attributes of God, when this is in perfect accordance with His teaching?'*[53] she was entranced. The message was certainly not delivered in the way Frend would have phrased it, but it was 'the sort of sentiment'[54] she would have expected from him. It was hard for her to resist the evidence that Mrs. Hayden had the power to communicate with the dead. Sophia was so impressed that she invited Mrs. Hayden to Camden Street so that Augustus could meet her.[55] After Alice died, the issues became much more immediate, and Sophia plunged into an investigation of the spirit world that would occupy her for the rest of her life.

Sophia's sense of the universe that Alice had entered was much more familiar than her husband's thoughts about *uses, trusts*, and *responsibilities*. She never claimed to be a 'sensitive', which meant that her personal glimpses into Alice's world were rare and fleeting, but she energetically engaged with those who claimed that capacity. In autumn 1857 she started a diary in which she followed her daughter through a visionary world that included a rich cast of characters from Cupid

---

52 For an overview of the English experience of table-turning and mesmerism in the 1850s and beyond, see Alison Winter, *Mesmerized: Powers of Mind in Victorian Britain* (Chicago: University of Chicago Press, 1998), pp. 276–306.
53 S.E. De Morgan, *From Matter to Spirit*, pp. 13–14.
54 S.E. De Morgan, *From Matter to Spirit*, p. 14.
55 Sophia De Morgan was among the minority in being impressed by Mrs Hayden. See Maurice Leonard, *People from the Other Side: The Enigmatic Fox Sisters and the History of Victorian Spiritualism* (Stroud: History Press, 2008), p. 82. For a full but, at the time of writing, not very accessible account of Mrs Hayden, see Sharon DeBartolo Carmack, *In Search of Maria B. Hayden: The American Medium Who Brought Spiritualism to the UK* (Salt Lake City: Scattered Leaves Press, 2020).

to glimpses of God and Christ.⁵⁶ It is difficult to make narrative sense of the place Sophia found Alice to be living in. What is clear is that Alice's mother was doing everything she could not to lose touch with her beloved child.

A decade after Alice's death, in 1863, the De Morgans presented their visions in *From Matter to Spirit: The Result of Ten Years' Experience in Spirit Manifestations Intended as a Guide to Enquirers*. The title page of the book stated simply that it was 'By C.D. with preface by A.B.' but those pretensions of anonymity were fleeting. Within days, everyone knew that the book was written by Sophia, the 'Preface' by Augustus. In it she followed the model of reason she had learned from her father and his many visitors into the wider nineteenth-century world in which she lived. Sophia was typical of her age in regarding spiritualism as a science.⁵⁷ One of the changes that had occurred in the twenty-five years since she tried to make sense of the death of her sister Harriet lay in the variety of sources she found relevant. She devoted nine pages to a close reading of the discussion of death in 1 Corinthians 15.35–57 that had supported her earliest conviction that Harriet still lived, but those pages are embedded in a fifty-five-page chapter that included sources from Plato to Swedenborg. The Christianity she learned from her father remained Sophia's touchstone, but he had also taught her to be open to other perspectives. Her book was the product of a Victorian who gave credence to many sources beyond the Bible. It remains a lasting testimony to a particular brand of spiritualism of which she, together with various other middle-class intellectuals and professionals including William and Mary Howitt and Royal Physician John Ashburner, were early propagators.⁵⁸

It was not easy for Sophia to carry her father and husband's views of reason beyond the sheltered classrooms of Cambridge and UCL into her woman's world of deathbed scenes, near-death experiences

---

56 Senate House Library, University of London, MS 913B/2/2.
57 See Richard Noakes, 'The Sciences of Spiritualism in Victorian Britain: Possibilities and Problems', in *The Ashgate Research Companion to Nineteenth-Century Spiritualism and the Occult*, ed. by Tatiana Kontou and Sarah Wilburn (Farnham: Ashgate, 2012), pp. 25–54, especially pp. 29–32, and Tatiana Kontou and Sarah Wilburn, 'Introduction', in the same volume, pp. 1–16 (pp. 1–4).
58 Alex Owen, *The Darkened Room: Women, Power and Spiritualism in Late Victorian England* (London: Virago, 1989), p. 21.

and ghost stories. She knew that she was not a medium, which meant that she had to collect information from others. Sophia was like her father in her willingness to listen seriously to those around her. But his strict standards of linguistic rationality did not work in her world of 6-year-olds, nursemaids, neighbours and mediums. She could, and did, routinely screen her informants for truth-telling, but she could not guarantee that they either spoke or wrote precisely and properly. It was very difficult for her to imagine her father saying things like *'we long to clasp you in our arms in this bright world of glory'*[59] and simply impossible that he would make spelling errors that turned 'Beautiful' into *'butiful'*, 'writing' into *'riting'*, and so on.[60] The power of reason was thus transmogrified as it moved from the masculine world of its usual defenders into the predominantly female world that existed by its side.

Sophia was nonetheless determined to use reason to identify the basic structures that underlay not only spelling mistakes, but a bewildering array of blowing curtains, turning tables, spirit writings and trance descriptions. She began by organising spiritual experiences on a hierarchical scale of materiality. The least exalted experiences were those like table-turning, which occurred on a material level that even she could observe. Somewhat higher up the scale was 'spirit writing', in which someone holding a pen was guided by spirit power. In this case the act of writing could be observed by many, but the force behind it was experienced only by one (or sometimes two) people. Highest of all were visions, dreams and voices perceived directly in the mind, because these had no inter-subjective material manifestations at all.

An elaborate theory of human development underlay this hierarchy of experiences. All people, Sophia explained, are made of a material body, an animating spirit, and an ever-developing soul. At the moment of death, the soul 'passes away' from the material realm, 'and, animated by the spirit, becomes the body of the next life'.[61] In this new form, the process of development continues; the spirits move ever closer to God and farther from the material world. Spirits who communicated through material manifestations like table-turning were at the lowest

---

59 S.E. De Morgan, *From Matter to Spirit*, p. 15.
60 S.E. De Morgan, *From Matter to Spirit*, p. 23.
61 S.E. De Morgan, *From Matter to Spirit*, p. 268.

level of spirit development, whereas those who communicated without such manifestations were higher on the developmental scale.

Sophia distilled these hierarchies of experience and development from her conversations, readings, and experiences, but she needed a rational ground for her spiritual theorising. She rested hers on what she called 'the Principle of Correspondence', defined as 'the law by which the external of one state agrees with the internal of that below it'.[62] She acknowledged that this principle might at first glance appear 'mystical and imaginary', but she insisted that it was 'intelligible enough' to render any conclusions drawn from it 'as certain as any branch of knowledge which can be deduced by well-marked steps from indisputable principles'.[63] Sophia's principle of correspondence provided a stable platform from which to evaluate a set of otherwise confusing and untethered phenomena. Over the course of years of contemplation, the aspects of the principle that seemed at first obscure became ever clearer until they were incontrovertible. The process of insight was the same as that experienced by students studying geometry. In the end the self-evidence of her principle supported Sophia's work in the same way that self-evident postulates supported geometry. Her effort to build a theory upon a principle that could be understood clearly and distinctly reflected the reasoned approach that De Morgan was teaching his students year after year in mathematics classes.

In 1859, the De Morgans moved to a new house on Chalcot Villas, soon renamed Adelaide Road, in Primrose Hill that was well suited to their growing brood. As Augustus put it, in the immediate aftermath of Alice's death, he could *'understand'* but not *'feel* that six left made any set-off against one gone',[64] but now both he and Sophia began again to enjoy their children's company. It was not difficult to do; their offspring were a sophisticated and fun-loving group. At the time of the move, William was 20 and, having spent three years at University College School followed by a year at UCL, was veering off to study at the Royal Academy of Arts. At 18 and 16 respectively, George and Edward were following the early stages of their own trajectories. A contemporary portrait captures the three of them before a house, with

---

62 S.E. De Morgan, *From Matter to Spirit*, p. 274.
63 S.E. De Morgan, *From Matter to Spirit*, p. 267.
64 Augustus De Morgan to a Friend, Jan. 19, 1861. S.E. De Morgan, *Memoir*, p. 304.

William playing the role of rakish art student, George a stolid scholar, and Edward a somewhat impish younger brother. Not pictured are their sisters—Annie aged 14, Chrissy aged 12 and Mary aged 9—but the girls were equal members of the tight group of De Morgan children. They enjoyed musical evenings in which Annie played the piano, Augustus his flute and Edward his violin.[65] They played elaborate games in which some would draw pictures and challenge the rest to write the stories to accompany them. They shared an interest in anagrams; 'Great gun, do us a sum!' is just one of a list of over two thousand created from Augustus's name, which Sophia carefully preserved.[66] As time went by and the boys began to venture out on their own, they still returned all but daily for time with the family on Adelaide Road.

William's decision to leave UCL after only one year confused and disappointed his father, but George thrived there. After graduating, George looked for other ways to pursue mathematics in London, and by 1866 had secured the position of mathematics teacher at University College School that he and his brothers had attended. This was not enough to support the development of his mathematical ideas, however, so in 1864, George and his friend Arthur Cowper Ranyard decided to form a society that focused on mathematics. At first they were thinking of a school group—either the London University Mathematics Society, or the 'University College Mathematical Society'—but by the time of their first regular meeting in January 1865 the group they co-founded had expanded its vision to become the London Mathematical Society, or LMS.[67] De Morgan was warmed by his son's enterprising spirit and thrilled to be named the new society's first president. The Society

---

65 Stirling, *William De Morgan*, p. 61.
66 Stirling, *William De Morgan*, pp. 64–65. Stirling attributes the list to William, but it was in fact created by De Morgan's fellow mathematician and professor of jurisprudence at UCL, John Thomas Graves. The collection is to be found in UCL Special Collections, MS Graves 36. A selection of these anagrams may also be found in Augustus De Morgan, *Budget*, pp. 82–83.
67 For the founding of the society see: S.E. De Morgan, *Memoir*, pp. 281–86; Adrian C. Rice, Robin J. Wilson and J. Helen Gardner, 'From Student Club to National Society: The Founding of the London Mathematical Society in 1865', *Historia Mathematica*, 22 (1995), 402–21; Adrian Rice, 'London Mathematical Society Historical Overview', in Susan Oakes, Alan Pears, and Adrian Rice, *The Book of Presidents 1865-1965* (London: London Mathematical Society, 2005), available on the London Mathematical Society's website at: https://www.lms.ac.uk/sites/default/files/About_Us/history/lms_full_history.pdf.

continues to flourish as the United Kingdom's premier learned society for mathematics in De Morgan House on Russell Square, with over 2,700 members worldwide. It confers triennially the De Morgan Medal, Britain's highest mathematical honour, and influenced the formations of similar bodies abroad, such as the French and American Mathematical Societies.[68]

Ever since his triumphant graduation from UCL, George had been in declining health. He died in October 1867. Soon Chrissy began to exhibit the delicacy that had presaged Alice and George's early deaths. She too died and was buried in Bournemouth in August 1870. Seven months later, on 18 March 1871, Augustus De Morgan died at home. He was buried with Alice and George in the family plot in Kensal Green Cemetery.

In the years surrounding Chrissy and Augustus's deaths, the remaining members of the family reorganised themselves. Edward, who was concerned about his own health, moved to South Africa for its weather, and Annie married Dr Reginald Edward Thomson in 1872.[69] In the same year Mary and Sophia moved to a yet smaller house on Cheyne Row, Chelsea, where both women turned their attention to writing. Sophia had already gathered De Morgan's notes and articles into *A Budget of Paradoxes*, a popular book which was published in 1872.[70] In Cheyne Row she turned her attention to establishing the righteousness of Augustus's causes in a *Memoir of Augustus De Morgan*. Finding the best way to present her husband's life's work entailed considerable negotiation with family and friends, but by 1882 she was finally satisfied that she had succeeded in explaining his life of reason. Sophia then turned to writing her own reminiscences in *Threescore Years and Ten*. Working with Mary to draw her memories together sustained her through the last decade of her life.

---

68 See London Mathematical Society, https://www.lms.ac.uk; Adrian C. Rice and Robin J. Wilson, 'From National to International Society: The London Mathematical Society 1867–1900', *Historia Mathematica*, 25 (1998), 185–217.

69 Their son was the archaeologist and Assyriologist Reginald Campbell Thompson, on whom see Clyde Curry Smith, 'Thompson, Reginald Campbell (1876–1941), Assyriologist and Archaeologist', *Oxford Dictionary of National Biography*, online edn (Oxford: Oxford University Press, 2004-).

70 See Chapter 7 of this volume.

Mary was also hard at work at her own writing. Her first work, *Six by Two*, was a collection of fictional stories about schoolgirls, co-edited with Edith Helen Dixon.[71] It was published in 1873, thirteen years before L.T. Meade's *A World of Girls*, the girls' school story seen as the starting point of what was to become an exceedingly popular genre. She was more interested in writing fairy tales, however, and published *On a Pincushion and Other Fairy Tales* in 1877. In the next decade she published two other collections: *The Necklace of Princess Fiorimonde* and *A Choice of Chance*.[72] In the same way that her mother had used imaginary characters to teach Alice good behaviour, Mary's fairy tales carried social messages. But whereas her mother's goal was to encourage Alice to cooperate, Mary wrote with an iconoclastic feminist slant: 'The Toy Princess', for example, was a satirical dig at the prim Victorian young lady expected to hold no opinions.

Like her brother William's writings, Mary's stories were popular in their own time, described by her obituarist in *The Times* as being 'of very distinguished quality, and ... the delight of more than one generation of children'.[73] Unlike William's, they have stood the test of time, republished sporadically throughout the twentieth century: as single anthologised stories and as collections, both illustrated by the original artists and interpreted by newer illustrators.[74] Scholarly interest

---

71   Edith Helen Dixon and Mary De Morgan, *Six by Two, Stories of Old School Fellows* (London: Virtue, 1873); repr. as *The French Girl at Our School and Other Stories* (London: Virtue, 1887).
72   Mary De Morgan, *Complete Fairy Tales* (New York: F. Watts, 1963).
73   'Obituary', *The Times*, 10 June 1907, p. 6. Repr in Marilyn Pemberton, *Out of the Shadows: The Life and Work of Mary De Morgan* (Newcastle upon Tyne: Cambridge Scholars Publishing, 2012), pp. 243–44 (p. 243).
74   Posthumous editions are: *On a Pincushion and Other Fairy Tales; The Necklace of Princess Fiorimonde and Other Stories*, introd. by Charity Chang (New York and London: Garland, 1977; facsimile reprints of the first editions); *On a Pincushion and Other Fairy Tales*, ill. by Jean Walmsley Heap (London: R. Ingram, 1950); *The Necklace of Princess Fiorimonde and Other Stories*, ed. by Roger Lancelyn Green and ill. by William De Morgan, Walter Crane and Olive Cockerell (London: Gollancz, 1963); ill. by Sylvie Monti (London: Hutchinson, 1990); and, most recently (Dinslaken: anboco, 2016). Anthologies with a story by Mary De Morgan (usually 'A Toy Princess') include: *A Staircase of Stories*, ed. by Louey Chisholm and Amy Steedman (New York: Putnam's, 1920); *A Book of Princesses*, ed. by Sally Patrick Johnson (Harmondsworth: Penguin, 1965); *Beyond the Looking Glass, Extraordinary Works of Fairy Tale and Fantasy: Novels and Stories from the Victorian Era*, ed. by Jonathan Cott (New York: Stonehill, 1973; 'Through the Fire'); *The Revolt of the Fairies and Elves*, ed. by Jack Zipes (London: Routledge, 1987); *The Oxford Book of*

in female writers and in Mary's chosen genre have assisted survival, although a monograph was devoted to her only in 2012.[75]

Mary was clearly a product of her family. Even if one discounts the statement in her *Times* obituary that she 'inherited from both parents very considerable literary power' as a value judgement and a very general one at that, her exposition of the position of women marks her as Sophia's daughter, while her satirical strain, independent thought and inventiveness are reminiscent of Augustus. From Augustus, too, could come astronomical awareness, shown in her 'The Story of the Opal', and from use of scientific phenomena in her tales.[76]

In 1872 William moved into a house just three doors down from his mother and sister on Cheyne Row where he began to set up a pottery studio. It took him several years and many mistakes to master the complicated processes involved in making pots and tiles, but when he succeeded in taming his medium, the ebullient imagination that was relegated to the margins of his father's reasoned world exploded in images of adventuring ships, fire-breathing dragons, contemplative mermaids and imps peering out from among flowers. The colours he finally mastered flashed in exuberant peacocks' tails, deep blue oceans and bees that positively buzz with redness. He easily matched the playfulness his father expressed in his doodles; in fact, Charles Dodgson, better known as Lewis Carroll, is said to have written *The Hunting of the Snark* in response to the De Morgan tiles he had installed in his college rooms.[77]

In the years that surrounded their father's and siblings' deaths, William and Mary grew close. Mary enjoyed making tiles in her brother's studio, while William developed a novel technique for etching the illustrations for *On a Pincushion*. William was a rather sociable person, whose range of artistic friends included Edward Burne-Jones

---

       Modern Fairy Tales, ed. by Alison Lurie (Oxford: Oxford University Press, 1993); *The Oxford Book of Children's Stories*, ed. by Jan Mark (Oxford: Oxford University Press, 1994; 'Nanina's Sheep').

75 Pemberton, *Out of the Shadows*.

76 Discussed in Laurence Talairach-Vielmas, *Fairy Tales, Natural History and Victorian Culture* (Houndmills: Palgrave Macmillan, 2014), pp. 65–79.

77 June Barrow-Green, 'Euclid, William De Morgan and Charles Dodgson', in *The London Mathematical Society and Sublime Symmetry* (London: London Mathematical Society, 2018), p. 17.

and William Morris. Mary was a more mercurial character, who either charmed or repelled those who encountered her.

Outside of Cheyne Row, the family's troubles continued. In 1877 Edward died after falling from a horse in South Africa, and in 1884 Annie died of the same disease that had taken Alice, George and Chrissie. Sophia's communion with the dead sustained her through all of these losses. Until the very end of her life, she remained as warmly interested in everyone around her as was the father she so adored. She became actively involved in the movement against vivisection, and from the time of its founding in 1882 continued her explorations of the afterlife as a member of the Society of Psychical Research in London. In 1895, at the age of 87, Sophia De Morgan died in her sleep and was buried with Augustus, Alice and George.

In 1887, at the age of 48, William married Evelyn Pickering, who was fifteen years his junior and had already established herself as a painter. The talented new member of the family was in many ways a suitable heiress to Sophia's world of strong women. On her canvases, resilient sea maidens stand united among ocean waves, and powerful women direct thunderstorms. Sophia's ways of understanding pain are reflected in Evelyn's later works, where spirits pull away from exhausted bodies, and Christ rises from a graveyard supported by angels. In her work, Evelyn expanded upon the reflections on matter and spirit that her mother-in-law had begun.

Despite decades of hard work William's studio was a commercial failure, and in the early twentieth century he followed his sister and turned to writing. The novels he wrote were in their time a runaway success, and William was compared in complimentary terms with Dickens and Thackeray.[78] According to A.C. Ward, his fiction 'amused, touched, consoled, and inspired a widespread multitude as hardly any English novelist had done since Dickens died in 1870'.[79] His writings are clearly fictional, but it is not difficult to detect signs of his family background in them. The portrayal of the life of the young artist in

---

78   For evidence of complimentary reception, with quotations, see Mark Hamilton, *Rare Spirit: A Life of William De Morgan 1839–1911* (London: Constable, 1997), pp. 119–23.

79   William De Morgan, *Joseph Vance: An Ill-Written Autobiography*, ed. by A.C. Ward (London: Oxford University Press, 1954), p. xiii.

Alice-for-Short reflects a warm experience with siblings coming and going through a parental home with a welcoming spiritualist matriarch.[80] The childhood pictured in *Joseph Vance* is a darker reflection of the struggles to be found in Sophia's nursery journals, with a demanding mother and emotionally distant father.[81] As historical sources, William's novels are opaque at best, but clearly William spent his final years ruminating on his experiences growing up as a De Morgan.

The success of William's novels supported him well financially until his death in 1917. They dated quickly, however, such that even *Joseph Vance*, reprinted as an Oxford University Press World's Classic in 1954, has been out of print for more than half a century. Despite public revulsion against Victorian art in the first half of the twentieth century, it is William's ceramics that have stood the test of time. He has been described since as having 'as conspicuous a mastery as did William Morris in his undertakings in design'; 'an artist in the true sense of the word'; 'perhaps the greatest of all English ceramic artists', with unique achievements and imaginative powers, and credited as the re-inventor of lustre.[82] In 1968 the De Morgan Foundation was established to preserve his and his wife's work.[83] It continues to flourish, and William's works are displayed in several English galleries—most strikingly, the Victoria and Albert Museum—as well as Cardiff Castle Museum in Wales and, in continental Europe, the International Museum of Ceramics at Faenza.

William did not claim to be practising reason when he created his designs, but the intense experiences of insight that inspired his father's and mother's research leap directly from his pieces. The symmetries that supported his father's logical thinking hold his imps in place, control the rolls of his dolphins, shape the flights of his dragons, and structure whole walls covered with carnations, roses, daisies and swans.

---

80 William De Morgan, *Alice-for-Short: A Dichronism* (London: Heinemann, 1907).
81 William De Morgan, *Joseph Vance: An Ill-Written Autobiography* (London: Heinemann, 1906).
82 William Gaunt and M.D.E. Clayton-Stamm, *William De Morgan* (London: Studio Vista, 1971), pp. 150–51; *Joseph Vance* (1954), p. ix; Hamilton, pp. 184–85.
83 For the continuing activities of the Foundation, see De Morgan Foundation, De Morgan Collection (2019), https://www.demorgan.org.uk/. The date of the Foundation's establishment is taken from Alan Crawford, 'Morgan, William Frend De (1839–1917)', 2009 version, *Oxford Dictionary of National Biography* (Oxford: Oxford University Press, 2004-); Gaunt and Clayton-Stamm (p. 155) give the date of establishment as 1969.

In his and Evelyn's world the nature, power, and limits of reason were being actively redefined, and they responded in and through his work. William's ceramic pieces and Evelyn's canvases stand as portals into the transcendent understandings that the De Morgans had entered through the reason they had learned from William's namesake, William Frend.

As literary studies move beyond the major canon, rediscovery of William De Morgan's novels is feasible. Following a surmise that they might be republished, Mark Hamilton ends his study of William with the words: 'It is very rare that someone can be a successful novelist ... an artist and an inventor, and, perhaps, equally rare to find someone of such varied talents who is also altogether an admirable and likeable human being.'[84] The areas of achievement differ from Augustus De Morgan's, but in the possession of variety of talent and human attraction, William clearly resembles his father.

Bibliography

Ashton, Rosemary, *Victorian Bloomsbury* (New Haven and London: Yale University Press, 2012). https://doi.org/10.12987/yale/9780300154474.001.0001

Carmack, Sharon DeBartolo, *In Search of Maria B. Hayden: The American Medium Who Brought Spiritualism to the UK* (Salt Lake City: Scattered Leaves Press, 2020).

De Morgan, Augustus, *A Budget of Paradoxes*, 2nd edn, ed. by David Eugene Smith, 2 vols. (Chicago: Open Court, 1915).

De Morgan, Mary, *Complete Fairy Tales* (New York: F. Watts, 1963).

De Morgan, Sophia Elizabeth, *From Matter to Spirit. The Result of Ten Years' Experience in Spirit Manifestations. Intended as a Guide to Enquirers* (London: Longman, Green, Longman, Roberts & Green, 1863).

—, *Memoir of Augustus De Morgan* (London: Longmans, Green, 1882).

—, *Threescore Years and Ten: Reminiscences of the Late Sophia Elizabeth De Morgan to which are Added Letters to and from her Husband the Late Augustus De Morgan, and Others*, ed. by Mary De Morgan (London: Bentley, 1895).

De Morgan, William, *Alice-for-Short: A Dichronism* (London: Heinemann, 1907).

—, *Joseph Vance: An Ill-Written Autobiography* (London: Heinemann, 1906).

---

84 Hamilton, p. 185.

—, *Joseph Vance: An Ill-Written Autobiography*, ed. by A.C. Ward (London: Oxford University Press, 1954).

—, *The Old Man's Youth and the Young Man's Old Age* (London: Heinemann, 1921).

Edgeworth, Maria, and Richard Lovell Edgeworth, *Practical Education*, 2 vols. (London: J. Johnson, 1798).

Fisch, Audrey, 'Uncle Tom and Harriet Beecher Stowe in England', in *The Cambridge Companion to Harriet Beecher Stowe*, ed. by Cindy Weinstein (Cambridge: Cambridge University Press, 2004), pp. 96–112. https://doi.org/10.1017/ccol052182592x.006

Frend, William, *Evening Amusements, or, The Beauty of the Heavens Displayed* (London: J. Mawman, 1804–22).

Fritsch, Rudolf and Gerda Fritsch, *The Four-Color Theorem: History, Topological Foundations, and Idea of Proof*, trans. by Julie Peschke (New York: Springer, 1998). https://doi.org/10.1007/978-1-4612-1720-6

Gaunt, William, and M.D.E. Clayton-Stamm, *William De Morgan* (London: Studio Vista, 1971).

Godwin, Joyce, *The Theosophical Enlightenment* (Albany: State University of New York Press, 1994).

Graves, Robert Perceval, *Life of Sir William Rowan Hamilton* (Dublin: Hodges, Figgis and London: Longmans, Green, 1882-9).

Hamilton, Mark, *Rare Spirit: A Life of William De Morgan 1839–1911* (London: Constable, 1997).

Hedrick, Joan D., *Harriet Beecher Stowe: A Life* (New York: Oxford University Press, 1994).

Knight, Frida, *University Rebel: The Life of William Frend (1757–1841)* (London: Gollancz, 1971).

Leonard, Maurice, *People from the Other Side: The Enigmatic Fox Sisters and the History of Victorian Spiritualism* (Stroud: History Press, 2008).

Leopold, W.F., 'The Study of Child Language and Infant Bilingualism', *Word*, 4 (1948), 1–17.

Noakes, Richard, 'The Sciences of Spiritualism in Victorian Britain: Possibilities and Problems', in *The Ashgate Research Companion to Nineteenth-Century Spiritualism and the Occult*, ed. by Tatiana Kontou and Sarah Wilburn (Farnham: Ashgate, 2012), pp. 25–54. https://doi.org/10.4324/9781315613352-9

Oakes, Susan, Alan Pears, and Adrian Rice, *The Book of Presidents 1865–1965* (London: London Mathematical Society, 2005).

Owen, Alex, *The Darkened Room: Women, Power and Spiritualism in Late Victorian England* (London: Virago, 1989).

Pemberton, Marilyn, *Out of the Shadows: The Life and Work of Mary De Morgan* (Newcastle upon Tyne: Cambridge Scholars Publishing, 2012).

Rice, Adrian C., Robin J. Wilson, and J. Helen Gardner, 'From Student Club to National Society: The Founding of the London Mathematical Society in 1865', *Historia Mathematica*, 22 (1995), 402–21. https://doi.org/10.1006/hmat.1995.1032

Rice, Adrian C., and Robin J. Wilson, 'From National to International Society: The London Mathematical Society 1867–1900', *Historia Mathematica*, 25 (1998), 185–217. https://doi.org/10.1006/hmat.1998.2198

Richards, Joan L., *Generations of Reason: A Family's Search for Meaning in Post-Newtonian England* (New Haven: Yale University Press, 2021). https://doi.org/10.12987/yale/ 9780300255492.001.0001

Shuttleworth, Sally, *The Mind of the Child: Child Development in Literature, Science and Medicine, 1840–1900* (Oxford: Oxford University Press, 2010). https://doi.org/10.1093/ acprof:oso/9780199582563.001.0001

Stirling, A. M. W., *William De Morgan and His Wife* (London: Butterworth, 1922).

Talairach-Vielmas, Laurence, *Fairy Tales, Natural History and Victorian Culture* (Houndmills: Palgrave Macmillan, 2014). https://doi.org/10.1057/9781137342409

Tucker, Robert, 'Augustus De Morgan', *Nature*, 4 Jan. 1883, pp. 217–20.

Tuke, Margaret J., *A History of Bedford College for Women* (London: Oxford University Press, 1939).

Whewell, William, *Of the Plurality of Worlds: An Essay*, ed. by Michael Ruse (Chicago: University of Chicago Press, 2001).

Wilson, Robin, *Four Colors Suffice: How the Map Problem Was Solved* (Princeton: Princeton University Press, 2002). https://doi.org/10.1515/9780691237565

Winter, Alison, 'A Calculus of Suffering: Ada Lovelace and the Bodily Constraints on Women's Knowledge in Early Victorian England', in *Science Incarnate: Historical Embodiments of Natural Knowledge*, ed. by Christopher Lawrence and Steven Shapin (Chicago: University of Chicago Press, 1998), pp. 202–39.

—, *Mesmerized: Powers of Mind in Victorian Britain* (Chicago: University of Chicago Press, 1998).

Fig. 13 Augustus De Morgan pictured in the 1860s. (Public domain, via Wikimedia Commons, https://commons.wikimedia.org/wiki/File:Augustus_De_Morgan.jpg)

PART III

THE BIBLIOGRAPHIC RECORD

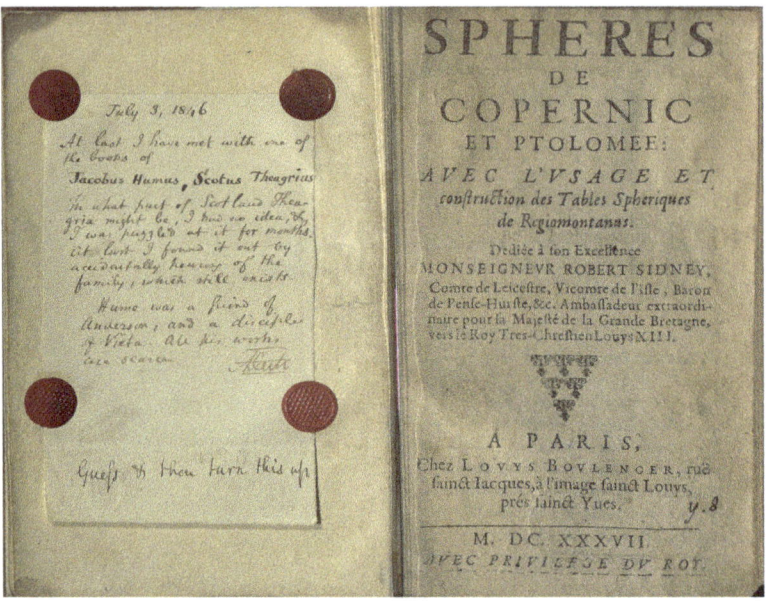

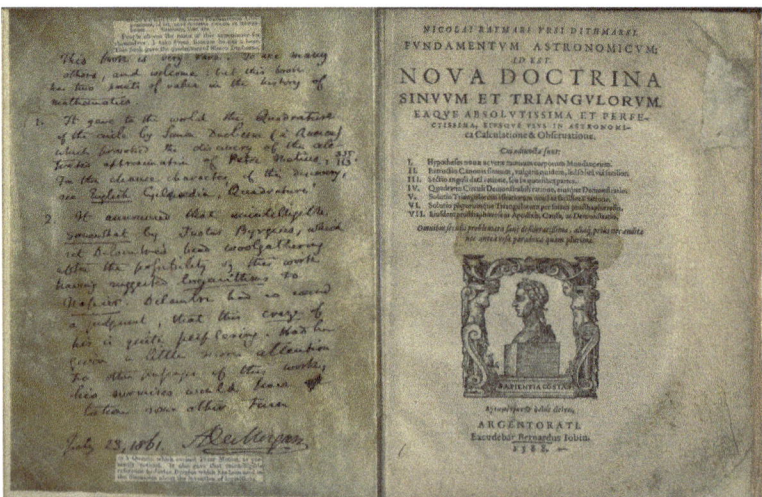

Fig. 14 Two typical examples of the sorts of notes De Morgan made in his books: on the identity of the author (above); and on the rarity and content of the book (below). ([DeM] M [Hume] SSR and [DeM] M [Ursus] SSR, reproduced by permission of Senate House Library, University of London.)

10. Augustus De Morgan's Library Revisited: Its Context and Its Afterlife

Karen Attar

> Professor De Morgan's unique mathematical library ... probably contains the most curious books on the history of mathematics to be found in England.
>
> — *The Spectator*[1]

Introduction

No study of Augustus De Morgan's life and work can be complete without a discussion of his mathematical library: the library which was the basis for some of his work, and which has been lauded in both the nineteenth and the twentieth centuries as one of the best mathematical libraries in the United Kingdom.[2] The library has been described elsewhere, such that a bare outline here will suffice to provide information about its content.[3] What has not been studied is the

---

1   'News of the Week', *The Spectator*, 1 Apr. 1871, p. 371.
2   The quotation in the epigraph reflects a nineteenth-century opinion, while in the twentieth century, the library was called 'one of the best surviving collections of early scientific books formed at this date' (A.N.L. Munby, *The History and Bibliography of Science in England: The First Phase, 1833–1845* (Berkeley: University of California, 1968), p. 12) and 'one of the finest accumulations of books on the history of mathematics in the country.' (Adrian Rice, 'Augustus De Morgan: Historian of Science', *History of Science*, 34 (1996), 201–40 (p. 222)).
3   See K.E. Attar, '"The Establishment of a First-Class University Library": The Beginnings of the University of London Library', *History of Universities*, 28 (2014),

contemporary context of the library, nor how it fared after De Morgan's death and why it continues to stand out. This chapter will seek to address these questions.

The library consists of approximately four thousand titles dating from 1474 until 1870. It holds material on all branches of mathematics, including astronomy, with arithmetic being especially strongly represented. There are associated titles of mathematical biography and bibliography and some philosophy. Important works are present in multiple editions. Euclid's *Elements* provides a particularly salient example, with editions ranging from the *editio princeps* of 1482 to Isaac Todhunter's 1862 edition for schools and colleges: editions abound in different languages, for different audiences, with different commentaries. Titles read to an extent as a roll call of significant mathematicians and works encapsulating landmark mathematical innovations: Michael Stifel, Niccolò Tartaglia, Albert Girard, François Viète and Thomas Harriot on symbolic algebra; Bonaventura Cavalieri on the geometry of 'indivisibles'; John Napier and Adriaan Vlacq on logarithms; Simon Stevin on decimal fractions; Johannes Kepler and Tycho Brahe on astronomy; Pierre de Fermat on number theory and analytic geometry; Jakob and Johann Bernoulli on probability and calculus; and so forth. Alongside these are popularising works and textbooks (by Luca Pacioli, Robert Recorde, William Oughtred, James Hodder, Edward Cocker, John Bonnycastle and others) and a host of relatively obscure works which De Morgan collected to contextualise trailblazing titles, on the principle that:

> The most worthless book of a bygone day is a record worthy of preservation. Like a telescopic star, its obscurity may render it unavailable for most purposes; but it serves, in the hands which know how to use it, to determine the places of more important bodies.[4]

Publications range from substantial tomes and multi-volume works to pamphlets, with numerous offprints and journal extracts among the nineteenth-century holdings.

44–65 (pp. 51–52); Karen Attar, 'Augustus De Morgan (1806–71), His Reading and His Library, in *The Edinburgh History of Reading: Modern Readers*, ed. by Mary Hammond (Edinburgh: Edinburgh University Press, 2020), pp. 62–82 (pp. 64–65).

4   Augustus De Morgan, *Arithmetical Books from the Invention of Printing to the Present Time* (London: Taylor & Walton, 1847), p. ii.

De Morgan annotated a significant minority of his books with notes about their rarity, their place in mathematical history, their quality in his opinion, their connection with him, or with anecdotes about their authors, ranging in length between a phrase and a paragraph.[5] Examples are: 'All the notes were made when I was a student at Cambridge. May 19/49'; 'Found in the *threepenny box* at a bookstall by A De Morgan'; 'Watt mentions no edition earlier than 1799'; and:

> The Royal Society published this paper, I may say, with avidity: I had not the least idea that it would be inserted in the Transactions, for which it was never intended. But they refused to publish the account of the manner in which their predecessors had falsified the second edition of the *Commercium Epistolicum*, for which I exposed them in the *Philosophical Magazine* for June 1848. But it is easier to make ink blush than philosophers, and I cannot say that they ever appeared ashamed of themselves. So I blush for them.[6]

Annotations appear on all sorts of books, from pamphlets to incunabula and to such iconic works as the first edition of Copernicus's *De Revolutionibus* (1543). Occasional volumes have notable provenance: for example, the sixteenth-century German Jesuit mathematician and astronomer Christoph Clavius and the seventeenth-century English poet and politician Edmund Waller.

The Contemporary Context

Mathematics was not a major or fashionable subject in which to collect in De Morgan's time, which explains how a professor with a large family and an annual income which seldom reached five hundred pounds could acquire a fine collection, independently of review and presentation

---

5 For a preliminary analysis of annotations, see Attar, 'Augustus De Morgan (1806–71)'.

6 On De Morgan's copies of the following books respectively: Giuseppe Venturoli, *Elements of the Theory of Mechanics*, trans. by D. Cresswell (Cambridge: Nicholson, 1822), Senate House Library (henceforth SHL) [DeM] N.1 [Venturoli] SSR; James Bradley, *A Letter to the Right Honourable George, Earl of Macclesfield, Concerning an Apparent Notion Observed in Some of the Fixed Stars* (London: [s.n.], 1747), SHL [DeM] M [Bradley]; George Alexander Stevens, *A Lecture on Heads*, new edn (London: G. Kearsley, 1787), SHL [DeM] (XVIII) BBc [Stevens]; Augustus De Morgan, 'On a Point Connected with the Dispute between Keil and Leibnitz about the Invention of Fluxions' [offprint] (London: printed by R. and J.E. Taylor, 1846), SHL [DeM] L° (B.P. 12).

copies among the more modern books and gifts of books of all vintages: books acquired from second-hand dealers (let alone street barrows) were generally cheap. Even auction prices could be modest, especially as De Morgan and other collectors bid directly without impediment, rather than employing agents. In 1809, Thomas Frognall Dibdin both encapsulated and canonised the contemporary attitudes of book collectors when he summarised the subjects on which bibliophiles collected as large paper copies (equated with limited editions), illustrated (i.e. extra-illustrated) copies, unique copies, copies printed upon vellum, first editions (and specifically Shakespeare's First Folio and Greek and Latin classics), true editions (i.e. editions with variants), unopened copies, black letter books, and books printed by Caxton, Wynkyn de Worde and the Manutius family. Little differentiates his list from those of Andrew Lang and Henry Wheatley at the end of the nineteenth century.[7] Discussions of specific collectors reinforce the message conveyed by general studies. De Ricci's seminal work on book collectors mentions nobody whose subject of interest was mathematics.[8] The sole collector featured in *The Dictionary of Literary Biography*'s *Nineteenth-Century British Book-Collectors and Bibliographers* who amassed mathematical works was the youthful James Orchard Halliwell (later Halliwell-Phillipps), before he moved on to Renaissance literature, and the chapter is silent about his mathematical collecting.[9] The auctions of the mathematical collections of De Morgan's Cambridge teacher and lifelong friend George Peacock and of James Orchard Halliwell demonstrate the overall indifference

---

7 Thomas Frognall Dibdin, *The Bibliomania or Book Madness, Containing Some Account of the History, Symptoms and Cure of this Fatal Disease*, ed. by P Danckwerts (Richmond: Tiger of the Stripe 2007), pp. 56–74; Andrew Lang, *The Library* (London: Macmillan, 1881), pp. 19–21, 76–122; Henry B. Wheatley, *Prices of Books: An Inquiry into the Price of Books which has Occurred in England at Different Periods* (London: G. Allen, 1898), p. 179.

8 Seymour de Ricci, *English Collectors of Books & Manuscripts (1530-1930) and their Marks of Ownership* (Cambridge: Cambridge University Press, 1930); see also William Younger Fletcher, *English Book Collectors* (London: K. Paul, Trench, Trübner, 1902).

9 Richard Maxwell, 'James Orchard Halliwell-Phillipps', in *Nineteenth-Century British Book-Collectors and Bibliographers*, ed. by William Baker and Kenneth Womack, *Dictionary of Literary Biography*, 184 (Detroit, Washington DC: Gale Research, 1997), pp. 202–18. Even a monograph devoted to Halliwell-Phillipps ignores his mathematical collecting: see Marvin Spevack, *James Orchard Halliwell-Phillipps: The Life and Works of the Shakespearean Scholar and Bookman* (New Castle, DE: Oak Knoll and London: Shepheard-Walwyn, 2001).

to mathematical books clearly: Peacock's sale was unusually sketchily described, and Halliwell's fetched exceptionally low prices.[10]

This does not mean that no general interest in mathematical books existed, but that interest was based on extra-textual factors and applied only to a small proportion of books. A few mathematical books fell into desirable categories for being printed on vellum (typically some copies of the 1482 Euclid) or on large paper, or for fine bindings. Collectible early printers produced small amounts of mathematics: four on Ptolemaic astronomy from the Aldine Press, all postdating the death of Aldus Manutius,[11] and a tiny part of both the Plantin and the Elzevir output. De Morgan possessed all of these works.[12] Books produced by early English printers—such as Henry Billingsley's first English edition of Euclid (printed by John Day, 1570), and especially those in black letter, such as Cuthbert Tunstall's *De arte supputandi* (Richard Pynson, 1522), the first English printed work to be devoted exclusively to mathematics—provided a major competing interest.[13] Auction catalogues of mathematical books, as of others, highlight these features. That of the book collector John Bellingham Inglis, who gave De Morgan the run of his library, provides particularly good examples of books falling in this last category.[14] The catalogues sometimes drew attention to 'rarity' or 'scarcity', terms which could be aimed at either the general collector or the mathematical connoisseur.

Further mathematical books were wanted insofar as they constituted part of general knowledge. From at least the Stuart period onwards, mathematics comprised one of several branches of knowledge which

---

10 See Munby, pp. 7 and 18–19.
11 See *The Aldine Press: Catalogue of the Ahmanson-Murphy Collection of Books by or Relating to the Press in the Library of the University of California, Los Angeles, Incorporating Works Recorded Elsewhere* (Berkeley: University of California Press, 2001), nos. 141, 266, 541 and 681. Ahmanson-Murphy provides a brief history of Aldine collecting, p. 13.
12 See Leon Voet, *The Plantin Press (155-1589): A Bibliography of the Works Printed and Published by Christopher Plantin at Antwerp and Leiden*, 6 vols. (Amsterdam: Van Hoeve, 1980-1983), vol. 6 (1983): Indices, p. 2633; Alphonse Willems, *Les Elzevier: histoire at annales typographiques* (Brussels: van Trigt, 1880), nos. 52, 244, 413, 503, 624 and 800.
13 Cuthbert Tunstall, *De arte supputandi libri quattuor* (London: R. Pynson, 1522).
14 Sotheby, Wilkinson & Hodge, *Catalogue of the Principal Portion of the Singularly Curious and Valuable Library of the late J.B. Inglis ...* (London: [Sotheby, Wilkinson & Hodge, 1871]).

would typically be represented in a gentleman's library, if not by a high proportion of books.[15] The desirable books were modern: when title pages of auction catalogues named specific titles for sale, these focused at least partly on comparatively recently published books, such as Francis Maseres' *Scriptores logarithmici* (1791–1807), Leonhard Euler's *Institutiones calculi differentialis* (1755) and his twelve-volume works, and titles by Francis Baily, Charles Hutton and Jean-Baptiste Biot among others.[16]

Supply by the book trade implies a certain demand. The bookseller Samuel Maynard made a living from selling specifically mathematical books, advertised as having been sourced from the libraries of deceased mathematicians, at his shop in Fleet Street and later Leicester Square. He was the foremost mathematical bookseller, whose stock after his death was sold by Sotheby's, but at least five others were congregated nearby.[17] Furthermore, auctioneers in De Morgan's lifetime happily sold mathematical collections, which they more often than not described on their title pages as 'valuable'. Sotheby catalogues dominate, possibly as the best preserved, but other examples stem from Hodgson, Lewis, Tait, and Southgate & Barrett, all in London, and Ballantyne in Edinburgh.

The mere fact that specifically mathematical libraries were sold proved that they were also consciously amassed. Giles Mandelbrote has detailed the presence of scientific (sometimes mathematical) books

---

15   See, for example, David Pearson, *Book Ownership in Stuart England: the Lyell Lectures, 2018* (Oxford: Oxford University Press, 2021), p. 21; Mark Purcell, *The Country House Library* (New Haven and London: Yale University Press for the National Trust, 2017), pp. 125–31.

16   Examples are taken from *A Catalogue of the Valuable Mathematical Library of the late Thomas Leybourn* (London: Hodgson, [1840]) and *Catalogue of the Valuable Mathematical & Miscellaneous Library of the late Reverend John Toplis, Late Fellow of Queens College, Cambridge, and Rector of South Walsham, Norwich* ([London: S. Leigh Sotheby & J. Wilkinson, 1858]).

17   S.L. Sotheby and John Wilkinson, *Catalogue of the Stock of Books of Mr. Samuel Maynard ... which will be Sold ... on Wednesday, 7th January [sic] 1863 and Two Following Days ...* ([London]: J. Davy and Sons, [1863]). For details of other mathematical booksellers frequented by De Morgan, see Karen Attar, 'Augustus De Morgan's Incunabula', in *Spotlights on Incunabula*, ed. by Anette Hagan, Library of the Written Word, 118 (Leiden: Brill, 2024), pp. 194–211 (p. 202). They were not the first: John Nichols described John Nourse, a bookseller on the Strand who died in 1780, as 'a man of science, particularly in the mathematical line' (W. Roberts, *The Book-Hunter in London: Historical and Other Studies of Collectors and Collecting* (London: Elliot Stock, 1895), p. 236).

in libraries up until the early eighteenth century, and Judith Overmier has continued the overview to the twentieth century.[18] The words 'mathematics' or 'mathematical' occur for twenty-two libraries between 1810 and 1870 in the SCIPIO database of art auction and rare book catalogues, often as one subject among several in a general library.[19] De Morgan trod a comparatively narrow path, as Overmier states cogently,[20] but he was not alone.

As marked-up sale catalogues and the occasional inscriptions in his books show, De Morgan purchased from the libraries of earlier collectors—chiefly mathematicians—sometimes at a remove. Sale catalogues available for the period 1820–1858 which specify mathematics (as opposed to general science) in their title are those of the astronomer Francis Baily; the clergyman George Burgess Wildig; a gentleman identified by De Morgan as the judge William Fuller Boteler; the mathematical teacher, headmaster and rector John Toplis; amateur scientist and mathematical enthisiast Abigail Baruh Lousada; and the mathematicians Thomas Galloway, William Wallace, Thomas Stephen Davies, Thomas Leybourn and John Playfair. The catalogue of the antiquary and literary scholar James Orchard Halliwell is also relevant, as De Morgan bought heavily from it.[21]

---

18   Giles Mandelbrote, 'Scientific Books and their Owners: A Survey to c. 1720', in *Thornton and Tully's Scientific Books, Libraries, and Collectors: A Study of Bibliography and the Book Trade in Relation to the History of Science*, 4th edn, ed. by Andrew Hunter (Aldershot: Ashgate, 2000), pp. 333–66; Judith Overmier, 'Scientific Book Collectors and Collections, Public and Private, 1720 to Date', in *Thornton and Tully's Scientific Books*, pp. 367–91. Of the collectors covered in this chapter, Overmier devotes paragraphs to Charles Babbage, Augustus De Morgan, and Guglielmo Libri.

19   OCLC, SCIPIO: Art and Rare Books Sales/Auction Catalogs. I am indebted to Karen Limper-Herz for searching SCIPIO for me. Ellen G. Wells locates twelve mathematical collectors for the same period: see Ellen G. Wells, 'Scientists' Libraries: A Handlist of Printed sources', *Annals of Science*, 40 (1983), 317–89.

20   Overmier, p. 368.

21   *Catalogue of the Valuable Astronomical, Mathematical, and General Library of the Late Francis Baily* ... ([London: S. Leigh Sotheby, 1845]); *Catalogue of a Selected Portion of the Scientific, Historical, and Miscellaneous Library of James Orchard Halliwell* ... ([London: S. Leigh Sotheby, 1840); *Catalogue of the Mathematical, Classical, and Miscellaneous Library of a Gentleman, Deceased* ... ([London: S. Leigh Sotheby, 1846]); *Catalogue of the Library of the Late Rev. George Burgess Wildig ... Together with Another Valuable Collection, Including all the Best English and French Mathematical Works* ... ([London: L.A. Lewis, 1854); *Catalogue of the Valuable Mathematical & Miscellaneous Library of the Late Reverend John Toplis; Catalogue of the Valuable Astronomical & Mathematical Library of the Late Thomas Galloway* ... ([London: S.

Of these libraries, Francis Baily's was evidently utilitarian. Imprints are chiefly from the nineteenth century (of a sample, 72 per cent) followed by eighteenth-century imprints (27 per cent), leaving just one percent of earlier books. Works on astronomy, annuities and probability predominate, in accordance with Baily's professional interests. Other libraries are more clearly collections. Among the owners, Thomas Galloway has been noted for his 'small but valuable library', primarily for his nineteen Keplers; the reference, which appears in his entry in the *Oxford Dictionary of National Biography*, stems ultimately from De Morgan's note of Galloway's books in his obituary for Galloway.[22] All the people named owned some mouth-watering items: even Baily, the most pedestrian amasser of mathematical books, owned the first edition of Copernicus's *De revolutionibus*, and a 'fine copy, in vellum' of Tycho Brahe's *Historia coelestis* (1666).[23]

What instantly differentiates De Morgan's library from the others is its sheer size. The quantity of lots is admittedly a crude guide to the size of collections because some catalogues bunch several items into one lot and, quite commonly, titles considered insignificant are noted merely as 'and others'. Yet even allowing for this propensity, De Morgan's library remains markedly larger than the others, the number of whose lots range from 228 (Wildig) to 1384 (Playfair) before one subtracts lots unconnected with mathematics or astronomy.

Leigh Sotheby & J. Wilkinson, 1852]); *Catalogue of the Mathematical, Philosophical, and Miscellaneous Library ... of Miss A.B. Lousada ...* ([London: Sotheby, 1834]); *Catalogue of the Valuable Library of the Late William Wallace, LL.D. Professor of Mathematics in the University of Edinburgh ...* ([London: C.B. Tait, 1843]); *A Catalogue of the Valuable Mathematical and Scientific Library of the Late T.S. Davies ... Professor of Mathematics at the Royal Military Academy, Woolwich ...* ([London: Southgate & Barrett, 1851]); *A Catalogue of the Valuable Mathematical Library of the Late Thomas Leybourn*); *Catalogue of the Library of the Late John Playfair ... Comprising a Valuable Collection of Mathematical, Philosophical, and Miscellaneous Books, Maps, &c &c ...* (Edinburgh: J. Ballantyne, 1820).

22 David Gavine, 'Galloway, Thomas (1796–1851)', *Oxford Dictionary of National Biography Online* (Oxford: Oxford University Press, 2004), https://doi.org/10.1093/ref:odnb/10312; For Augustus De Morgan's obituary of Galloway, see *Monthly Notices of the Royal Astronomical Society* (1851–52), 87–89. The ODNB entry for Thomas Leybourn notes 'his library of nearly a thousand books', without indicating its content; see Niccolò Guicciardini, 'Leybourn, Thomas (c.1769–1840), Mathematician', in *Oxford Dictionary of National Biography Online*, https://doi.org/10.1093/ref:odnb/16622.

23 Baily sale catalogue, lots 452 and 227 respectively.

The chronological range of titles is a further differential. Leybourn, Halliwell, Toplis, and Boteler each owned the occasional incunable, but none as many as De Morgan's twenty-one.[24] The presence of antiquated elementary textbooks, such as Cocker and Hodder on arithmetic, in De Morgan's library is a third; others barely own such books. The extent to which De Morgan owned multiple editions of books is a fourth. To have possessed more than one Euclid, edited by different editors, is common, and examples are present of owners having two or more editions of another particular work, whereby Galloway's five editions of Newton's *Principia* stand out.[25] Yet nobody owned as many multiple editions as Augustus De Morgan with his three editions of Cuthbert Tunstall's *De arte supputandi*, six of Newton's *Principia* (including the first French edition), eight editions of Sacrobosco's *Sphaera mundi*, nine of Robert Recorde's *Ground of Arts*, plus two of his *Whetstone of Witte*, and so forth.

Of mathematical libraries auctioned during De Morgan's lifetime, there remains that of Guglielmo Libri, which far surpassed De Morgan's library numerically, yet has probably received less attention.[26] Libri's collection has been tarnished by the method by which he acquired some of his books: he was a known thief.[27] Possibly the mathematical section was diluted by the strengths of other areas of Libri's library, just as Halliwell's mathematical collection pales beside his Shakespeareana. Much of the explanation for its lower profile is that, like the other libraries noted above, it has been dispersed. A dispersed library is generally harder to research than an intact one, although Benjamin Wardhaugh's scholarly examination of Charles Hutton's library (considered by De Morgan to be the best of its time), dispersed in 1816, proves that it can be done.[28]

---

24 Listed in Attar, 'Augustus De Morgan's Incunabula', pp. 195–96.
25 Galloway catalogue, lots 470–02 and 474–75.
26 *Catalogue of the Mathematical, Historical, Bibliographical and Miscellaneous Portion of the Celebrated Library of M. Guglielmo Libri* ... (London: S. Leigh Sotheby & J. Wilkinson, [1861]).
27 For a full account of Libri and his books, see P. Alessandra Maccioni Ruju and Marco Mostert, *The Life and Times of Guglielmo Libri (1802–1869), Scientist, Patriot, Scholar, Journalist and Thief: A Nineteenth-Century Story* (Hilversum: Verloren, 1995).
28 Leigh and Sotheby, *A Catalogue of the Entire, Extensive and Very Rare Mathematical Library of Charles Hutton* ... (London: Leigh & Sotheby, [1816]); a marked-up copy is in De Morgan's library at Senate House Library, [DeM] Z (B.P.34); Benjamin Wardhaugh, 'Collection, Use, Dispersal: The Library of Charles Hutton and the Fate of Georgian Mathematics', in *Beyond the Learned Academy: The Practice of*

Two further mathematical libraries remained undispersed. Charles Babbage's library of 2,581 lots, described as the best of the time apart from De Morgan's, is at the Royal Observatory in Edinburgh. That of De Morgan's London friend and colleague John Thomas Graves, larger than De Morgan's and sharing its characteristics of breadth, school textbooks, multiple editions and a goodly number of fifteenth-century imprints, is at University College London.[29] Neither of Babbage's biographers refer to his library, presumably because it has no bearing on his scientific achievements.[30] Graves's library is noted in his entry in the *Oxford Dictionary of National Biography*.[31] Its comparative obscurity is harder to explain. One reason could be that it made less impact on University College London than De Morgan's library did on the central University of London Library: whereas De Morgan's library was the University of London's founding collection, Graves's, albeit indubitably significant, joined an already respectable library.[32]

The remainder of the explanation lies in factors which set De Morgan's library apart from all the others. The first of these is that De Morgan was a leading mathematician who wrote for various audiences, and whose bibliography of arithmetical books (1847) was cited as an authority in contemporary auction catalogues and continued to be respected long afterwards. Secondly, some of De Morgan's writings were closely connected with his own books and made the fact of his

---

Mathematics, 1600–1850, ed. by Philip Beeley and Christopher Hollings (Oxford: Oxford University Press, 2024), pp. 158–84.

29 Described respectively in M.R. Williams, 'The Scientific Library of Charles Babbage', *Annals of the History of Computing*, 3 (1981), 235–40, and in Alison R. Dorling, 'The Graves Mathematical Collection in University College London', *Annals of Science*, 33 (1976), 307–09. Babbage's books were listed in *Mathematical and Scientific Library of the late Charles Babbage of No. 1, Dorset street, Manchester Square* ([London: Sotheby, Wilkinson & Hodge, 1872]); Graves's in *Catalogue of Books in the General Library and in the South Library of University College, London*, 3 vols. (London: Taylor and Francis, 1879), where they are identified as coming from him.

30 See Maboth Moseley, *Irascible Genius: A Life of Charles Babbage, Inventor* (London: Hutchinson, 1964); H.W. Buxton, *Memoir of the Life and Labours of the Late Charles Babbage, Esq. F.R.S.*, Charles Babbage Institute Reprint Series for the History of Computing, 13 (Cambridge, MA: MIT Press, 1988).

31 Adrian Rice, 'Graves, John Thomas (1806-1870)', *Oxford Dictionary of National Biography Online*, https://doi.org/10.1093/ref:odnb/11311.

32 See Attar, 'Establishment', p. 44, and especially *Catalogue of Books in the General Library*, pp. iii–v.

library known—above all his *Budget of Paradoxes*, devoted entirely to books he possessed.[33] The third factor, as stated above, is that De Morgan annotated a significant minority of his books, for his own amusement or edification, for posterity, or both: he both enjoyed and discernibly used his collection, putting it to work for the benefit of mathematical history and bibliography. By immortalising some of his books in his writings, especially the *Budget* and his arithmetical bibliography (based entirely on works he had seen and partly on works that he owned),[34] De Morgan raised awareness of his ownership of old mathematical books. The resulting prestige of his collection ultimately ensured its long-term preservation and hence an afterlife.

The Afterlife

The afterlife of De Morgan's library began a fortnight after his death with a speculation in *The Spectator* on 1 April 1871:

> Would it be impossible, by the way, to secure for the University the late Professor de Morgan's unique mathematical library, which probably contains the most curious collection of books on the history of mathematics to be found in England? ... if it could be obtained, there would be a special fitness in securing it for the University of London, which would then have a really good start towards the formation of a fine classical and scientific library.[35]

The timing was propitious. The University of London had only recently acquired its own building and thereby space for a library, and Sir Julian

---

33 Augustus De Morgan, *A Budget of Paradoxes* (London: Longmans, Green, 1872). See also Chapter 7 in this volume.

34 Augustus De Morgan, *Arithmetical Books from the Invention of Printing to the Present Time* (London: Taylor & Walton, 1847). In 1908 David Eugene Smith called this 'one of our best single sources' (David Eugene Smith, *Rara Arithmetica: A Catalogue of the Arithmetics Written before the Year MDCI, with a Description of those in the Library of George Arthur Plimpton, of New York* (Boston and London: Ginn, 1908), p. xii); this bibliography refers several times to De Morgan's. As late as 1967, A. Rupert Hall noted the continuing use of De Morgan's bibliography and described it as a 'minor classic' (Augustus De Morgan, *Arithmetical Books from the Invention of Printing to the Present Time*, introd. by A. Rupert Hall (London: Hugh K. Elliott, [1967]), p. [vii]).

35 'News of the Week', *The Spectator*, 1 Apr. 1871, p. 371; see also 'Miscellaneous', *Birmingham Daily Post*, 7 Apr. 1871, p. 6; 'Multiple News Items', *The Sheffield and Rotherham Independent*, 7 Apr. 1871, p. 3.

Goldsmid had noted the importance for the University of having a library, so that the possible provision met a newly revealed requirement.[36] On 10 May, the University Chancellor, Lord Granville, appealed at the annual University of London degree ceremony for books to fill the empty bookshelves of the new University of London Library. That afternoon, a meeting took place to determine how to honour Augustus De Morgan, at which, again according to *The Spectator*, 'There was also a great desire to purchase his rare mathematical library (valued at something like £1,200) on behalf of the University of London'.[37] Samuel Loyd, First Baron Overstone, was present. He and De Morgan had moved in similar orbits: Overstone served from 1828 until 1844 on the council of University College London, where De Morgan was a professor, and as a banker could have known De Morgan's actuarial work and his views on decimal currency (upon the introduction of which the two men held opposing views). Whether he knew De Morgan and his activity as a collector personally is a matter for speculation.[38] But he was a wealthy member of the University Senate who heard the plea, and he purchased the collection for the University. In June 1871, following receipt of the books, he wrote to the Senate:

> It is a source of satisfaction to me to have been the means of preventing the disperssion [sic] of this remarkable collection of mathematical Works; and I gladly present it to the London University, as a testimony of my appreciation of the service which that Body has rendered to the extension and improvement of Education in all its branches throughout the United Kingdom, and in the hope that it may prove the first fruits of a Library which shall ere long become such in all respects as the London University ought to possess.[39]

De Morgan's library not only remained physically under one roof, but its contents were readily identifiable by the note '[D.M.]' at the end of titles

---

36 Attar, 'Establishment', p. 50.
37 'News of the Week', *The Spectator*, 13 May 1871, p. 563.
38 No correspondence between the two men is present in Samuel Jones Loyd, Baron Overstone, *The Correspondence of Lord Overstone*, ed. by D.P. O'Brien, 3 vols. (Cambridge: Cambridge University Press, 1971).
39 Senate minute 156, 14 June 1871 (p. 49); letter from Lord Overstone to W.B. Carpenter, Registrar, 10 June 1871; cited in *Catalogue of the Library of the University of London, Including the Libraries of George Grote and Augustus De Morgan* (London: Taylor & Francis, 1876), p. [iv].

from his library in the University of London's first printed catalogue, from 1876. The catalogue's very title emphasised the presence of De Morgan's books within the University's library: *Catalogue of the Library of the University of London, Including the Libraries of George Grote and Augustus De Morgan*.[40] It remains a lasting record: indeed, the only printed record, as provenance was omitted when the catalogue was updated.[41] A specially designed bookplate was inserted into each volume.

As neither the first catalogue nor the books themselves record shelfmarks, we know neither the order in which the books were originally kept, nor whether they were integrated with, or kept apart from, other library holdings. While Reginald Arthur Rye described them in 1908 as a 'special collection', this could refer to provenance rather than treatment.[42] Certainly many books were borrowable; at the time of electronic cataloguing in the early twenty-first century, circulation labels remained on several eighteenth-century books (some of which had been borrowed as late as the 1970s).

The books were initially held in some disarray. In his arithmetical bibliography of 1908, David Eugene Smith noted of De Morgan's library: 'While some of the books were sold ... most of them were purchased by Lord Overstone and presented to the University of London.'[43] Although Smith's phraseology strongly suggests that these were books discarded by De Morgan—for example in a large cull when the family moved from Adelaide Road to 6 Merton Road, near Primrose Hill, in 1868[44]—University of London Librarian Reginald Arthur Rye chose to assume theft from the University. He explained 'that the University Library was at one time uncared for and stacked in the rooms of the Central Building at South Kensington, to which access was readily obtainable for the

---

40 The Classical historian George Grote, Vice-Chancellor of the University of London, bequeathed his library to the University within months of De Morgan's death.
41 University of London, *Hand-Catalogue of the Library, Brought Down to the End of 1897* (London: HMSO, [1900])
42 Reginald Arthur Rye, *The Libraries of London: A Guide for Students* (London: University of London, 1908), p. 24. George Grote's books were similarly designated as a special collection, despite definitely being dispersed throughout the collections.
43 Smith, p. [498].
44 Sophia Elizabeth De Morgan, *Memoir of Augustus De Morgan* (London: Longmans, Green, 1882), p. 364.

purpose of meetings, examinations, and the like.' Rye listed 172 missing titles, many 'of value and interest', including 'a few early printed books'.[45]

Had Rye's enumeration been accurate, University negligence would have been dire. The list included three quarto incunabula: Giorgio Chiarini's *Libro che tratta di mercanzie ed usanze dei paesi* (Florence: Francesco di Dino, 1481; Rye attributed authorship to the printer); Joannes de Sacrobosco's influential astronomical work *Sphaera mundi*, with Gerardus Cremonensis, *Theorica planetarum* (Venice: Franz Renner, 1478), and Paolo Veneto's *Logica parva* (Milan, Christoph Valdarfer, 1474).[46] It also included an obscure arithmetic manual, entitled *In desem boechelge[n] uyne men ein kurtze[m] wech vn[n] güde manier bald tzo leren rechen mit zyfferen nae der konst algorismi* ('*Algorismus*') (Cologne, 1526), one of the rarest titles in the collection.

Fortunately, Rye overestimated the loss. The vast majority of items listed as lost, including those listed above, remain safely at the University of London. Many are pamphlets within bound volumes and are hence easy to overlook, especially as the printed catalogues do not state the contents of such *sammelbände*. Minor works by Jacques Philippe Marie Binet, Augustin-Louis Cauchy, Michel Chasles, Sir John Herschel, Charles Hutton and William Whewell all fall into this category, which includes offprints and extracts and in which items, by virtue of their comparatively ephemeral nature, lack title pages. The occasional more substantial item is part of a *sammelband* and therefore also possible to overlook: Petrus Ramus's 336-page quarto *Mathematicarum libri unus et triginta*, edited by Lazarus Schöner (Frankfurt, 1599) and bound with Ramus's *Arithmeticae libri duo* of the same year, is an example. But other items are hefty volumes, multi-volume sets, or both: John Bale's *Index Britanniae scriptorium* (1657) is a two-volume folio of 1,128 pages; Sir John Hill's *Review of the Works of the Royal Society of London* (2nd edn, 1780) comprises 686 pages, Johann Jacob Grynaeus's folio *Adagia* (Frankfurt, 1646) is a folio of 946 pages, and Daniel Neal's *History of the Puritans* (London, 1822) is in five volumes. Inability to find them implies either wilfulness or disorder.

---

45 Senate House Library, Library Committee minutes, 1901-13, UoL/.UL/1/1/1: minute 113, meeting of 27 Apr. 1908, and minute 123, meeting of 29 June 1908.
46 ISTC ic00449000, ij00402000 and ip00220000, respectively.

However, Rye's accusation bore some substance. Most items wanting can be inferred to be slight from the designation 'n.p., n.d' ('no place, no date'), or are reports on subjects ranging from Henry Toynbee's *Report to the Committee of the Meteorological Office on the Use of Isobaric Curves* (1869) to a *Sydney College Report of the Syndicate appointed to Consider Whether it is Expedient to Afford Greater Encouragement to the Pursuit of those studies for the Cultivation of Which Professorships Have Been Founded in the University, and, if so, by What Means that Object May be Best Accomplished* (1848). T. T. Wilkinson's *On Some Points in the Restoration of Euclid's Porisms*, listed as an octavo without place or date of publication, appears to be a ghost. Perceived duplicates are salient, when Rye located only one of two copies of a book recorded in the catalogue, and only one copy remains today: the eighth edition of John Hawkins's *The Young Clerk's Tutor Enlarged* (1675), based on the work of Edward Cocker, and Gregor Reisch's *Margarita Philosophica* ([Strasbourg], 1504). Collections did not arrive in the nineteenth century with an obligation to retain all items, and the loss of the second copy in both instances could have resulted from a discreet sale of duplicates.[47]

In the early twentieth century the De Morgan library, along with those in the University Library generally, were classified, following a scheme devised by Reginald Arthur Rye based on the Dewey Decimal Classification.[48] Dewey had divided mathematics into eight main classes from 511 to 519 (with general mathematical works in class 510): arithmetic, algebra, geometry, conic sections, trigonometry, descriptive geometry, analytic geometry and quaternions, calculus, and probability. Class 518 remained empty. Each class was subdivided: whereby arithmetic and trigonometry, for example, each had nine subdivisions, algebra had 21, while geometry had 41 in the basic geometry section alone. Astronomy occupied numbers 521 to 529: theoretic, practical and spherical, descriptive, maps and observations, earth, geodesy, navigation, ephemerides, and chronology, again sub-divided, and

---

47 No such sale is recorded in Library Committee minutes. That some books were sold as duplicates is apparent from annotations in the archive copy of the Library's 1900 catalogue (University of London Archive, UoL/UL/8/1).

48 Melvil Dewey, *Decimal Classification and Relative Index: for Libraries, Clippings, Notes, etc.*, 6th edn (Boston: Library Bureau, 1899). For Rye's system, see R.A. Rye, 'Table of Shelf Classification and Arrangement for the General Library', University of London Archive, UoL/UL/8/2.

with general works at 520. Rye followed Dewey precisely for Class M, Astronomy, minus the subdivisions. He adapted mathematics (Class L), changing Dewey's order, reducing the emphasis on geometry, relegating probability to a subdivision, and adding a class to the end, as the table below shows. Both systems placed mathematical bibliography within bibliography.

UL classmark	Subject	Dewey number
L	Mathematics (general)	510
L1	Arithmetic	511
L1.1	Book-keeping	
L1.2	Mensuration	
L2	Algebra	512
L2.1	Probabilities / Group theory	519
L3	Calculus	517
L4	Annuities: insurance	519.5
L5	Trigonometry	514
L6	Geometry: conic sections	513; 515
L7	Analysis and functions	516
L8	Mathematical tables	511.9; 512.9
L9	Weights and measures	n/a

Any classification scheme imposes a certain order on a collection. But with its failure to subdivide, Rye's system (now termed 'Old Classification') was clearly far cruder than Dewey. A large collection devoted to a single subject exposes this drawback mercilessly; why Rye ignored the possibility for greater distinction in a library which began with a rich subject-based special collection is a mystery. Moreover, Old Classification depended heavily on names, verbal descriptions, and dates for further arrangement. The De Morgan Library, with several editions of a single work, suffered. Many books were either undifferentiated, or were differentiated by means of such long, unwieldy classmarks as [DeM] L6 [Apollonius. Pergaeus] or, in an extreme example, [DeM] L6 [Euclid – Elements – English - 1834]. One hundred years later, retrieval remains challenging.

Although De Morgan's library is not distinguished by fine bindings, any large collection is likely to have a smattering of them. Of the eighty-three bindings exhibited from various countries and centuries featured in the exhibition of bookbindings with which the University of London Library celebrated its move from the Imperial Institute in South Kensington to the new Senate House building in London's Bloomsbury area in 1937, ten were from De Morgan's library.[49] This was the first time the University had promoted printed material in the library in a publication, beyond descriptions in library directories. Manuscripts fared better. Receipt of the Black Prince manuscript in 1921, a medieval manuscript providing an eye-witness account of some battles of the Hundred Years War, provided the incentive to produce a catalogue of the Library's manuscripts. This was published in the same year and made known manuscript material generated by De Morgan and the few other manuscripts he owned.[50]

Some twenty years later, misfortune struck the collection. At about 4.00 a.m. on Saturday, 16 November 1940, a high-explosive German bomb exploded in the Library's strong-room, damaging, among others, several De Morgan books. Only two were considered irrevocably damaged and were not replaced, Bonaventura Cavalieri's *Exercitationes geometricae sex* (Bologna, 1647) and the 23rd, very rare, edition of Francis Walkingame's *The Tutor's Assistant* (London, 1787). Both remain in the Library, with De Morgan's note on the half-title verso of the Cavalieri. Another thirteen books were described as 'books badly damaged, but which could be repaired if they cannot be replaced'; most were sixteenth, seventeenth and eighteenth-century imprints from the De Morgan Library. The Library was able to purchase three books to 'replace' bomb-damaged copies: Francesco Maurolico's *Cosmographia* (Venice, 1543), John Wilkins's *A Discourse Concerning a New World & Another Planet* ([London], 1640) and Blaise Pascal's *Traitez de l'equilibre des liqueurs* (Paris, 1698). The damaged books remain and have attained

---

49 Reginald Arthur Rye and Muriel Sinton Quinn, *Historical and Armorial Bookbindings Exhibited by the University Library: Descriptive Catalogue* (London: University of London, 1937).

50 Reginald Arthur Rye, *Catalogue of the Manuscripts and Autograph Letters in the University Library at the Central Building of the University of London* (London: University of London Press, 1921). See Chapter 11 in this volume.

a new meaning as memorials to the London Blitz and to the role played by the University of London in the Second World War.

After the war, De Morgan's library was accessible to the extent that any collection was: catalogued in the University of London's card catalogue, and described in standard library directories.[51] Its items of economic interest, for example concerning finance, were included in the printed catalogue of the Goldsmiths' Library of Economic Literature, with notes recording De Morgan's ownership and the presence where applicable of his manuscript notes.[52] Subsequently, Maxine Merrington of University College London's School of Library, Archive and Information Studies trawled through those of the De Morgan books which had been gathered together at the time to produce a published listing of letters inserted in books in De Morgan's library, enlarging awareness of the circles in which he moved and the respect in which he was held.[53] Yet the library's impact was modest. The emphasis of early printed books before the digital era remained their content, and the history of mathematics was a niche subject. When in the 1980s the Library decided to stop acquiring books in the sciences, De Morgan's library was on a limb.

The collection gained a new life when scholarly interest turned to post-production copy-specific features of books. De Morgan's annotations, the distinguishing feature lauded in the nineteenth century, were well suited to late-twentieth-century academic fashion with its burgeoning scholarly interest in the history of books (emphasising post-production history) and of reading.[54] The London Rare Books School, established in 2007, exploited this, using De Morgan's books for their bindings, evidence of provenance (including text deliberately obliterated) and the nature of mathematical illustrations for courses on historic

---

51 See J.H.P. Pafford, 'The University of London Library', in *The Libraries of London*, 2nd edn, ed. by Raymond Irwin and Ronald Staveley (London: Library Association, 1961), pp. 140–56; *A Directory of Rare Book and Special Collections in the United Kingdom and the Republic of Ireland*, ed. by Moelwyn I. Williams (London: Library Association, 1985).

52 Margaret Canney and David Knott, *Catalogue of the Goldsmiths' Library of Economic Literature*, Vol. 1, *Printed Books to 1800* (Cambridge: Cambridge University Press for the University of London Library, 1970).

53 Maxine Merrington, *A List of Certain Letters Inserted in Books from the Library of Augustus De Morgan (1806–1871) now in the University of London Library* (London: University of London Library, 1990). See Chapter 11 in this volume.

54 Pioneered in the English-speaking world by D.F. McKenzie, *Bibliography and the Sociology of Texts* (London: British Library, 1986).

bookbindings, provenance and the history of publishing, respectively. A single book, valued by De Morgan for its pioneering content, could be of interest for courses on bibliography, incunabula and the history of collections in addition to those named immediately above.[55]

Censuses, formerly focused on traditional collectable books like Caxtons and Shakespeare quartos and First Folios, are now extending to scientific books, such as Vesalius's *De humani corporis fabrica* (1543). De Morgan owned a copy of both the first and second edition of Copernicus's *De Revolutionibus*, which are included in Owen Gingerich's *An Annotated Census of Copernicus' De revolutionibus (Nuremberg, 1543 and Basel, 1566)* (2002), complete with De Morgan's note on the title page of his copy of the first edition: 'Aug. 4. 1864. I have this day entered all the corrections required by the Congregation of the Index (1620) so that any Roman Xtian may read the book with a good conscience'; De Morgan has annotated his copy accordingly.[56] All incunabula are being gathered into a census in the CERL database *Material Evidence in Incunabula*,[57] such that information about De Morgan's fifteenth-century books can feed into general knowledge of how the earliest editions of Euclid, Pacioli, Regiomontanus and others have been read and treated over the ages.

De Morgan's incunabula were recorded on the International Short Title Catalogue.[58] Online cataloguing of the entire collection in 2004/5, enabled by a grant from the Vice-Chancellor's Development Fund, was a major advance. Sometime before 1998, a De Morgan rare book sequence had been set apart. As a forerunner to online cataloguing, the remainder of the collection was collocated physically.[59] Bookplates revealed one

---

55 See Karen Attar, 'Senate House Library and the London Rare Books School' (London: Institute of English Studies, University of London, 2020), about Johannes Widmann, *Behende vnd hubsche Rechenung auff allen Kauffmanschafft* (Leipzig: Konrad Kachelofen, 1489), http://englishstudies.blogs.sas.ac.uk/2020/06/15/senate-house-library-and-the-london-rare-books-school.

56 Repeated with an illustration in David Pearson, *Books as History* (London: British Library, 2008), p. 131.

57 Consortium of European Research Libraries, *Material Evidence in Incunabula* (2021), https://www.cerl.org/resources/mei/main.

58 British Library, *International Short Title Catalogue (ISTC): The International Database of 15th-Century European Printing* (2016), https://data.cerl.org/istc/_search.

59 Excepted were a couple of De Morgan's own books, interleaved and with substantial insertions and annotations, which had been classified as archival material; these remained *in situ*, with catalogue records in the book catalogue.

half of the books in the mathematical section of the Library's 'old classification' sequence, every one of which was opened, to have been De Morgan's. Further books were found across several sequences. Some were instantly identifiable from a uniform binding of half calf and drab brown cloth, even before the book was opened. Others were not, and books continued to emerge long after the conclusion of the project. Currently 3,852 titles on Senate House Library's online catalogue index De Morgan as a former owner, which tallies well with the four thousand titles estimated by 1908 to be in the collection.[60] De Morgan's books are all recorded on Library Hub Discover (formerly Copac), the union catalogue of British national and research libraries.[61] Books printed up to the year 1830 have records in CERL's Heritage of the Printed Book Database, raising their profile nationally and internationally.[62]

As all electronic catalogue records code the date, language and country of publication of each item, online cataloguing of De Morgan's library instantly enabled accurate analysis of his library in terms of these features, singly or in combination. The printed and card catalogues had recorded author, title, edition, format and the place and date of publication. Online cataloguing followed standard rare book cataloguing rules of the time.[63] Electronic records routinely included details of publishers and pagination and added notes. They indexed printers and publishers for books printed up to 1700, indexed subjects, noted salient bindings, and described all evidence of provenance and all imperfections. They indexed all former owners, including De Morgan, and the best way to gain an overview of the collection is to search the name index for De Morgan as a former owner.

For the workings of De Morgan's mind and his library, provenance and other copy-specific notes were the most important features. De Morgan himself had several volumes bound and took care about what went into each for uniformity of content and format. It is possible to

---

60 See University of London, Catalogue, https://catalogue.libraries.london.ac.uk
61 Jisc, *Library Hub Discover*, https://discover.libraryhub.jisc.ac.uk.
62 Consortium of European Research Libraries, Heritage of the Printed Book Database (2018), https://www.cerl.org/resources/hpb/main.
63 Association of College and Research Libraries, American Library Association and Office for Descriptive Cataloging Policy, Library of Congress, *Descriptive Cataloging of Rare Books*, 2nd edn (Washington, D.C.: Cataloging Distribution Service, Library of Congress, 1991).

see the results of his organisation through a note like: 'SHL copy is no. 1 of 8 items on philosophy and economics, 1813-1841, in vol. with binder's spine title: "De Morgan Tracts". (B.P.304)', and then to browse virtually the list of contents of the volume. Provenance notes at their most basic read: 'SHL copy is from the library of Augustus De Morgan'. The fact and date of inscriptions is noted, as is their location: for example, 'SHL copy is from the library of Augustus De Morgan, with his note, 3 Dec. 1857, tipped on p. [3]' (L (B.P.21) SSR). Occasionally the subject of the annotation is noted, as for Thomas Keith's *Complete practical arithmetician* (1788): 'SHL copy is bound with the author's Key to The complete practical arithmetician (1790). Notes on Bonnycastle's Arithmetic and Algebra written on front flyleaf.' Even brief notes serve a purpose, flagging points of interest for potential researchers. Electronic cataloguing further helped to shed light on the circulation of mathematical books in the Victorian era by indicating books which had belonged to mathematicians or astronomers before De Morgan: Francis Baily, Olinthus Gregory, Thomas Galloway and others. The information is incomplete because it relies on either former owners having recorded their names in books, or on De Morgan having noted the method of acquisition, and comparison of his books with marked-up Sotheby sale catalogues shows that he did this only sporadically. However, it opens the field for further research.

Following cataloguing came digitisation, with the facilitation of more research. In 2020 Brill released the first segment of De Morgan's library digitally, 'The Augustus De Morgan Collection'. Digitisation of each book from cover to cover took advantage of the togetherness of a broad mathematical library. De Morgan's annotations provided a unique selling point to justify the reproduction of books which as texts were frequently available from other sources.

With some exceptions such as early editions of Euclid, Newton and Copernicus, the history of mathematics can be a hard sell for outreach purposes. The first book on logarithms or the earliest reference to the decimal point lacks the general allure of the first edition of a familiar literary work, a travel book, a work on magic, or an exquisitely illustrated volume. De Morgan's books have been used for blog posts and displays about the holdings of Plantin and Elzevir imprints, connected with anniversaries of both publishers. A twenty-first-century library initiative focused on the importance of De Morgan's books as Senate House Library's founding collection: a virtual exhibition of 150 items to celebrate the library's 150th anniversary in 2021 concentrated on post-1871 imprints to emphasise the library's dynamism, but devoted one of its nine sections to the founding collections.[64] Nine of the fifteen books in that section were from De Morgan's library: primarily obvious treasures, but also Ramchundra's *A Treatise on Problems of Maxima and Minima, Solved by Algebra* (1850), to show De Morgan's interest in Indian mathematics and his support for, and connections with, other mathematicians (letters from Ramchundra to De Morgan are tipped in). Some of De Morgan's books appeared in two treasures volumes. The thirty items in *Director's Choice*, a handy gift for library visitors, included two from De Morgan's library, selected for his annotations: Clairaut's *Théorie de la lune* (1765) and John Leland's *Commentarii de Scriptoribus Britannicis* (1709).[65] A more substantial volume showcasing sixty items included three from De Morgan's library.[66] One was the only known complete copy of Bernard de Granollachs's *Lunarium ab anno 1491 ad annum 1550* (Lyon: Johannes Siber, 1491).[67] Another, John Bonnycastle's *The Scholar's Guide to Arithmetic*, edited by Edwin Colman Tyson (1828), is also scarce, but was more important for De Morgan's annotation of 1857 identifying it as 'the book which suggested the existence of the deficiency to supply which I wrote my own arithmetic in 1830.'

De Morgan's library has been used to meet interest in the history of women, through Abigail Lousada's former ownership of books and through Ada Lovelace's translation and edition of Luigi Menabrea's *Sketch of the Analytical Engine Invented by Charles Babbage Esq.* (1843). The latter proves especially valuable when choosing material to display to honour women computer scientists at the University's annual conferral of honorary degrees. Charles Dodgson's *An Elementary Treatise on Determinants* (1867) or his *Formulae of Plane Trigonometry* (1861; copy

---

64 Senate House Library, University of London, 'Senate House Library in 150 Items' (2021), https://www.london.ac.uk/senate-house-library/exhibitions-and-events/exhibitions/senate-house-library-150/senate-house-library-in-150-items.

65 Christopher Pressler, *Director's Choice: Senate House Library, University of London* (London: Scala in association with Senate House Library, 2012).

66 *Senate House Library, University of London*, ed. by Christopher Pressler and Karen Attar (London: Scala in association with Senate House Library, 2012).

67 ISTC ig00340700.

stamped 'Presented by the publisher') come into their own displayed together with early editions of the *Alice* books Dodgson wrote as Lewis Carroll. The *Scholia in Euclidis elementorum geometriae* by Monachus Isaacus (1573), *Nova de universis philosophia* by Francesco Patrizi (1593) and Simon Stevin's *Oeuvres mathématiques* (1634) are of interest not merely for themselves, but for showing three different ownership stamps of the French statesman and historian Jacques-Auguste de Thou: one when he was a bachelor, and one for each of his two marriages.

To an extent, De Morgan's library is valued today for reasons which he, too, regarded as important. De Morgan was ahead of his time in following up provenance quite broadly, regarding scientific provenance as being as valuable as ownership by renowned bibliophiles. His pioneering bibliographical work demonstrated his interest in the history of the book. Twenty-first-century scholarship centres on these aspects. While widespread availability of titles, primarily through digitisation, has weakened the interest in the content of the books which mattered to De Morgan, their combined value as the library of a major scholar in the subject retains meaning. For the modern scholar, the way De Morgan, as a leading Victorian mathematician and bibliographer, used and interacted with his books has gained a fascination, which even those who shortly after his death commented on the enhancement of his annotations could not have envisaged. To say that the retention of his library as a unit, the multiplicity of levels of interest and the individuality stamped upon it by De Morgan's annotations have kept and increased its significance in a digital age may seem a trite and obvious conclusion. Yet any historic book collection that can increase its relevance over 150 years is impressive. A note De Morgan made on one of his copies of the sale catalogue of Charles Hutton's library, mourning its dispersal, may suggest that he wanted his own to remain intact after his death. One hopes that, could he see it today, he would be pleased.

Bibliography

Sale catalogues

Unless otherwise stated, sale catalogues are published by an iteration of Sotheby (e.g. Sotheby, Wilkinson & Hodge; S.L. Sotheby and John Wilkinson) and the place of publication is London.

Catalogue of a Selected Portion of the Scientific, Historical, and Miscellaneous Library of James Orchard Halliwell ... (1840).

A Catalogue of the Entire, Extensive and Very Rare Mathematical Library of Charles Hutton ... (1816).

Catalogue of the Library of the Late John Playfair ... Comprising a Valuable Collection of Mathematical, Philosophical, and Miscellaneous Books, Maps, &c &c ... (Edinburgh: J. Ballantyne, 1820).

Catalogue of the Library of the Late Rev. George Burgess Wildig ... Together with Another Valuable Collection, Including all the Best English and French Mathematical Works ... (L.A. Lewis, 1854).

Catalogue of the Mathematical, Classical, and Miscellaneous Library of a Gentleman, Deceased ... (1846).

Catalogue of the Mathematical, Historical, Bibliographical and Miscellaneous Portion of the Celebrated Library of M. Guglielmo Libri ... (1861).

Catalogue of the Mathematical, Philosophical, and Miscellaneous Library ... of Miss A.B. Lousada ... (1834).

Catalogue of the Principal Portion of the Singularly Curious and Valuable Library of the late J.B. Inglis ... (1871).

Catalogue of the Stock of Books of Mr. Samuel Maynard ... which will be Sold ... on Wednesday, 7th January [sic] 1863 and Two Following Days ... (1863).

Catalogue of the Valuable Astronomical & Mathematical Library of the Late Thomas Galloway ... (1852).

Catalogue of the Valuable Astronomical, Mathematical, and General Library of the Late Francis Baily ... (1845).

Catalogue of the Valuable Library of the Late William Wallace, LL.D. Professor of Mathematics in the University of Edinburgh ... (C.B. Tait, 1843).

A Catalogue of the Valuable Mathematical Library of the late Thomas Leybourn (Hodgson, 1840).

Catalogue of the Valuable Mathematical & Miscellaneous Library of the late Reverend John Toplis ... (1858).

A Catalogue of the Valuable Mathematical and Scientific Library of the Late T.S. Davies ... Professor of Mathematics at the Royal Military Academy, Woolwich ... (Southgate & Barrett, [1851]).

Mathematical and Scientific Library of the late Charles Babbage of No. 1, Dorset street, Manchester Square (1872).

Other

The Aldine Press: Catalogue of the Ahmanson-Murphy Collection of Books by or Relating to the Press in the Library of the University of California, Los Angeles, Incorporating Works Recorded Elsewhere (Berkeley: University of California Press, 2001). https://doi.org/ 10.1525/9780520328563

Attar, K.E., '"The Establishment of a First-Class University Library": The Beginnings of the University of London Library', *History of Universities*, 28 (2014), 44–65.

Attar, Karen, 'Augustus De Morgan (1806–71), His Reading and His Library', in *The Edinburgh History of Reading: Modern Readers*, ed. by Mary Hammond (Edinburgh: Edinburgh University Press, 2020), pp. 62–82. https://doi.org/10.3366/edinburgh/ 9781474446112.003.0004

—, 'Augustus De Morgan's Incunabula', in *Spotlights on Incunabula*, ed. by Anette Hagan, Library of the Written Word, 118 (Leiden: Brill, 2024), pp. 194–211. https://doi.org/10.1163/9789004681378_011

Buxton, H.W., *Memoir of the Life and Labours of the Late Charles Babbage, Esq. F.R.S.*, Charles Babbage Institute Reprint Series for the History of Computing, 13 (Cambridge, MA: MIT Press, 1988).

Canney, Margaret and David Knott, *Catalogue of the Goldsmiths' Library of Economic Literature*, vol. 1, *Printed Books to 1800* (Cambridge: Cambridge University Press for the University of London Library, 1970).

Catalogue of Books in the General Library and in the South Library of University College, London, 3 vols. (London: Taylor & Francis, 1879).

Catalogue of the Library of the University of London, Including the Libraries of George Grote and Augustus De Morgan (London: Taylor & Francis, 1876).

De Morgan, Augustus, *A Budget of Paradoxes* (London: Longmans, Green, 1872).

—, *Arithmetical Books from the Invention of Printing to the Present Time* (London: Taylor & Walton, 1847).

—, *Arithmetical Books from the Invention of Printing to the Present Time*, introd. by A. Rupert Hall (London: Hugh K. Elliott, [1967]).

—, [Obituary of Thomas Galloway], *Monthly Notices of the Royal Astronomical Society* (1851–52), 87–89.

De Morgan, Sophia Elizabeth, *Memoir of Augustus De Morgan* (London: Longmans, Green, 1882).

De Ricci, Seymour, *English Collectors of Books & Manuscripts (1530–1930) and their Marks of Ownership* (Cambridge: Cambridge University Press, 1930).

Dewey, Melvil, *Decimal Classification and Relative Index: for Libraries, Clippings, Notes, etc.*, 6th edn (Boston: Library Bureau, 1899).

Dibdin, Thomas Frognall, *The Bibliomania or Book Madness, Containing Some Account of the History, Symptoms and Cure of this Fatal Disease*, ed. by P Danckwerts (Richmond: Tiger of the Stripe, 2007).

Dorling, Alison R., 'The Graves Mathematical Collection in University College London', *Annals of Science*, 33 (1976), 307–09.

Fletcher, William Younger, *English Book Collectors* (London: K. Paul, Trench, Trübner, 1902).

Gavine, David, 'Galloway, Thomas (1796–1851)', *Oxford Dictionary of National Biography Online* (Oxford: Oxford University Press, 2004). https://doi.org/10.1093/ref:odnb/10312

Hand-Catalogue of the Library [of the University of London], Brought Down to the End of 1897 (London: HMSO, [1900]).

Lang, Andrew, *The Library* (London: Macmillan, 1881).

Loyd, Samuel Jones, Baron Overstone, *The Correspondence of Lord Overstone*, ed. by D.P. O'Brien, 3 vols. (Cambridge: Cambridge University Press, 1971).

Mandelbrote, Giles, 'Scientific Books and their Owners: A Survey to c. 1720', in *Thornton and Tully's Scientific Books, Libraries, and Collectors: A study of Bibliography and the Book Trade in Relation to the History of Science*, 4th edn, ed. by Andrew Hunter (Aldershot: Ashgate, 2000), pp. 333–66.

Maxwell, Richard, 'James Orchard Halliwell-Phillipps', in *Nineteenth-Century British Book-Collectors and Bibliographers*, ed. by William Baker and Kenneth Womack, *Dictionary of Literary Biography*, 184 (Detroit, Washington DC: Gale Research, 1997), pp. 202–18.

Merrington, Maxine, *A List of Certain Letters Inserted in Books from the Library of Augustus De Morgan (1806–1871) now in the University of London Library* (London: University of London Library, 1990).

Moseley, Maboth, *Irascible Genius: A Life of Charles Babbage, Inventor* (London: Hutchinson, 1964).

Munby, A.N.L., *The History and Bibliography of Science in England: The First Phase, 1833–1845* (Berkeley: University of California, 1968).

Overmier, Judith, 'Scientific Book Collectors and Collections, Public and Private, 1720 to Date', in *Thornton and Tully's Scientific Books, Libraries, and Collectors: A Study of Bibliography and the Book Trade in Relation to the History of Science*, 4th edn, ed. by Andrew Hunter (Aldershot: Ashgate, 2000), pp. 367–91.

Pafford, J.H.P., 'The University of London Library', in *The Libraries of London*, 2nd edn, ed. by Raymond Irwin and Ronald Staveley (London: Library Association, 1961), pp. 140–56.

Pearson, David, *Book Ownership in Stuart England: the Lyell Lectures, 2018* (Oxford: Oxford University Press, 2021). https://doi.org/10.1093/oso/9780198870128.001.0001

Pearson, David, *Books as History* (London: British Library, 2008).

Pressler, Christopher, *Director's Choice: Senate House Library, University of London* (London: Scala in association with Senate House Library, 2012).

Pressler, Christopher and Karen Attar (ed.), *Senate House Library, University of London* (London: Scala in association with Senate House Library, 2012).

Purcell, Mark, *The Country House Library* (New Haven and London: Yale University Press for the National Trust, 2017).

Rice, Adrian, 'Augustus De Morgan: Historian of Science', *History of Science*, 34 (1996), 201–40. https://doi.org/10.1177/007327539603400203

—, 'Graves, John Thomas (1806–1870)', *Oxford Dictionary of National Biography Online* (Oxford: Oxford University Press, 2004). https://doi.org/10.1093/ref:odnb/11311

Roberts, W., *The Book-Hunter in London: Historical and Other Studies of Collectors and Collecting* (London: Elliot Stock, 1895).

Ruju, P. Alessandra Maccioni and Marco Mostert, *The Life and Times of Guglielmo Libri (1802–1869), Scientist, Patriot, Scholar, Journalist and Thief: A Nineteenth-Century Story* (Hilversum: Verloren, 1995).

Rye, Reginald Arthur, *Catalogue of the Manuscripts and Autograph Letters in the University Library at the Central Building of the University of London* (London: University of London Press, 1921).

—, *The Libraries of London: A Guide for Students* (London: University of London, 1908).

Rye, Reginald Arthur and Muriel Sinton Quinn, *Historical and Armorial Bookbindings Exhibited by the University Library: Descriptive Catalogue* (London: University of London, 1937).

Smith, David Eugene, *Rara Arithmetica: A Catalogue of the Arithmetics Written before the Year MDCI, with a Description of those in the Library of George Arthur Plimpton, of New York* (Boston and London: Ginn, 1908).

Spevack, Marvin, *James Orchard Halliwell-Phillipps: The Life and Works of the Shakespearean Scholar and Bookman* (New Castle, DE: Oak Knoll and London: Shepheard-Walwyn, 2001).

Voet, Leon, *The Plantin Press (155–1589): A Bibliography of the Works Printed and Published by Christopher Plantin at Antwerp and Leiden*, 6 vols. (Amsterdam: Van Hoeve, 1980–1983), vol. 6 (1983).

Wardhaugh, Benjamin, 'Collection, Use, Dispersal: The Library of Charles Hutton and the Fate of Georgian Mathematics', in *Beyond the Learned Academy: The Practice of Mathematics, 1600–1850*, ed. by Philip Beeley and Christopher Hollings (Oxford: Oxford University Press, 2024), pp. 158–84. https://doi.org/10.1093/oso/9780198863953.003.0007

Wells, Ellen G., 'Scientists' Libraries: A Handlist of Printed sources', *Annals of Science*, 40 (1983), 317–89.

Wheatley, Henry B., *Prices of Books: An Inquiry into the Price of Books which has Occurred in England at Different Periods* (London: G. Allen, 1898).

Willems, Alphonse, *Les Elzevier: histoire at annales typographiques* (Brussels: van Trigt, 1880).

Williams, M.R., 'The Scientific Library of Charles Babbage', *Annals of the History of Computing*, 3 (1981), 235–40.

Williams, Moelwyn I. (ed.), *A Directory of Rare Book and Special Collections in the United Kingdom and the Republic of Ireland* (London: Library Association, 1985).

I told Airy of your topsy turvy views about Orion: he declares he will stand upon his head the first fine night and look at it. I think I shall suggest to him a generalisation upon the Greenwich method of observing. Let an observer take transits in a direct and reversed position on successive nights, and there must be a ± destruction of personal equation, and his determinations would be very valuable and effective, particularly that of blood to the head.

And now a very good night to you, and finer weather than we have had. Such a succession of rain has not occurred for a long time.

Pray remember me kindly to Maclear and C Smith

Yours very sincerely,
A De Morgan

5 Upper Gower S[t]
Feb. 20/37

Fig. 15 This letter, now housed in the archives of the Royal Society, was sent by De Morgan to the astronomer Sir John Herschel in 1837, while Herschel was in South Africa surveying the skies of the southern hemisphere. The cartoon is De Morgan's tongue-in-cheek suggestion as to how the Astronomer Royal, George Airy, might obtain similar 'topsy turvy views' of the Orion constellation from Greenwich. (HS/6/183, reproduced by permission of the Royal Society.)

11. Augustus De Morgan: The Archival Record

Karen Attar, Alexander Lock, Katy Makin, Jane Maxwell, Virginia Mills and Diana Smith

> The history of science is almost entirely the history of books and manuscripts.
>
> — Augustus De Morgan[1]

Introduction

There is no single De Morgan archive, and, tantalisingly, much of Augustus De Morgan's activity is unrecorded. Few drafts of publications remain, for example, and no documentation about his library in terms of accessions records—if he ever kept a list of acquisitions—or invoices. However, as the *Oxford Dictionary of National Biography* notes, De Morgan 'was a prolific correspondent, often adorning his letters with well-drawn caricatures and sketches'.[2] His correspondence with a wide range of scientific luminaries and other acquaintances, together with some mathematical manuscripts, grace libraries in England, Scotland, the Republic of Ireland, and the United States of America, easily locatable from the list of repositories ending the entry for De Morgan in the *Oxford Dictionary of National Biography*.

---

1 Augustus De Morgan, 'On the Earliest Printed Almanacs', in *Companion to the Almanac for 1846* (London: Charles Knight, 1846), pp. 1–31 (p. 1).
2 Leslie Stephen & I. Grattan-Guinness, 'De Morgan, Augustus (1806–1871)', in *The Oxford Dictionary of National Biography* (Oxford: Oxford University Press, 2004–), online edn., https://doi.org/10.1093/ref:odnb/7470.

Electronic catalogues and websites of the holding institutions provide further details of what precisely is available and how to access it. Helpful as they are, in themselves these are arid. They cannot give a flavour of the correspondence or, through it, the man. They indicate the fact of De Morgan's broad scholarly network, but not its nature. Publications of De Morgan's letters do achieve this, but cover only a fragment of the extant material.[3] This chapter looks more closely at the archival records pertaining to Augustus De Morgan held in his chief areas of residence, Cambridge (as a student) and London.[4] It complements these with two sets of records held elsewhere, in Edinburgh and in Dublin, and attempts to show how the archives reveal the man.

The British Library (BL)

The British Library holds some sixty manuscripts by or about Augustus De Morgan across several collections. These can be identified by examining the printed *Index of Manuscripts in the British Library* or by searching the Library's online manuscripts catalogue.[5] The manuscripts range from private correspondence on both personal and professional matters through to a biographical note in the 'Original, letters, with corrected proofs of memoirs, etc., addressed to Charles Griffin,

---

3   See *The Boole-De Morgan Correspondence*, ed. by G.C. Smith (Oxford: Clarendon Press, 1982); Sophia Elizabeth De Morgan, *Memoir of Augustus De Morgan* (London: Longmans, Green, 1882); Sophia Elizabeth De Morgan, *Threescore Years and Ten: Reminiscences of the Late Sophia Elizabeth De Morgan, to which are Added Letters to and from her Husband, the Late Augustus De Morgan, and Others*, ed. by Mary A. De Morgan (London: Bentley, 1895); James Smith, *The Quadrature of the Circle: Correspondence between an Eminent Mathematician and James Smith* (London: Simpkin, Marshall, 1861).

4   Only the Royal Astronomical Society (RAS) is excepted, because De Morgan's astronomical activity, which mines the RAS records, is covered thoroughly in Chapter 3 of this volume; moreover, its letters are primarily to, not from De Morgan. The catalogue of the RAS archives is available at: https://ras.ac.uk/library/archives/introduction-to-archives. The RAS archive includes correspondence with multiple astronomers, of whom Richard Sheepshanks (68 letters, 1842-1852) stands out. Particularly distinctive in the RAS archives are twelve De Morgan-authored biographies of scientists reprinted from Charles Knight's *Gallery of Portraits* (1833–1837), with De Morgan's annotations and drawings.

5   *Index of Manuscripts in the British Library* (Cambridge: Chadwyck-Healey, 1984–1986); British Library, 'Explore Archives and Manuscripts', http://searcharchives.bl.uk/.

publisher of the "Handbook of Contemporary Biography," (London, 1861)'.[6] Most of the material is correspondence, and the largest number of letters are found among the papers of De Morgan's friend and fellow mathematician Charles Babbage. Other letters by De Morgan can be found in the collections of several distinguished nineteenth-century figures, including the social reformer Sir Rowland Hill; the antiquary, book collector and university administrator Philip Bliss; the clergyman and campanologist Rev. Henry Thomas Ellacombe; the army officer Sir Charles Pasley; and the naturalist Alfred Russel Wallace.[7]

Letters include a number of De Morgan's playful pictures alongside more serious mathematical sketches. The surviving correspondence is largely good-humoured and the topics discussed are wide-ranging and well informed, demonstrative of the breadth of De Morgan's expertise and interests. As one biographical note held in the British Library confirms, De Morgan 'was a voluminous writer on branches of mathematics ... history and bibliography' and these are largely the topics discussed in the letters.[8] In his letters to Babbage, De Morgan frequently commented on Babbage's published work and on occasion—having 'read the book through (God's blessing be on you for writing a short book)'—had cause to question Babbage's logic (with equations) or point out a 'mistake' in his 'figures'.[9]

Elsewhere De Morgan corresponded with the Principal Librarians at the British Museum, writing to Sir Henry Ellis, regarding the Domesday Book and the length of 'the old English mile'. With Ellis's successor, Anthony Panizzi, of whom De Morgan was a friend and

---

6 British Library, London (henceforth BL), Add MS 28509, f. 421r.
7 See BL, Add MS 31978, Letters to Rowland Hill, 1837–1879; Add MS 33206, Correspondence with the Rev. H. T. Ellacombe, 1824-1882; Add MS 28509, Letters with corrected proofs of memoirs, etc. addressed to Charles Griffin, publishers, 1860; Add MS 22786*, Letters from Augustus De Morgan to Antonio Panizzi; Add MS 34578, Correspondence of Rev. Philip Bliss, 1806–185; Add MS 36717 and 36724, Correspondence and papers, literary, political and general, of Antonio Panizzi, 1823–1877; Add MS 37185, 37186-94 and 37199-200, Correspondence of Charles Babbage, 1806–1871; Add MS 38514, Correspondence and papers of Sir Henry Ellis, 1757–1850; Add MS 41964; Correspondence, letter-books, diaries, and note-books of Gen. Sir Charles William Pasley, 1764–1861; Add MS 46439, Alfred Russel Wallace Papers, 1848–1914.
8 BL, Add MS 28509, f. 421r.
9 BL, Add MS 37190, f. 167 and Add MS 37189, ff. 142r.-142v, Letters from Augustus De Morgan to Charles Babbage, n.d.

staunch supporter,[10] De Morgan discussed the activities of the Italian mathematician and manuscript thief Guglielmo Libri as well as the authorship and provenance of manuscripts, including one by Galileo.[11] When unfamiliar with a subject, De Morgan would correspond with experts to learn more, and was willing to share his own perspective. This was the case with the 'subject of the oscillation of buildings' that housed and rang bells, a topic which De Morgan had 'never turned my attention to' but which he found 'curious in a mathematical point of view'.[12] He discussed the matter over a series of letters with Henry Thomas Ellacombe, suggesting they compose 'a collection of facts ... for very little is known about the matter.'[13]

The Royal Society (RS)

Augustus De Morgan never sought to become a Fellow of the Royal Society. His name appears on no Candidates lists, and he never attended a meeting,[14] for he felt that, despite steps by Fellows to reform the Society in the 1840s, it continued to favour privilege over scientific

---

10 Their acquaintance had begun in 1828 when they were among the first professorial appointees at the new London University, later University College London: De Morgan as professor of mathematics and Panizzi as professor of Italian. For evidence of friendship with Panizzi, see De Morgan's note in his copy of J. Rogg, *Handbuch der mathematischen Literatur vom Anfange der Buchdruckerkunst bis zum Schlusse des Jahrs 1830* (Tübingen, 1830; Senate House Library, University of London, [DeM] CC4L [Rogg]) about him and Panizzi examining the book together. Note also De Morgan's support of Panizzi's arrangement of the British Museum catalogue, *Report of the Commissioners Appointed to Enquire into the Constitution and Government of the British Museum* (London: William Clowes and Sons, 1850), pp. 377–78; noted in P.R. Harris, *A History of the British Museum Library, 1753–1973* (London: British Library, 1998), p. 170, and described in Edward Miller, *Prince of Librarians: The Life and Times of Antonio Panizzi of the British Museum* (London: British Library, 1988), p. 181.
11 BL, Add MS 38514, f. 159, Letter from Augustus De Morgan to Sir Henry Ellis, 15 Oct. 1838; Add MS 36724, ff. 287v. –288v, Letter from Augustus De Morgan to Anthony Panizzi, 18 Aug. 1867; Add MS 22786*, Three letters from Augustus De Morgan to Antonio Panizzi, 18 Apr., 27 Apr., and 2 May 1859.
12 BL, Add MS 33206, ff. 66r.–67v, Letter from Augustus De Morgan to Rev. H. T. Ellacombe, 2 Nov. 1863.
13 BL, Add MS 33206, f. 69v, Letter from Augustus De Morgan to Rev. H. T. Ellacombe, 4 Nov. 1863.
14 As a non-Fellow his name would have been recorded in the ordinary meeting minutes as a 'stranger' given leave to be present at a weekly meeting, or in the visitors' book if he had attended on another occasion.

attainment and was therefore at odds with his own professionalising principles.[15] Yet some valuable records of De Morgan entered the archive of the organisation he held in such low regard, largely through the correspondence and agency of his friends Sir John Herschel and Francis Baily.

Some 250 letters from De Morgan form a small but significant part of the Herschel papers, a vast collection of letters received by the astronomer and polymath Sir John Herschel from his extensive network of learned correspondents.[16] One hundred and thirty-four letters from Herschel to De Morgan also survive in the collection in the form of amanuensis copies (HS/23-25), and very occasionally as unsent drafts (HS/6): the presence of both sides of the correspondence arises from the fact that after De Morgan's and Herschel's deaths, the widows copied and sent each other many of the letters that passed between their husbands.[17] The correspondence spans the 1830s to the 1860s without significant breaks, concluding in 1870 shortly before both men died. It covers primarily the two men's shared interests in astronomy and mathematics, but also extends into personal matters.

De Morgan wrote his earliest letters to Herschel (1831) in his capacity as Secretary of the Royal Astronomical Society (RAS), requesting papers for their publications and Herschel's opinions on Society business. (Herschel was an RAS founder, oftentimes Council Member and

---

15 For further discussion see Rebekah Higgitt, 'Why I don't FRS my Tail: Augustus De Morgan and the Royal Society', *Notes and Records*, 60 (2006), 253–59, available at http://doi.org/10.1098/rsnr.2006.0150.

16 Royal Society, London (henceforth RS), HS/6/174-434. See the Royal Society, 'Search Archives', https://catalogues.royalsociety.org/. Summaries of Herschel's correspondence can be found at the Adler Planetarium, Calendar of the Correspondence of Sir John Herschel, http://historydb.adlerplanetarium.org/herschel and on Cambridge University Library, Epsilon [Nineteenth-Century Scientific Correspondence], https://epsilon.ac.uk. The digitised letters are available through the Royal Society archive catalogue and through the Royal Society, *Science in the Making*, https://makingscience.royalsociety.org.

17 Several of these letters are reproduced in Sophia De Morgan's *Memoir* of her husband. For correspondence between Sophia De Morgan and Lady Herschel and one of her daughters, see RS, HS/6/437-443. The discrepancy with the number of letters he received from De Morgan suggests that not all Herschel's outgoing correspondence was copied or retained for this collection though there are no obvious gaps identified. The flow of correspondence between them may have been originally unbalanced, with De Morgan as the more prolific letter writer, quite possible as Herschel had several protracted periods of illness which made him less able to write.

sometime President, or, in De Morgan's words, one of its 'wisest heads'.) Initially formal and administrative, the letters soon began to display De Morgan's trademark wit, neat turn of phrase and what Herschel called his 'punning humour'. When Herschel was in the Cape, De Morgan sent self-confessed 'gossipy letters' about their scientific acquaintances. Their letters are an exchange both of intellectual equals and of friends. Topics include Herschel's work in astronomy, contemporary astronomical discoveries by others, further aspects of Herschel's broad-ranging work and publications such as his treatise on sound and noise, and his theory of perspective, and, briefly, Herschel's photographic processes (1846).[18] In 1844-1845 Herschel, then engaged in establishing a benefit society for the mutual insurance of artisans, sought and received extensive actuarial and policy advice from De Morgan on running and financing friendly societies. The two men also commented on politics, psychology and recommended reading, including the occasional novel.

The letters became increasingly familiar in tone, and De Morgan sometimes included responses in illustration form. For example, Herschel's comments on having made some 'topsy turvy observations of the constellation Orion' whilst in the southern hemisphere were answered with an illustration of an acrobatic astronomer looking through a telescope whilst standing on his head.[19] In another letter De Morgan replies tongue in cheek to Herschel's comments on the need to reform stellar nomenclature by suggesting that he might like to name any newly mapped nebulae observed in the Southern Hemisphere after the president and officers of the RAS. Others are written in verse, contain thanks in sonnet form, riddles (why is Sir John Herschel as good as two astronomers – because he's a double star gazer), mathematical puzzles such as magic squares, equations 'fattened' for the festive season and gentle jibes about 'the great moon hoax', a series of articles published in an American newspaper in 1835 claiming Herschel had discovered civilisation on the moon.[20]

De Morgan also shared his own chiefly mathematical interests and current research with Herschel, covering algebra, probability, functions,

---

18 For letters on Herschel's treatise on sound and noise, see, e.g., RS, HS/6/184, 284 and 420; for his theory of perspective, see RS, HS/6/424.
19 RS, HS/6/183.
20 See e.g. RS, HS/6/188–89.

calculus, trigonometry, differential equations, partial differentials, infinity in physics, Euclid and resolving fractions. He sometimes included lengthy explanations of his theorems, equations and puzzles for comment, or notified Herschel that he was sending him his latest publications or critiquing Herschel's. Other notable subjects include logic and reasoning and the nature of absolute truth, the Augustan and Julian calendars and new- and old-style dates. Only Augustus and Sophia De Morgan's investigations into spiritualism seem to be lacking.

In the 1850s De Morgan's 'plot regarding the introduction of the decimal coinage' was a recurring topic, for which he sought and gained Herschel's advice and support. The subject dropped after the government decided in 1859 against introducing decimal coinage. De Morgan's interest in history of mathematics and scientific biography is also evident in his correspondence with Herschel. It also brought him into limited direct contact with the Royal Society and some accounts of this are retained in the Society's administrative records.

De Morgan was one of the first writers to draw attention to the seventeenth-century dispute between Isaac Newton and Gottfried Leibniz over the priority for the calculus, and to Newton's use of cronyism to secure a decision in his favour. In preparing an article on the dispute in 1845–1846 De Morgan corresponded with the Secretary of the Royal Society, Charles Weld, requesting access to documents and details of the case against Leibniz as presented by Newton's friends in the *Commercium Epistolicum* (1712). The Royal Society's outgoing letter books contain copies of Weld's replies.[21]

The Royal Society further holds drafts of two papers De Morgan submitted to be considered for publication in one of its journals:

---

21 See RS, MS/426. For De Morgan's publication on the matter, see 'On the Additions Made to the Second Edition of the *Commercium Epistolicum*', *London, Edinburgh and Dublin Philosophical Magazine and Journal of Science*, 3rd ser., 2 (1848), 446–56; 'A Short Account of Some Recent Discoveries in England and Germany Relative to the Controversy on the Invention of Fluxions', *Companion to the Almanac for 1852*, pp. 5–20. Weld's original letters are bound into De Morgan's copy of 'On the Additions Made' and related articles: see Senate House Library, University of London, [DeM] L (B.P.1) SR. For a discussion of De Morgan's contribution to the dispute, see especially Rebekah Higgitt, *Recreating Newton: Newtonian Biography and the Making of Nineteenth-Century History of Science* (London: Pickering & Chatto, 2007), and Adrian Rice, 'Vindicating Leibniz in The Calculus Priority Dispute: The Role of Augustus De Morgan', in *The History of the History of Mathematics*, ed. by Benjamin Wardhaugh (New York: Lang, 2012), pp. 89–114.

'Description of a Calculating Machine, invented by Mr Thomas Fowler of Torrington in Devonshire' (1840) and 'Comparison of the First and Second Editions of the *Commercium Epistolicum*' (submitted in 1846).[22] Although these papers were ultimately rejected, the Society did publish one paper by him, entitled 'On a Point Connected with the Dispute between Keil and Leibnitz about the Invention of Fluxions', in its *Philosophical Transactions* in 1846. These were De Morgan's only communications with the Royal Society, aside from the donation of five eighteenth-century letters of historical significance, found in Baily's collection after his death.[23]

Senate House Library, University of London (SHL)

The archival material pertaining to De Morgan at Senate House Library falls into three distinct categories. Firstly, there is the correspondence: one archival box, chiefly of letters written by Augustus De Morgan. Secondly, there are letters written primarily to De Morgan connected with the books he owned, filed within the books. Thirdly are manuscripts by De Morgan, or printed works of his so heavily annotated by him as to occupy a grey area between archives and printed books.

The letters inside books came to the University of London in 1871 with the books themselves. Some are listed in a booklet by Maxine Merrington.[24] The online book catalogue notes the presence of letters in books in the records for the relevant books, with writer and date, and books with letters in them are brought up most easily by doing a keyword search on 'Augustus De Morgan' in conjunction with 'ALS', for

---

22 See RS, AP/23/24 and AP/29/2 respectively. A review of the latter by George Peacock, recommending that the paper be printed only in abstract, is also present, at RR/1/57. See also Rice, 'Vindicating Leibniz', p. 104.

23 See RS, EL/M3/60a. The letters donated are Royal Society, EL/I1/183, Letter from Pierre Maupertuis to James Bradley concerning the figure of the earth, the first announcement confirming Newton's theory from the Lapland expedition; EL/I1/183–86, Four letters from William Jones and James Hodgson regarding the determination of longitude.

24 Maxine Merrington, *A List of Certain Letters Inserted in Books from the Library of Augustus De Morgan (1806–1871) now in the University of London Library* ([London]: University of London Library, 1990). As the De Morgan library was partially dispersed at the time of compilation and later reconstituted, some classmarks and details of location are no longer applicable.

'autograph letter(s) signed'.[25] To establish the content of any letter, it is necessary to read the letters themselves.

Correspondents include the mathematical and other scientific luminaries whose letters to and/or from De Morgan are preserved in other repositories: Sir George Biddell Airy, Charles Babbage, Sir William Rowan Hamilton, Sir John Herschel, William Whewell, as well as people whose letters are not held elsewhere, among them the mathematicians Thomas Galloway and George Salmon. A full-text database, *The Augustus De Morgan Collection*, is currently in the process of online publication by Brill; by capturing the content of all books within De Morgan's library from cover to cover, it will publish all letters in the context of the books in which they are contained.

The letters show above all the widespread respect in which De Morgan and his knowledge were held. Some also show his kindness, such as a letter from Thomas Weddle inserted in his *Solving Numerical Equations of All Orders* informing De Morgan of his successful application for a mathematical professorship at Sandhurst and thanking him for his reference, and one from Carel J. Matthes of Amsterdam giving De Morgan Willem Kersseboom's *Kort Bewys* ... (1738) as 'a slight mark of my gratefulness for the many kindnesses you showed me during my stay in London'.[26] Other writers ask De Morgan to recommend or publicise their writings: 'Your good opinion of my work will so much enhance its value, that I must ... excuse the liberty I am taking', explained William Henry Oakes, in a letter in his *Table of the Reciprocals of Numbers* (1865).[27] Still others request De Morgan's elucidation of mathematical matters, as when Isaac Todhunter, a former student of De Morgan and fellow early member of the London Mathematical Society, requested elucidation on a couple of points in a paper about least squares, in a letter in Todhunter's *A History of the Mathematical Theory of Probability* (1865).[28]

---

25 Senate House Library catalogue, https://catalogue.libraries.london.ac.uk/search~S1.
26 Senate House Library, University of London (henceforth SHL), [DeM] L.2 [Weddle] fol. and [DeM] L.4 [Kersseboom] SSR respectively.
27 SHL, [DeM] L.8 [Oakes]. David Bierens de Haan, in a letter in his *Nouvelles tables d'intégrales définies* (1867), wrote in a similar vein ([DeM] L.3 [Haan] fol. SSR).
28 SHL, [DeM] L.2.1 [Todhunter].

The De Morgan manuscripts are catalogued in Senate House Library's archival database.[29] The only pure manuscript, which entered the University of London with his library, is the text of his *Elements of Arithmetic*, 169 leaves in an oblong (landscape-format) book.[30] Although numerous words and phrases have been crossed out, De Morgan's handwriting in it is unusually large and neat. He later annotated it: 'This is the MSS [sic] of the first separate work I ever wrote. A. De Morgan. May 10, 1853'. Interleaved printed copies are present of his *Formal Logic* (1847), with De Morgan's notes and diagrams, press cuttings and letters on the blank leaves and his pencilled annotations within the text,[31] *Syllabus of a Proposed System of Logic* (1860) and *Arithmetical Books from the Invention of Printing to the Present Time* (1847).[32] Correspondence, sometimes drawing De Morgan's attention to additional works, corrections to the text and verifications make clear that *Arithmetical Books* is work in progress towards a second edition, never published. Some correspondence also entered the University, possibly with the library, and definitely by 1921, when the first catalogue of manuscripts and autograph letters was published. Most copious is a group of letters to De Morgan, bound together, mostly from the French mathematician Jean-Baptiste Biot, dating from 1855 to 1858.[33] Several of these pertain to Biot's article on Isaac Newton in the *Biographie Universelle* and his controversy with Sir David Brewster, about whose scholarship on Newton De Morgan had reservations.[34] From an educational viewpoint, the most remarkable is a draft of De Morgan's letter to the council of the University of London (i.e. University College London) of 24 July 1831, resigning his Chair in protest against Granville Sharp Pattison's removal from the Chair of Anatomy: '... I should think it discreditable to hold a professorship under you one moment longer.'[35]

---

29   SHL archival catalogue, https://archives.libraries.london.ac.uk/home.
30   SHL, MS165.
31   SHL, MS776/1-2. See Joan Richards, 'Augustus De Morgan, *Formal Logic: or, The Calculus of Inference, Necessary and Probable*', in *Senate House Library, University of London*, ed. by Christopher Pressler and Karen Attar (London: Scala, 2012), no. 40.
32   SHL, MS776/3.
33   SHL, AL6.
34   See Paul Theerman, 'Unaccustomed Role: The Scientist as Historical Biographer: Two Nineteenth-Century Portrayals of Newton', *Biography*, 8 (1985), 145–62.
35   SHL, AL45. The autograph letters are described in Reginald Arthur Rye, *Catalogue of the Manuscripts and Autograph Letters in the University Library at the Central*

The University acquired further miscellaneous correspondence—letters both to and from De Morgan, apparently gathered by De Morgan's great-niece Joan Antrobus—from an auction at Christies in November 1990 (catalogued as MS913). Sophia De Morgan had solicited at least some of it for her memoir of her husband, as shown by a letter to her from Lord Brougham, dated 24 August 1874, accompanying letters from De Morgan to him from 1851 to 1857, stating: 'I send herewith all the letters of the late Mr De Morgan I have been able to find. You are perfectly welcome not only to publish them, but to keep them'.[36] Correspondents include various mathematicians or other scientists, among them George Airy, Peter Hardy, Olinthus Gregory, George Boole, John Lubbock, George Peacock, and John Graves. The letters mix personal and mathematical matters, as De Morgan writes about what absorbs him, including his work, while the correspondents send best wishes to each other's wives and refer to their health. Some letters concern books, such as one to W.H. Smyth about the merits of quarto versus octavo volumes, concluding: 'Except for expence [sic] I am myself a quartist'.[37]

What distinguishes the De Morgan correspondence at Senate House Library from most of that elsewhere is the inclusion of family letters: letters from De Morgan to his father-in-law, William Frend, and to his mother, correspondence with his wife before and during their marriage, and letters to and from his children. They reveal an affectionate father, as in a scrap from De Morgan addressed to 'Miss Alice De Morgan, Upstairs' and saying:

> I want to see if you can read writing; so mind you try and remember every word of this, and find it out by yourself. Your Mamma cannot write to night, so I write instead. Mr. Baily has sent us two wood pigeons.[38]

The affection was clearly reciprocated, as shown, for example, by a letter from De Morgan's daughter Christiana telling him about a fracas in which her brothers were involved at school and siding with them against 'old Keys', namely headmaster Thomas Hewitt Key.

Building of the University of London ... (London: University of London Press, 1921).
36   SHL, MS913A/2/10.
37   SHL, MS913A/2/5. Letter dated May 1852.
38   SHL, MS913A/1/3.

University College London (UCL)

Material relating to various aspects of De Morgan's life and work is scattered throughout University College London's special collections, mainly classified under the umbrella of 'additional manuscripts', and partly among the first archives the College acquired.[39]

The presence of lecture notes is a unique feature of the UCL archives, most notably De Morgan's handwritten copy of his introductory lecture delivered at the opening of classes in mathematics at UCL in November 1828.[40] This is especially important because, unlike several of his professorial colleagues, De Morgan chose not to publish his inaugural lecture soon after delivery. Rough notes also survive for a couple of his other lectures, such as one he gave in October 1862 at the opening of UCL's academic year.[41] Again, since this lecture was never published, these notes in De Morgan's hand give the only clues as to its content.

Lecture notes taken by De Morgan's students complement his own. In the original college notebook of John Golch Hepburn, a UCL student from the late 1840s, are transcribed details of mathematics classes as they happened.[42] The manuscript contains notes from twenty-one of De Morgan's lectures on algebraic geometry and differential calculus, delivered between 11 March and 13 May 1847, thus giving us a rare insight into what De Morgan's students would have experienced in his lecture room more than a century and a half ago.

One of the most interesting mathematical manuscripts is 'Elements of Statics', a draft of an unfinished, early book De Morgan began at the request of the Society for the Diffusion of Useful Knowledge (SDUK).[43] De Morgan recorded, surely for posterity, opposite the first page:

> Elements of Statics
>
> Written in the Summer of 1827

---

39 See University College London (henceforth UCL) archives catalogue, https://archives.ucl.ac.uk/.
40 UCL, MS ADD 3.
41 UCL, MS ADD 2; see Chapter 6 in this volume.
42 UCL, MS ADD 5.
43 UCL, MS ADD 27. See Adrian Rice, 'Inspiration or Desperation? Augustus De Morgan's Appointment to the Chair of Mathematics at London University in 1828', *British Journal for the History of Science*, 30 (1997), 257–74 (pp. 270–71).

> This is the first attempt I ever made at writing for publication. It was commenced at the proposal of the Useful Knowledge Society in its earliest days - but was never published, nor even completed. I sent it in with my testimonials when I was a Candidate for the Maths chair in the Univ. of London in 1827, and I think it was as useful as the testimonials.
>
> A De Morgan
>
> May 10/53

The date is the one on which De Morgan annotated his manuscript copy of *Elements of Arithmetic* in a similarly nostalgic fashion (see above); comparison of the annotation of the two manuscripts, now in two separate repositories, sheds light on De Morgan's habit of browsing in his study, marking books long after the time of creation or acquisition.[44] The volume consists of around one hundred folios in De Morgan's hand, interspersed throughout with his neat diagrams. Additions and amendments on spare pieces of paper are attached to the relevant page with red sealing wax. As the College acquired the manuscript before records of archive donations were routinely kept, the depositor is unknown.

UCL is also a major repository of material pertaining to the De Morgan family, holding in particular a volume of family history and a folder of associated ephemera.[45] Entitled 'Memorandums on the Descendants of Captain John De Morgan' it was compiled by Augustus De Morgan in the 1850s and 1860s, and opens in his usual witty style:

> Such account as I can give of my family is contained in two books. The first is well known by the name of Genesis, ascribed by Jewish tradition to Moses. The second is this book itself, which my own handwriting will identify as compiled by me. Moses gave no account of his materials: I have given what I could. Moses wrote in Hebrew: I in English. Moses was a public writer, I am a private one. Many are the oppositions between me and Moses ...

---

44 On De Morgan's marking of books, see Karen Attar, 'Augustus De Morgan (1809–71), His Reading and His Library', in *The Edinburgh History of Reading: Modern Readers*, ed. by Mary Hammond (Edinburgh: Edinburgh University Press, 2020), pp. 62–82, especially p. 69.

45 UCL, MS ADD 7.

Beginning with his paternal great-grandfather, who served in the British Army in India, the book goes on to explore numerous subsequent branches and generations of the family, including its connection with the eighteenth-century mathematician James Dodson, Augustus De Morgan's maternal great-grandfather. The ephemera folder contains a note and letter by Dodson, while the volume contains a host of information on the various strands of De Morgan's family, featuring family trees, genealogical information, stories, anecdotes, articles, witticisms and drawings, including two frequently reproduced cartoons of De Morgan, one of him lecturing in 1865 and an undated 'Sketch of Professor De Morgan in the Pillory'.

UCL additionally holds a small number of letters of Augustus De Morgan and Sophia Frend (subsequently his wife), to and from family and friends.[46] It includes one undated letter from De Morgan to Miss Frend and one from her to him from 1836, the year before their wedding. Miscellaneous notes and papers include the scripts of two comic plays by an unknown author (possibly William Frend), in the hand of his daughter Sophia, perhaps performed by friends and family in the mid-1830s. The first manuscript, bearing the title *The Comet*, is a witty play written in verse, set amidst the excitement of the British astronomical community during the re-appearance of Halley's Comet in 1835. Centred around the Royal Astronomical Society, of which De Morgan was a prominent member, it gently mocks both him and his scientific fellows. The second manuscript, probably written slightly later, contains another satirical sketch, this time a short play in prose sending up De Morgan's views on algebra and the style of teaching at UCL.[47]

Like other repositories, UCL holds correspondence between De Morgan and other intellectuals: the logician George Boole, philosopher John Stuart Mill, fellow mathematician and book-collector John Graves and others.[48] UCL also holds the archive of the Society for the Diffusion of Useful Knowledge (SDUK), with which De Morgan was heavily involved, and this archive includes over one hundred letters from De Morgan written between 1827 and 1844. The college also owns a

---

46   UCL, MS ADD 163.
47   Helena M. Pycior, 'Early Criticism of the Symbolical Approach to Algebra', *Historia Mathematica*, 9 (1982), 392–412.
48   UCL, MS ADD 97. The letters date from 1842 to 1871.

substantial collection of the papers of the lawyer, statesman, and UCL co-founder, Lord Henry Brougham, which contains several letters from De Morgan on scholarly matters, such as their common interest in the life and work of Isaac Newton. Also housed at UCL are manuscripts from the early years of the London Mathematical Society (LMS). Among them are a few letters from De Morgan (its first president) to fellow LMS member and vice-president, Thomas Archer Hirst, who was to succeed him as UCL's professor of mathematics in 1867.

The College's own records provide valuable information about De Morgan's professional life at UCL, including his initial application for the mathematics professorship in 1827 and the circumstances leading to his two resignations.[49] These documents show his strained relationship with the College Council throughout his tenure at UCL, culminating in his final resignation in 1866, which left De Morgan feeling so let down by the College that he never returned. It therefore seems unlikely that he would have sanctioned the subsequent donation of his manuscripts to the college library—a circumstance seemingly forgotten half a century later, when at least one item in the archives was 'Presented to University College London by Mrs. William De Morgan'.[50]

Cambridge University Library (CUL)

'Airy is the prince of *method*ists.... My theory is that when he tries his pen on blotting-paper, he makes a duplicate by the pressing machine, files, and indexes it', wrote Augustus De Morgan to William Rowan Hamilton in 1852.[51] De Morgan exaggerates, but not by much. The papers of his friend and colleague George Biddell Airy, Astronomer Royal from 1835 to 1881, at Cambridge University Library occupy twelve cubic metres: a voluminous collection which, in fairness to Airy, is a fine record of his life and work rather than a frustrating mix of important papers obscured by unworthy blotting papers.[52] The copy press letters are in fact there,

---

49 UCL, UCLA/CORR/1912; UCLA/CORR/729.
50 UCL, MS ADD 2.
51 Trinity College, Dublin, TCD MS 1493/610; Quoted in Anne van Weerden, *A Victorian Marriage: Sir William Rowan Hamilton* (Stedum: J. Fransje van Weerden, 2017), p. 333.
52 Cambridge University Library (henceforth CUL), GBR/0180/RGO 6. The papers are described in Cambridge University Library, ArchiveSearch: https://

representing Airy's outgoing letters. De Morgan's papers at Cambridge University Library are primarily to be found in this collection, which is part of the archive of the Royal Greenwich Observatory. Airy's archive retains his original arrangement by subject matter, with catch-all miscellaneous correspondence files divided into short date spans, such that finding all instances of a correspondent's letters and works in his collection is not straightforward. De Morgan is represented by an estimated two hundred letters or more, housed in seventy-eight different boxes. One scholar is of the opinion that 'due to the extensive nature of the collection, large parts of it have remained unexplored by historians' thus far.[53]

The papers initially appear to be of a narrow cast, of a mathematician corresponding with an astronomer. Airy's headings include: 'Papers relating to the Astronomical Society', 'Miscellaneous astronomical papers', 'Correspondence on scientific institutes', 'Papers on engineering and inventions' and 'Mathematical theories and calculations'. Both men excelled in their fields, and there are indeed pleasingly intricate mathematical workings and explanations, discussing partial differential equations, probabilities, mathematical principles and current and historical mathematicians and their works. However, just like the De Morgan-Herschel letters at the Royal Society, the correspondence extends beyond their shared vocations and Airy's headings to reflect their close friendship and wide range of interests. De Morgan moves easily from mathematics, astronomy and logic to history, bibliography and literature: he describes his articles for the *Penny Cyclopaedia*, advises on bell tuning, confers on writing memorials for their mutual friend Richard Sheepshanks, shares doggerel inspired by Longfellow and Tennyson, 'irate with their transcendental egoism',[54] refers to the 'aereals' and 'aerunculae' (the Airy children), shares puns and wordplay that delight him, and discusses the work of others. De Morgan's close friendship with Airy is not always unquestioning, and in a letter from 31 December 1857 he sounds particularly forward-thinking: 'Some friend of Babbage sneers at you in the Lit: Gaz: for not wanting Scheutz's

---

archivesearch.lib.cam.ac.uk/.
53 Daniel Belteki, 'Papers of George Biddell Airy', Cambridge Digital Library page: https://cudl.lib.cam.ac.uk/collections/rgo6/1
54 CUL, RGO 6/376, 15 Nov. 1855.

[difference] machine at the observatory. Why do you not want it? You are to remember the people who did not want to see Jupiter's satellites & those who did not want to be vaccinated &c &c'.[55]

Three other collections at the library feature De Morgan's letters, of which two mainly concern De Morgan's writing of reviews and articles for journals. The William Thomson, Lord Kelvin papers include ten letters written from 1845 to 1849 about contributions to *The Cambridge and Dublin Mathematical Journal*. The papers also include De Morgan's testimonial for Thomson from June 1846 for the chair of Natural Philosophy at Glasgow.[56] The letters from De Morgan to the journalist and writer William Hepworth Dixon feature nearly forty letters to Dixon on his reviews and articles for *The Athenæum* from the period 1856–1868, with one other letter concerning Dixon standing for Parliament in 1868.[57] Additionally, the papers of William Christie include a duplicate copy of a letter from De Morgan to Airy dated 1 January 1847 in a collection of letters regarding the existence of Neptune and the controversy over the name of the planet between 1846 and 1878.[58]

Trinity College, Cambridge (TCC)

De Morgan researchers consulting the William Whewell papers bequeathed to Trinity College, Cambridge follow in De Morgan's own footsteps. In early March 1847, De Morgan asked Whewell to look in his papers for a particular letter of his which might provide useful evidence in his argument with the Scottish philosopher Sir William Hamilton. On 16 March he wrote again, 'Your practice of keeping letters is most praiseworthy—and you shall rank next to Airy for extreme method....'[59]

Of the seventy-seven letters written by Augustus De Morgan in the Library's Modern Manuscripts collection, all but two have a connection to the William Whewell papers.[60] Whewell spent most of his life at

---

55 CUL, RGO 6/433, 31 Dec. 1857.
56 CUL, MS Add.7342 (William Thomson papers generally); MS Add.7342 Tm3 (testimonial).
57 CUL, MS Add 9428.
58 CUL, RGO 7/247. For the dispute, refer to Chapter 3 in this volume.
59 Trinity College, Cambridge (henceforth TCC), Add.MS.a.202/111.
60 The papers have been catalogued in the Library's archival database, AtoM, which may be found linked from the Library's main page, https://trin.cam.ac.uk/library.

Trinity College Cambridge, from his entrance as a sub-sizar in 1812 to his death in his twenty-fifth year as Master in 1866. His papers include over five thousand letters covering five decades of Whewell's varied mathematical, philosophical, scientific and literary career. This correspondence contains much information on the shared intellectual world inhabited by Whewell and De Morgan, and is anchored by two particularly large runs of letters from their mutual friends Sir George Airy and Sir John Herschel.

The De Morgan letters date from 1832 to 1866, mainly the period from 1845 to 1863. All but seven are addressed to Whewell.[61] The letters document a long friendship and shared interests in mathematics, philosophy, etymology and the work of various scientific organisations, and are notable for their deep dives into mathematics and logic, particularly syllogistic reasoning and the concept of infinity. De Morgan often writes at length about his work in these fields, sharing his latest findings, commenting on the works of Sir William Hamilton and others and requesting Whewell's help in finding appropriate words to express concepts in syllogism. He also actively engages with Whewell's texts, with comments ranging from a proposal that a mathematical problem Whewell thinks he has solved could in fact have an infinite number of answers, to his own theories on the pluralities of worlds and the concept of multipresence.[62] De Morgan's interest in historical research also surfaces, with queries relating to the life of Newton and Bacon, often referring to his own library and asking for information from books in Trinity College Library. There is whimsy, too: the depth and focus of many of De Morgan's letters is often leavened by instances of his love of puns. Concerning a marginal note of Whewell's on a Smith's Prize paper, 'Hast thou appealed unto Caesar &c', De Morgan points out that Whewell has spelled Caesar's name wrongly: 'when an experimental geometer appeals to his geometry it is to "See, sir!" not to Caesar'.[63]

Sixteen of De Morgan's letters to Whewell are printed in Sophia De Morgan's *Memoir*, including one from April 1863 discussing Aristotle on

---

Searching for WHWL in the database will find the collection-level record for the papers.
61 See especially TCC, Add.MS.a.202/95-156.
62 TCC, Add.MS.a.202/115, 126-7 and 147 respectively.
63 TCC, Add.MS.a.202/99.

the infinite, which she regrets is incomplete but which sits complete in the collection.⁶⁴ Many of the printed letters have ellipses to indicate lacunae, but the letter of 12 July 1850 silently omits De Morgan's sharpest words about Sir William Hamilton: 'I sum up the argument as follows—Sir W., not knowing how obscure his programme was, took it for granted that I had copied from it—that is, he thought I was the Father of Lies because he did not know that he himself was the Prince of Darkness.'⁶⁵

A volume in Trinity College Library of copies of William Whewell's letters made by his biographer Isaac Todhunter includes excerpts from eighteen letters written to De Morgan between 1841 and 1864. Many of these letters relate to the De Morgan letters in the Whewell collection, and only one of them is published in Todhunter's memoir.⁶⁶ The collection also includes two letters to Robert Leslie Ellis concerning the four-colour problem and spherical triangles, and four letters about Ellis to his sister, Whewell's wife, Lady Affleck.⁶⁷

Two letters from De Morgan appear in two other collections in the Modern Manuscripts collection. A congratulatory letter to George Peacock on his marriage demonstrates De Morgan's playful nature, with a logical argument in favour of ladies listing both maiden and married names on wedding cards:

> Let AB represent the duration of the lady's life and M the point of marriage ... now, because by common courtesy, a lady is not a discontinuous function, it follows that what is true up to the limit is true at the limit: Therefore, at the instant M, her name is Selwyn. But, for a similar reason, her name at the same moment, is also Peacock. Therefore, at the moment M, she has both names, whence both ought to appear on the wedding cards, Q.E.D.⁶⁸

In addition to the letters, Trinity College has a volume of printed material and letters relating to Richard Sheepshanks created by Augustus De Morgan.⁶⁹ Sheepshanks's *A Letter to the Board of Visitors of the Greenwich*

---

64 S.E. De Morgan, *Memoir*, p. 319; cf. TCC, Add.MS.a.202/149.
65 TCC, Add.MS.a.202/120; cf S.E. De Morgan, *Memoir*, pp. 212–13.
66 Isaac Todhunter, *William Whewell: An Account of his Writings, with Selections from his Literary and Scientific Correspondence* (London: Macmillan, 1876; repr. Cambridge: Cambridge University Press, 2011).
67 TCC, Add.MS.c.67/111-112 and Add.MS.a.202/139-142 respectively.
68 TCC, Add.MS.b.49.65. See S.E. De Morgan, *Memoir*, pp. 202–03.
69 TCC, Adv.c.16.32.

Royal Observatory in Reply to the Calumnies of Mr Babbage ... is represented in the 1854 and 1860 editions, which are closely annotated by De Morgan and bound with letters sent to him by George Airy, Francis Baily, John Herschel, Sheepshanks, James South, and W.H. Smyth, with one letter written by De Morgan to Sheepshanks, retrieved after Sheepshanks' death.

Although De Morgan was an undergraduate at Trinity College, little in the Trinity College Archives relates to De Morgan, and Venn's *Alumni Cantabrigiensis* has gathered almost all available information about him, as for many men of this period.[70] This publication mines the information to be found in the Admissions Book, including father's name, previous school and headmaster, omitting only the name of the assigned tutor, J. P. Higman. More information about De Morgan may conceivably come to light relating to his participation in the musical society CAMUS as an accomplished flautist. It is more probable that further cataloguing of papers in the Modern Manuscripts collections will yield more mentions of De Morgan by his contemporaries.

Edinburgh University Library (EUL)

The De Morgan letters at Edinburgh University Library are a subset of the correspondence to the antiquary and literary scholar James Orchard Halliwell, later Halliwell-Phillipps, some of whose enviable Shakespearean book collection also went to Edinburgh. The two men had known each other from the early 1840s, and in 1845 De Morgan intervened to put an end to the short-lived Historical Society of Science, of which Halliwell was Secretary, without undue pressure on Halliwell.[71] De Morgan purchased copiously from Halliwell's sale of mathematical books in 1840.[72]

---

70 John Venn and J.A. Venn, *Alumni Cantabrigienses: A Biographical List of All Known Students, Graduates and Holders of Office at the University of Cambridge from the Earliest Times to 1900*, 10 vols. (Cambridge: Cambridge University Press, 1922–1954; repr. 2011), pt. 2, vol. 2. Also available online as University of Cambridge, *ACAD: A Cambridge Alumni Database*, https://venn.lib.cam.ac.uk.

71 S.E. De Morgan, *Memoir*, p. 124. For further examples of contact between the two men, see Marvin Spevack, *James Orchard Halliwell-Phillipps: The Life and Works of the Shakespearean Scholar and Bookman* (London: Shepheard-Walwyn, 2001).

72 *Catalogue of a Selected Portion of the Scientific, Historical and Miscellaneous Library of James Orchard Halliwell* ... ([London: S.L. Sotheby, 1840]); see De Morgan's copy of

The twelve letters at Edinburgh from De Morgan to Halliwell date chiefly from 1858 to 1867, with one from 1840. Whereas most of De Morgan's extant non-familial correspondence is with scientific figures, the Halliwell letters are interesting as an example of correspondence with a man who devoted most of his adult life to other areas. The correspondence is cordial. In a letter of 13 November 1859, De Morgan wrote: 'I am greatly pleased to see your handwriting after so many years'.[73] Halliwell may have written about family matters, to judge from a hope in a letter of 3 October 1867 that Halliwell had settled to his satisfaction with his father-in-law, the irascible Sir Thomas Phillipps.[74]

A couple of letters are introductions to Halliwell of other scholars, one of them the Italian mathematical historian Prince Baldassarre Boncompagni: an indication of the far-flung mathematical circles in which De Morgan, without leaving England, moved.[75] In some he thanks Halliwell for a book or tract; such letters add a dimension to the material evidence of the objects in De Morgan's library.[76] A couple are requests to Halliwell to delve into the use of particular words. A couple refer to the 'Macclesfield correspondence', namely *Contents and Index of the Correspondence of Scientific Men of the Seventeenth Century*, compiled by De Morgan (1842). Two mention book circulation, with the sale of Samuel Maynard's books (a few of which De Morgan bought), and books falling into the hands of the booksellers Davis & Dickson. Some letters, like those among De Morgan's correspondence elsewhere, make clear his broad and close reading. He refers to Shakespeare and specifically to Samuel Ayscough's 1790 glossary of Shakespeare, and compares a preface to an almanac with Chaucer's preface to *A Treatise on the Astrolabe*. Perhaps most interesting are comments which document De Morgan's attitude towards books: 'I suppose every book has a history,

---

the catalogue at Senate House Library, University of London, [DeM] Z (B.P.354).
73   Edinburgh University Library (henceforth EUL), LOA 70/8.
74   EUL, LOA 125/37.
75   EUL, LOA 46/15.
76   See Edward Cocker, *Cocker's Arithmetick*, ed. by John Hawkins, 20th edn (London: E. Tracey, 1700), with a letter from Halliwell to De Morgan, n.d., tipped in (SHL, [DeM] L.1 [Cocker] SSR); Christian Wurstisen, *The Elements of Arithmeticke Most Methodically Deliuered*, trans. by Thomas Hood (London: R. Field, 1596), with a letter from Halliwell to De Morgan, 4 Mar. 1865, tipped in (SHL, [DeM] L.1 [Wurstisen] SSR); this latter is the subject of De Morgan's letter to Halliwell of 6 Mar. 1865 (LOA 98/15).

if it was but looked for'; his belief in pasting cuttings into books, on the basis that 'Little things of this kind are often useful in history, in ways which cannot be conjectured until they arise in fact'; and his complaint about finding items within volumes of tracts: 'O these volumes of tracts! They keep safe—and so does the grave!'[77]

Trinity College Dublin (TCD)

De Morgan enters Trinity College Dublin through his correspondence with the Irish astronomer and mathematician Sir William Rowan Hamilton, the discoverer of quaternions. There are over 250 letters from De Morgan in the Hamilton collection in Trinity College, which includes one or two copies of letters sent to him by Hamilton.[78] In date they range from 1841 to 1865, the year of Hamilton's death; the first refers to the two men having met twelve years previously in London, the only time they met. De Morgan eventually wrote Hamilton's obituary.[79]

The scientifically most important texts among the De Morgan-Hamilton correspondence are well known, having been used frequently in publications about both men, their influence on one another, and the manner in which their thinking developed.[80] Reading the entire series of letters, without focusing on scientific nuggets only, reveals something of the sweep of De Morgan's conversational style to the reader. The letters show many of De Morgan's attractive personal traits. His clarity of expression can be understood as being spontaneous and not limited to prepared texts; his kindness and concern for Hamilton, whom he advises against overwork, shines out; his wit and humour, albeit sometimes rather leaden to modern ears, remains. These qualities have

---

77 EUL, LOA 98/15, 95/49 and 25/37, respectively.
78 Trinity College, Dublin, TCD MS 1493. See Manuscript and Archive Online Catalogue (MARLOC), https://manuscripts/catalogue.tcd.ie/CalmView.
79 Augustus De Morgan, 'Sir W.R. Hamilton', *Gentleman's Magazine and Historical Review*, n.s. 1 (1866), 128–34; accessible online at: https://www.maths.tcd.ie/pub/HistMath/People/Hamilton/Gentmag/GentMag.html. De Morgan helped Hamilton during his lifetime: see Charlotte Simmons, 'Augustus De Morgan Behind the Scenes', *College Mathematics Journal*, 42 (2011), 33–40.
80 See on the Hamilton side Robert Perceval Graves, *Life of Sir William Rowan Hamilton: Knt., LL.D., D.C.L., M.R.I.A., Andrews Professor of Astronomy in the University of Dublin, and Royal Astronomer of Ireland* (Dublin: Hodges, Figgis, 1882–89).

been attested to in print, but truly come alive when a reader immerses herself in the full and unedited experience of archival research. The energy in which De Morgan ranges from mathematical conundrum to literary quotation, to reported conversations with other scientists, to high-end gossip, combined with the buzz of seeing his *actual* ink on his *actual* paper, is a reading experience that remains attractive to even the most seasoned of researchers.

The De Morgan-Hamilton correspondence, like other epistolary exchanges, as a documentary genre, opens up various lines of research which do not depend completely on the biographies of the authors. The allusions used in personal, historical records reflect something of the cultural references shared among groups of individuals, and metaphorical agricultural references in the De Morgan letters provide an excellent example. Writing to Hamilton about avoiding anachronistic readings of pre-nineteenth-century scientists, De Morgan cautions Hamilton on 27 January 1853 that 'you have to see with his light' or 'plough with his heifer'.[81] Surprising perhaps to a modern reader, but probably unexceptional among the intelligentsia in what were still heavily agricultural countries.

A further line of research might focus on the editorial process. Right up until late in the twentieth century, the editorial approach to the publication of letters of important individuals favoured the omission of anything personal, as shown above for De Morgan in correspondence at Trinity College, Cambridge. Nowadays, that process itself is the subject of enquiry. In the De Morgan letters in Dublin, too, the blue pencil of the editor reveals what was considered not to be suitable for publication. For example, in the letter of 10 January 1854, following the death of De Morgan's 15-year-old daughter Alice, the reference to an autopsy, requested by De Morgan, has been crossed out. Such significant excisions highlight the possible value of going back to the archives rather than relying on value decisions taken in previous centuries.

Another increasingly important research strand, supported by historical epistolary exchange, is the role of letters in the dissemination of knowledge and in permitting the mapping of historical intellectual networks. Not only does De Morgan's reference to other people's work

---

81   See Graves, v. 3, p. 438. De Morgan further echoes biblical phraseology.

and conversations situate both himself and Hamilton in a constantly growing landscape of shared information, a virtual landscape in its own way, but he also shows his frustration when it does not work as efficiently as it might. In an important letter of 1844, responding to Hamilton's discovery of quaternions the previous year, De Morgan blames John Graves for his delay in engaging with Hamilton on the subject. Hamilton wrote to Graves first upon making his discovery but Graves, in speaking with De Morgan, failed to mention it. De Morgan wrote:

> He never dropped a hint about *imagining* imaginaries. On such little things do our thoughts depend. I do believe that had he said no more than "Hamilton *makes* his imaginary quantities" I should have got what I wanted.[82]

Conclusion

Within each institution and across the repositories, the De Morgan papers provide an important insight into the life, work and acquaintances of Augustus De Morgan. As letters in all the repositories show, they not only testify to his wide-ranging interests and expertise on a range of subjects from mathematics and logic to bibliography, but their style, at times playful, at times serious and questioning, provides a uniform key to his character. De Morgan's correspondents were some of the most prominent scholars in the nineteenth century in their chosen fields, and the nature of the letters reveals that he was on close terms with them. Thus the letters provide evidence of De Morgan's wider social networks and the respect in which his peers held him. They and other manuscript material left by him further demonstrate the importance of the archival record and the unreliability and incompleteness of print to convey it. The sheer quantity of the combined material makes one marvel, as for so many prolific Victorians, at De Morgan's prodigious productivity, especially as his epistolary activity was matched by his output of mathematical articles. The archival record furthermore illumines elements of Victorian culture that extend well beyond Augustus De Morgan, as well as pointing to further topics for research.

---

82 See Graves, v. 3, p. 256.

Bibliography

De Morgan, Sophia Elizabeth, *Memoir of Augustus De Morgan* (London: Longmans, Green, 1882).

—, *Threescore Years and Ten: Reminiscences of the Late Sophia Elizabeth De Morgan, to which are Added Letters to and from her Husband, the Late Augustus De Morgan, and Others*, ed. by Mary A. De Morgan (London: Bentley, 1895).

Graves, Robert Perceval, *Life of Sir William Rowan Hamilton: Knt., LL. D., D.C.L., M.R.I.A., Andrews Professor of Astronomy in the University of Dublin, and Royal Astronomer of Ireland* (Dublin: Hodges, Figgis, 1882–89).

Harris, P.R., *A History of the British Museum Library, 1753–1973* (London: British Library, 1998).

Higgitt, Rebekah, *Recreating Newton: Newtonian Biography and the Making of Nineteenth-Century History of Science* (London: Pickering & Chatto, 2007). https://doi.org/10.4324/ 9781315653068

—, 'Why I don't FRS my Tail: Augustus De Morgan and the Royal Society', *Notes and Records*, 60 (2006), 253–59. https://doi.org/10.1098/rsnr.2006.0150

Merrington, Maxine, *A List of Certain Letters Inserted in Books from the Library of Augustus De Morgan (1806–1871) now in the University of London Library* ([London]: University of London Library, 1990).

Miller, Edward, *Prince of Librarians: The Life and Times of Antonio Panizzi of the British Museum* (London: British Library, 1988).

Pycior, Helena M., 'Early Criticism of the Symbolical Approach to Algebra', *Historia Mathematica*, 9 (1982), 392–412.

Report of the Commissioners Appointed to Enquire into the Constitution and Government of the British Museum (London: William Clowes, 1850).

Rice, Adrian, 'Inspiration or Desperation? Augustus De Morgan's Appointment to the Chair of Mathematics at London University in 1828', *British Journal for the History of Science*, 30 (1997), 257–74. https://doi.org/10.1017/s0007087497003075

—, 'Vindicating Leibniz in the Calculus Priority Dispute: The Role of Augustus De Morgan', in *The History of the History of Mathematics*, ed. by Benjamin Wardhaugh (New York: Lang, 2012), pp. 89–114. https://doi.org/10.3726/978-3-0353-0261-5/7

Richards, Joan, 'Augustus De Morgan, *Formal Logic: or, The Calculus of Inference, Necessary and Probable*', in *Senate House Library, University of London*, ed. by Christopher Pressler and Karen Attar (London: Scala, 2012).

Rye, Reginald Arthur, *Catalogue of the Manuscripts and Autograph Letters in the University Library at the Central Building of the University of London …* (London: University of London Press, 1921).

Simmons, Charlotte, 'Augustus De Morgan Behind the Scenes', *College Mathematics Journal*, 42 (2011), 33–40. https://doi.org/10.4169/college.math.j.42.1.033

Smith, G.C., ed., *The Boole-De Morgan Correspondence* (Oxford: Clarendon Press, 1982).

Smith, James, *The Quadrature of the Circle: Correspondence between an Eminent Mathematician and James Smith* (London: Simpkin, Marshall, 1861).

Spevack, Marvin, *James Orchard Halliwell-Phillipps: The Life and Works of the Shakespearean Scholar and Bookman* (London: Shepheard-Walwyn, 2001).

Stephen, Leslie, and I. Grattan-Guinness, 'De Morgan, Augustus (1806–1871)', in *The Oxford Dictionary of National Biography* (Oxford: Oxford University Press, 2004–). https://doi.org/10.1093/ref:odnb/7470

Theerman, Paul, 'Unaccustomed Role: The Scientist as Historical Biographer: Two Nineteenth-Century Portrayals of Newton', *Biography*, 8 (1985), 145–62. https://doi.org/10.1353/bio.2010.0582

Todhunter, Isaac, *William Whewell: An Account of his Writings, with Selections from his Literary and Scientific Correspondence* (London: Macmillan, 1876; repr. Cambridge: Cambridge University Press, 2011). https://doi.org/10.1017/cbo9781139105415

Venn, John and J.A. Venn, *Alumni Cantabrigienses: A Biographical List of All Known Students, Graduates and Holders of Office at the University of Cambridge from the Earliest Times to 1900*, 10 vols. (Cambridge: Cambridge University Press, 1922–1954; repr. 2011); also available online as University of Cambridge, *ACAD: A Cambridge Alumni Database*, https://venn.lib.cam.ac.uk

Weerden, Anne van, *A Victorian Marriage: Sir William Rowan Hamilton* (Stedum: J. Fransje van Weerden, 2017).

12. Bibliography of the Works of Augustus De Morgan

William Hale

Previous attempts to list De Morgan's voluminous works can be found in Sophia De Morgan's *Memoir* of her husband (London: Longmans, Green, 1882) and in G.C. Smith's edition of *The Boole-De Morgan Correspondence* (Oxford: Clarendon Press, 1982). I have combined and corrected these, and supplemented them chiefly with De Morgan's signed contributions to *The Athenæum* and *Notes and Queries* to produce a bibliography of nearly 500 entries. Space and time precluded the inclusion of the more than 1,000 unsigned reviews which De Morgan contributed to *The Athenæum*; a list of these, extracted from the marked-up copies in the library of City University of London, can be found at https://athenaeum.city.ac.uk/reviews/contributors/contributorfiles/DEMORGAN,Augustus.html.

I have also omitted individual references to the roughly 700 articles he wrote for the *Penny Cyclopaedia* between 1833 and 1843. These are itemised in S.E. De Morgan's *Memoir* (1882), pp. 407–14.

One work notably absent from this bibliography is *On Probability* (*Library of Useful Knowledge*) (London: Baldwin & Cradock, [1830]). This book is still sometimes attributed to De Morgan in library catalogues; according to De Morgan himself, a reprint was mistakenly issued in a binding lettered 'De Morgan On Probability' around 1845.[1] He repeatedly disavowed authorship, notably in *A Budget of Paradoxes*

---

1   Augustus De Morgan, 'Authorship of the Treatise on Probability Published by the Society for the Diffusion of Useful Knowledge', *Assurance Magazine and Journal of the Institute of Actuaries*, 9 (1861), 238.

(1872), pp. 167–68. The book's authors were John William Lubbock and John Elliot Bethune.

The list below is divided into three sections: monographs written solely by De Morgan, contributions by De Morgan to other books and encyclopaedic works, and journal articles. Within sections the ordering is chronological by year of publication as far as possible, though editions of the same work and articles in series are grouped together. I have confined myself to editions published in De Morgan's lifetime, or immediately after his death under the editorship of his widow or daughter, and omitted unchanged reprints of books.[2]

Abbreviations Used in the Bibliography Below

CDMJ – *Cambridge and Dublin Mathematical Journal*

CMJ – *Cambridge Mathematical Journal*

Comp. Alm. – *Companion to the Almanac*

JIA – *Journal of the Institute of Actuaries*

Lond. Edin. Phil. Mag. – *London & Edinburgh Philosophical Magazine*

Lond. Edin. Dubl. Phil. Mag. – *London, Edinburgh & Dublin Philosophical Magazine*

Mon. Notices Royal Astron. Soc. – *Monthly Notices of the Royal Astronomical Society*

NQ – *Notes and Queries*

QJE – *Quarterly Journal of Education*

QJPAM – *Quarterly Journal of Pure and Applied Mathematics*

TCPS – *Transactions of the Cambridge Philosophical Society*

---

2   I am grateful for the sight of an unpublished list of De Morgan's articles by Olivier Bruneau, which allowed me to fill a number of lacunae in this bibliography.

Monographs

The Elements of Arithmetic (London: John Taylor, 1830).

——, 2nd edn, considerably enl. (London: John Taylor, 1832).

——, 3rd edn (London: John Taylor, 1835).

——, 4th edn (London: Taylor & Walton, 1840).

——, 5th edn, enl. (London: Taylor & Walton, 1846).

——, 6th edn (London: Edward Stanford, 1876).

Remarks on Elementary Education in Science: An Introductory Lecture, Delivered at the Opening of the Classes of Mathematics, Physics, and Chemistry in the University of London, November 2, 1830 (London: Thomas Davidson for John Taylor, 1830).

On the Study and Difficulties of Mathematics (*Library of Useful Knowledge. Mathematics*) (London: Baldwin & Cradock, 1831).

Examples of the Processes of Arithmetic and Algebra (*Library of Useful Knowledge. Mathematics*) (London: Baldwin & Cradock, [1831?]).

Elementary Illustrations of the Differential and Integral Calculus (*Library of Useful Knowledge. Mathematics*) (London: Baldwin & Cradock, 1832).

The Elements of Spherical Trigonometry (*Library of Useful Knowledge. Mathematics*) (London: Baldwin & Cradock, 1834).

The Elements of Algebra Preliminary to the Differential Calculus, and Fit for the Higher Classes of Schools in which the Principles of Arithmetic are Taught (London: John Taylor, 1835).

——, 2nd edn (London: Taylor & Walton, 1837).

The Connexion of Number and Magnitude: An Attempt to Explain the Fifth Book of Euclid (London: Taylor & Walton, 1836).

An Explanation of the Gnomonic Projection of the Sphere: And of Such Points of Astronomy As Are Most Necessary in the Use of Astronomical Maps (London: Baldwin & Cradock, 1836). 'Published under the superintendence of the Society for the Diffusion of Useful Knowledge.'

The Elements of Trigonometry and Trigonometrical Analysis Preliminary to the Differential Calculus (London: Taylor & Walton, 1837).

Thoughts Suggested by the Establishment of the University of London: An Introductory Lecture Delivered at the Opening of the Faculty of Arts in University College, Oct. 16, 1837 (London: Taylor & Walton, 1837).

An Essay on Probabilities and on their Application to Life Contingencies and Insurance Offices (*The Cabinet Cyclopaedia. Natural Philosophy*) (London: Longman, Orme, Brown, Green & Longmans, and John Taylor, 1838).

——, New edn (London: Longman, Orme, Brown, Green & Longmans, and John Taylor, 1841).

——, New edn (London, 1849).

Remarks on an Accusation Made by the Proprietors of the Encyclopaedia Metropolitana, Against the Author of An Essay on Probabilities... (London: for the author, and sold by Taylor & Walton, 1838).

First Notions of Logic Preparatory to the Study of Geometry (London: Taylor & Walton, 1839).

——, 2nd edn (London: Taylor & Walton, 1840).

The Differential and Integral Calculus (*Library of Useful Knowledge*) (London: Baldwin & Cradock, 1842). Published in 25 instalments between 1836 and 1842.

The Globes Celestial and Terrestrial (London: Malby, 1845). '... intended to accompany Malby's Globes, published under the superintendence of the Society for the Diffusion of Useful Knowledge'.

——, 2nd edn (London: William S. Orr, 1847).

——, 3rd edn (London: William S. Orr, 1854).

Arithmetical Books from the Invention of Printing to the Present Time (London: Taylor & Walton, 1847).

Formal Logic, or The Calculus of Inference Necessary and Probable (London: Taylor & Walton, 1847).

Statement in Answer to an Assertion Made by Sir William Hamilton, Bart., Professor of Logic in the University of Edinburgh (London: Richard and John E. Taylor for the author, 1847).

Trigonometry and Double Algebra (London: Taylor, Walton, & Maberly, 1849).

The Book of Almanacs: With an Index of Reference, By Which the Almanac may be Found for Every Year ... (London: Taylor, Walton, & Maberly, 1851).

——, 2nd edn (London: James Walton, 1871).

Observations in Favour of a Decimal Coinage ([London: M.S. Rickerby for the Decimal Association, 1854]). A reprint of the Introduction to Decimal Association, *Proceedings* (1854); see below.

Summary of the Decimal Coinage Question (London: M.S. Rickerby [for the Decimal Association], 1855). Another reprint of the Introduction to Decimal Association, *Proceedings* (1854); see below.

Reply to the Facetiae of the Member for Kidderminster in the Debate on Mr. Brown's Resolutions, June 12, 1855 (London: M.S. Rickerby [for the Decimal Association, 1855]). Also printed in *Debate on the Decimal Coinage Question* (1855); see below.

Answers to the Questions Communicated by Lord Overstone to the Decimal Coinage Commissioners. No. 1 (London: M.S. Rickerby [for the Decimal Association], Nov. 1857).

Syllabus of a Proposed System of Logic (London: Walton & Maberly, 1860).

A Budget of Paradoxes, ed. by S.E. De Morgan (London: Longmans, Green, 1872). See below for the original articles from *The Athenæum*.

De Morgan, S.E., *Memoir of Augustus De Morgan: With Selections From his Letters* (London: Longmans, Green, 1882).

Newton, His Friend, and His Niece, ed. by S.E. De Morgan and A.C. Ranyard (London: Elliot Stock, 1885).

De Morgan, S.E., *Threescore Years and Ten: Reminiscences... To Which are Added Letters To and From... A. De Morgan,* ed. by Mary A. De Morgan (London: Richard Bentley & Son, 1895).

Contributions to Other Works

Bourdon, L.P.M., *The Elements of Algebra,* trans. from the three first chapters [of *Élémens d'algèbre*] ... by A. De Morgan (London: John Taylor, 1828).

Approximately 700 articles in *The Penny Cyclopaedia* (London: Charles Knight, 1833–1843). Listed in S.E. De Morgan's *Memoir* (1882), pp. 407–14.

'Bradley', 'Delambre', 'Descartes', 'Dollond', 'Euler', 'Halley', 'Harrison', 'W. Herschel', 'Lagrange', 'Laplace', 'Leibniz', 'Maskelyne', in *The Gallery of Portraits: with Memoirs* (London: Charles Knight, 1833–37).

'Calculus of Functions', in *Encyclopaedia Metropolitana,* vol. 2 (London: Baldwin & Cradock [&c.], 1836), pp. 305–92. Also issued separately as *A Treatise on the Calculus of Functions.*

'Theory of Probabilities', in *Encyclopaedia Metropolitana,* vol. 2 (London: Baldwin & Cradock [&c.], 1837), pp. 393–490. Also issued separately as *A Treatise on the Theory of Probabilities.*

Rigaud, S.P. and S.J. Rigaud, eds, *Correspondence of Scientific Men of the Seventeenth Century,* Contents and Index by A. De Morgan (Oxford: University Press, 1842).

'Newton', in *Cabinet Portrait Gallery of British Worthies,* vol. XI (London: Charles Knight, 1846), 78–117.

'Mr Babbage's Calculating Machine', in Weld, Charles Richard, *The Eleventh Chapter of the History of the Royal Society* (London: Richard Clay [for Charles Babbage], 1848). Reprint of De Morgan's review in *The Athenæum,* 14 Oct. 1848, with notes by Babbage. A subsequent printing included De Morgan's response, also from *The Athenæum*, 16 Dec. 1848. Republished in Babbage's *The Exposition of 1851* (London: John Murray, 1851).

'Diophantus', 'Eucleides', 'Heron', 'Hipparchus', 'Sosigenes', 'Theon', 'Ptolomaeus', in Smith, William, ed., *A New Classical Dictionary of Biography, Mythology, and Geography* (London: John Murray; Taylor, Walton, & Maberly, 1850).

Decimal Association, *Proceedings*, with introd. by A. De Morgan (London: M.S. Rickerby [for the Association], 1854). Reprinted 1855. Introduction issued as *Summary of the Decimal Coinage Question* (1855); see above.

Debate on the Decimal Coinage Question in the House of Commons, June 12th, 1855. With Remarks [by A. De Morgan] on the Speech of the Hon. Member for Kidderminster (London: M.S. Rickerby [for the Decimal Association], 1855). *Remarks* also issued separately as *Reply to the Facetiae* (1855); see above.

Baily, Francis, *Journal of a Tour in Unsettled Parts of North America in 1796 & 1797*, ed. with preface, by A. De Morgan (London: Baily Brothers, 1856).

Ramchundra, *A Treatise on the Problems of Maxima and Minima, Solved by Algebra*, ed. with introd. by A. De Morgan (London: Wm. H. Allen, 1859).

'Logic', in Knight, Charles, ed., *English Cyclopaedia*, vol. 5 (London: Bradbury and Evans, 1861), cols. 340–54.

'Tables', in Knight, Charles, ed., *English Cyclopaedia*, vol. 7 (London: Bradbury and Evans, 1861), cols. 976–1016.

C. D. [i.e. S.E. De Morgan], *From Matter to Spirit: The Result of Ten Years Experience in Spirit Manifestations*, pref. by A.B. [i.e. A. De Morgan] (London: Longman, Green, Longman, Roberts & Green, 1863).

'Note on the Construction of the Table', in Oakes, W.H., *Table of the Reciprocals of Numbers from 1 to 100,000* (London: Charles & Edwin Layton, [1865]).

Schrön, Ludwig, *Seven-Figure Logarithms of Numbers From 1 to 108000 and of Sines, Cosines, Tangents, Cotangents to Every 10 Seconds of the Quadrant*, 5th edn, corr. and stereotyped, with a description of the tables added by A. De Morgan (London: Williams and Norgate, 1865).

Articles

'Life Assurance', *Comp. Alm.* (1831), 86–105.

'Polytechnic School of Paris', *QJE*, 1 (1831), 57–74.

'Notice of Some Tables of Different Species for Facilitating Calculation [review]', *QJE*, 1 (1831), 129–37.

'Pinnock's *Catechisms* [review]', *QJE*, 1 (1831), 179–83.

'Walker's *Theory of Mechanics* [review]', *QJE*, 1 (1831), 193–296.

'On Mathematical Instruction', *QJE*, 1 (1831), 264–79.

'Bayley's *Elements of Algebra* [review]', *QJE*, 2 (1831), 155–58.

'Darley's *System of Popular Geometry* [review]', *QJE*, 2 (1831), 336–44.

'Eclipses', *Comp. Alm.* (1832), 5–13.

'*A Plan for Conducting the Royal Naval School* [review]', *QJE*, 3 (1832), 42–49.

'Study of Natural Philosophy', *QJE*, 3 (1832), 60–73.

'*A Preparation for Euclid* [review]', *QJE*, 3 (1832), 129–32.

'Barlow's *Mathematical Tables* [review]', *QJE*, 3 (1832), 158–60.

'On Some Methods Employed for the Instruction of the Deaf and Dumb', *QJE*, 3 (1832), 203–19.

'Wood's *Algebra* [review]', *QJE*, 3 (1832), 276–85.

'Quetelet on Probabilities [review]', *QJE*, 4 (1832), 101–09.

'Young's *Elements of Mechanics* [review]', *QJE*, 4 (1832), 116–24.

'State of the Mathematical and Physical Sciences in the University of Oxford', *QJE*, 4 (1832), 191–208.

'Von Türk's *Phenomena of Nature* [review]', *QJE*, 4 (1832), 332–36.

'On Comets', *Comp. Alm.* (1833), 5–15.

'Demonstration of the Preceding Method' *annexed to* Zach, F.X. von 'A New Method of Reducing the Apparent Distance of the Moon', *Memoirs of the Royal Astronomical Society*, 5 (1833), 245–52.

'On Teaching Arithmetic', *QJE*, 5 (1833), 1–16.

'Cunningham's *Arithmetic* [review]', *QJE*, 5 (1833), 129–30.

'On the Method of Teaching Fractional Arithmetic', *QJE*, 5 (1833), 210–22.

'On the Method of Teaching the Elements of Geometry, pt. 1', *QJE*, 6 (1833), 35–49.

'——, pt. 2', *QJE*, 6 (1833), 237–51.

'*The School and Family Manual* [review]', *QJE*, 6 (1833), 108–12.

'Busby's *Catechism of Music* [review]', *QJE*, 6 (1833), 318–26.

'On the General Equation of Curves of the Second Degree', *TCPS*, 4 (1833), 71–78 (read 15 Nov. 1830).

'On the Moon's Orbit', *Comp. Alm.* (1834), 5–23.

'*Geometry Without Axioms* [review]', *QJE*, 7 (1834), 105–15.

'Ritchie's *Principles of Geometry* [review]', *QJE*, 7 (1834), 118–25.

'Elementary Works by M. Quetelet [review]', *QJE*, 7 (1834), 347–50.

'On the Notation of the Differential Calculus, Adopted in Some Works Lately Published at Cambridge', *QJE*, 8 (1834), 100–10.

'Airy's *Gravitation* [review]', *QJE*, 8 (1834), 316–25.

'Reports of the British Association. Vols. I. and II. London: 1833 and 1834 [review; sometimes cited as "English Science"]', *The British and Foreign Review* (1835), 1, 134–57.

'Halley's Comet', *Comp. Alm.* (1835), 5–15.

'On Taylor's theorem', *Lond. Edin. Phil. Mag.*, 3rd ser., 7 (1835), 188–92.

'Peacock's *Treatise on Algebra* [review], *QJE*, 9 (1835), 91–110, 293–311.

'Thompson's *Progress of Physical Science* [review]', *QJE*, 10 (1835), 122–34.

'Ecole polytechnique', *QJE*, 10 (1835), 330–40.

'On the General Equation of Surfaces of the Second Degree', *TCPS*, 5 (1835), 77–94 (read 12 Nov. 1832).

'Old Arguments Against the Motion of the Earth', *Comp. Alm.* (1836), 5–20.

'On the Relative Signs of Coordinates' *Lond. Edin. Phil. Mag.*, 3rd ser., 9 (1836), 249–54.

'Notices of English Mathematical and Astronomical Writers Between the Norman Conquest and the Year 1600', *Comp. Alm.* (1837), 21–44.

'Theorie Analytique des Probabilités. Par M. Le Marquis de Laplace. &c. 3ième edn. 1820 [review]', *The Dublin Review*, 2 (Apr. 1837), 338–54, 3 (July 1837), 237–48.

'The Mathematics: Their Value in Education', *Central Society of Education, Publication*, 1 (1837), 114–44.

'On Cavendish's Experiment', *Comp. Alm.* (1838), 26–43.

'On the Solid Polyhedron', *Lond. Edin. Phil. Mag.*, 3rd ser., 12 (1838), 323–24.

'Professional Mathematics', *Central Society of Education, Publication*, 2 (1838), 132–47.

'Sketch of a Method of Introducing Discontinuous Constants into the Arithmetical Expressions for Infinite Series, in Cases where they Admit of Several Values', *TCPS*, 6 (1838), 185–93 (read 16 May 1836).

'On a Question in the Theory of Probabilities', *TCPS*, 6 (1838), 423–30 (read 26 Feb. 1837).

'Notices of the Progress of the Problem of Evolution', *Comp. Alm.* (1839), 34–52.

'On the Rule for Finding the Value of an Annuity on Three Lives', *Lond. Edin. Phil. Mag.*, 3rd ser., 15 (1839), 337–39. Reprinted in *The Assurance Magazine and Journal of the Institute of Actuaries*, 10 (1861), 27–28.

'On the Calculation of Single Life Contingencies', *Comp. Alm.* (1840), 5–24.

'Prospectuses of New Life Assurance Companies [The Necessity of Legislation for Life Assurance]', *Dublin Review*, 9 (Aug. 1840), 49–88.

'Description of a Calculating Machine Invented by Mr. T. Fowler', *Lond. Edin. Phil. Mag.*, 3rd ser., 17 (1840), 385–86. Summary. Reprinted in *Proceedings of the Royal Society*, 4 (1843), 243–44.

'On the Perspective of the Coordinate Planes', *CMJ*, 2 (1841), 92–93 (Feb. 1840).

'On a Simple Property of the Conic Sections', *CMJ*, 2 (1841), 202–03.

'A Short Mode of Reducing the Square Root of a Number to a Continued Fraction', *CMJ*, 2 (1841), 239–40.

'On the Use of Small Tables of Logarithms in Commercial Calculations, and on the Practicability of a Decimal Coinage', *Comp. Alm.* (1841), 5–21.

'Jones' *Value of Annuities and Reversionary Payments* [review]', *Dublin Review*, 11 (Aug. 1841), 104–33.

'Peyrard's *Elements of Euclid* [review]', *Dublin Review*, 11 (Nov. 1841), 330–55.

'A Suggestion Relative to Barrett's Method of Computing the Values of Life Contingencies', *Lond. Edin. Dubl. Phil. Mag.*, 3rd ser., 18 (1841), 268–70.

'On the History of Fernel's Measure of a Degree', *Lond. Edin. Dubl. Phil. Mag.*, 3rd ser., 19 (1841), 445–47.

'——, Additional Note', *Ibid.*, 3rd ser., 20 (1842), 116–17.

'——, In Reply to Mr Galloway's Remarks', *Ibid.*, 3rd ser., 20 (1842), 230–33.

'——, [a further note on the subject]' *Ibid.*, 3rd ser., 20 (1842), 408–11.

'On Life Contingencies, No. 2' *Comp. Alm.* (1842), 1–19.

'*Report of the Commissioners Appointed to Consider the Steps to be Taken for Restoration of the Standards of Weights and Measures* [review]', *Dublin Review*, 12 (May 1842), 466–93.

'Science and Rank: London, 1842', *Dublin Review*, 13 (Nov. 1842), 413–48.

'On the Invention of the Signs + and –; And on the Sense in which the Former was Used by Leonardo da Vinci', *Lond. Edin. Dubl. Phil. Mag.*, 3rd ser., 20 (1842), 135–57.

'On the Foundation of Algebra', *TCPS*, 7 (1842), 173–87 (read 9 Dec. 1839).

'——, no. II', *TCPS*, 7 (1842), 287–300 (read 29 Nov. 1841).

'——, no. III', *TCPS*, 8 (1849), 139–42 (read 27 Nov. 1843).

'——, no. IV, On Triple Algebra', *TCPS*, 8 (1849), 241–54 (read 28 Oct. 1844).

'Remarks on the Binomial Theorem', *CMJ*, 3 (1843), 61–62 (Nov. 1841).

'References for the History of the Mathematical Sciences', *Comp. Alm.* (1843), 40–65.

'On the Invention of the Circular Parts', *Lond. Edin. Dubl. Phil. Mag.*, 3rd ser., 22 (1843), 350–53.

'On the Almost Total Disappearance of the Earliest Trigonometrical Canon', *Mon. Notices Royal Astron. Soc.*, 6, no. 15 (April 1843), 221–28. Reprinted with addition in *Lond. Edin. Dubl. Phil. Mag.*, 3rd ser., 26 (1845), 517–26.

'Easter-Day, 1845', *The Athenæum*, 872 (13 Jul. 1844), 646.

'On Arithmetical Computation', *Comp. Alm.* (1844), 1–22.

'On the Reduction of a Continued Fraction to a Series', *Lond. Edin. Dubl. Phil. Mag.*, 3rd ser., 24 (1844), 15–17.

'On the Equation $(D + a)^n y = X$', *CMJ*, 4 (1845), 60–62 (Feb. 1844).

'——, Addendum', *CMJ*, 4 (1845), 96 (May 1844).

'On a Law Existing in the Successive Approximations to a Continued Fraction', *CMJ*, 4 (1845), 97–99 (May 1844).

'On the Ecclesiastical Calendar', *Comp. Alm.* (1845), 1–36.

'Baily's Repetition of the Cavendish Experiment on the Mean Density of the Earth &c. [review of books]' *Dublin Review*, 18 (Mar. 1845), 75–112.

'Speculators and Speculations [review of books]', *Dublin Review*, 19 (Sept. 1845), 99–129.

'Book-keeping [review of books]', *Dublin Review*, 19 (Dec. 1845), 433–53.

'On Arbogast's Formulae of expansion', *CDMJ*, 1 (1846), 238–55.

'On a Point Connected with the Dispute Between Keil and Leibnitz About the Invention of Fluxions', *Philosophical Transactions of the Royal Society of London*, 136 (1846), 107–09 (read Jan. 29 1846).

'On the Earliest Printed Almanacs', *Comp. Alm.* (1846), 1–31.

'Mathematical Bibliography [review of Hain's *Repertorium Bibliographicum* and Panizzi's *Catalogue of the Scientific Books in the Library of the Royal Society*]' *Dublin Review*, 21 (Sept. 1846), 1–37.

'On the Derivation of the Word *Theodolite*', *Lond. Edin. Dubl. Phil. Mag.*, 3rd ser., 28 (1846), 287–89.

'On the First Introduction of the Words *Tangent* and *Secant*', *Lond. Edin. Dubl. Phil. Mag.*, 3rd ser., 28 (1846), 382–87.

'Professor De Morgan on the Syllogism', *Athenæum*, 1026 (26 Jun. 1847), 671.

'Recurrences of Eclipses and Full Moons', *Comp. Alm.* (1847), 53–55.

'On Helps to Calculation [review of 'Tables', in *Supplement to the Penny Cyclopaedia*]', *Dublin Review*, 22 (Mar. 1847), 74–92.

'On the Opinion of Copernicus with Respect to the Light of the Planets', *Mon. Notices Royal Astron. Soc.*, 7 (1847), 290–94.

'A New-Year's Puzzle', *Athenæum*, 1053 (1 Jan. 1848), 14; no. 1061 (26 Feb. 1848), 215.

'Punning Dates', *Athenæum*, 1084 (5 Aug. 1848), 772.

'On the effects of Competitory Examinations, Employed as Instruments in Education', *Athenæum*, 1096 (28 Oct. 1848), 1076–77. Reprinted in *The Educational Times* (1 Dec. 1848), 56–59.

'Suggestion on the Integration of Rational Fractions', *CDMJ*, 3 (Nov. 1848), 238–42.

'On Decimal Coinage', *Comp. Alm.* (1848), 5–21.

'An Account of the Speculations of Thomas Wright of Durham', *Lond. Edin. Dubl. Phil. Mag.*, 3rd ser., 32 (1848), 241–52.

'On the Additions Made to the Second Edition of the *Commercium Epistolicum*', *Lond. Edin. Dubl. Phil. Mag.*, 3rd ser., 32 (1848), 446–56.

'On a Property of the Hyperbola', *Lond. Edin. Dubl. Phil. Mag.*, 3rd ser., 33 (1848), 546–48.

'Short Supplementary Remarks on the First Six Books of Euclid's Elements', *Comp. Alm.* (1849), 5–20.

'On a New Species of Equations of Differences', *CDMJ*, 4 (1849), 87–90.

'On a Point in the Solution of Linear Differential Equations', *CDMJ*, 4 (1849) 137–39.

'On Anharmonic Ratio', *Lond. Edin. Dubl. Phil. Mag.*, 3rd ser., 35 (1849), 165–71.

'On Divergent Series and Various Points of Analysis Connected with Them', *TCPS*, 8 (1849), 182–203 (read 4 Mar. 1844).

'Methods of Integrating Partial Differential Equations', *TCPS*, 8 (1849), 606–13 (read 5 June 1848).

'On the Structure of the Syllogism, and on the Application of the Theory of Probabilities to Questions of Argument and Authority [On the Syllogism, no. I]', *TCPS*, 8 (1849), 379–408 (read 9 Nov. 1846).

'On the Symbols of Logic, the Theory of the Syllogism, and In Particular of the Copula, And the Application of the Theory of Probabilities to Some Questions of Evidence [On the Syllogism, no. II]', *TCPS*, 9 (1856), 79–127 (Read 25 Feb. 1850).

'On the Syllogism, no. III, and on Logic in General', *TCPS*, 10 (1864), 173–230 (Read 8 Feb. 1858).

'On the Syllogism, no. IV, and on the Logic of Relations', *TCPS*, 10 (1864), 331–58 (read 23 Apr. 1860).

'Appendix: On the Syllogism of Transposed Quantity', *TCPS*, 10 (1864), 355*–58*.

'On the Syllogism, no. V, and on Various Points of the Onymatic System', *TCPS*, 10 (1864), 428–88 (read 4 May 1863).

'Syllogistic Systems', *Athenæum*, 1192 (31 Aug.1850), 927.

'On Ancient and Modern Usage in Reckoning', *Comp. Alm.* (1850), 5–34.

'Extension of the Word "Area"', *CDMJ*, 5 (May 1850), 139–42.

'Remark on the General Equation of the Second Degree', *The Mathematician*, 3 (1850), 154–55.

'——, Supplement', *Ibid.*, 3, Suppl. (Nov. 1850), 4–5.

'An Organized Method of Making the Resolution Required in the Integration of Rational Fractions', *Mathematician*, 3 (1850), 242–46.

'Remark on Horner's Method of Solving Equations', *Mathematician*, 3 (1850), 289–91.

'The Geometrical Foot', *NQ*, 2 (1850), 133.

'Engelman's *Bibliotheca Scriptorum Classicorum*', *NQ*, 2 (1850), 296, 328.

'On the Equivalence of Compound Interest with Simple Interest Paid When Due', *The Assurance Magazine*, 1:4 (1851), 335–36.

'Application of Combinations to the Explanation of Arbogast's Method', *CDMJ*, 6 (1851), 35–37.

'On the Mode of Using the Signs + and – in Plane Geometry', *CDMJ*, 6 (1851), 156–60.

'On the Connexion of Involute and Evolute in Space', *CDMJ*, 6 (1851), 267–74.

'On Some Points in the History of Arithmetic', *Comp. Alm.* (1851), 5–18.

'[Remarks upon the Gregorian Calendar]', *Mon. Notices Royal Astron. Soc.*, 11 (1851), 147–48.

'Difficulty of Getting Rid of a Name [signed: M]', *NQ*, 4 (1851), 173.

'Spurious Edition of Baily's *Annuities*', *NQ*, 4 (1851), 19-20; 8 (1853), 242.

'Note on the Calendar', *NQ*, 4 (1851), 218.

'On a Method of Checking Annuity Tables at Different Rates of Interest by Help of One Another', *Assurance Magazine*, 2 (1852), 380–91.

'The Case of M. Libri', *Bentley's Miscellany*, 32 (1852), 107–15.

'On Partial Differential Equations of the First Order', *CDMJ*, 7 (Feb. 1852), 28–35.

'On the Signs + and – in Geometry (Continued) and On the Interpretation of the Equation of a Curve', *CDMJ*, 7 (Nov. 1852), 242–51.

'A Short Account of some Recent Discoveries in England and Germany Relative to the Controversy on the Invention of Fluxions', *Comp. Alm.* (1852), 5–20.

'On Indirect Demonstration', *Lond. Edin. Dubl. Phil. Mag.*, 4th ser., 4 (1852), 435–38.

'On the Authorship of the "Account of the *Commercium Epistolicum*" Published in the *Philosophical Transactions*', *Lond. Edin. Dubl. Phil. Mag.*, 4th ser., 3 (1852), 440–44.

'On the Early History of Infinitesimals in England', *Lond. Edin. Dubl. Phil. Mag.*, 4th ser., 4 (1852), 321–30.

'Query on the Controversy about Fluxions', *NQ*, 5 (1852), 103.

'James Wilson, M.D.', *NQ*, 5 (1852), 276, 399–400.

'Francis Walkinghame', *NQ*, 5 (1852), 441.

'Book of Almanacs', *NQ*, 5 (1852), 519.

'Mathematical Note', *CDMJ*, 8 (1853), 93-4.

'On the Difficulty of Correct Descriptions of Books', *Comp. Alm.* (1853), 5–19.

'Some Suggestions in Logical Phraseology', *Proceedings of the Philological Society*, 6 (1853), 27–30.

'Thomas Wright of Durham', *NQ*, 8 (1853), 218.

'Attainment of Majority', *NQ*, 8 (1853), 250–51, 372.

'Lord Halifax and Mrs. Catherine Barton', *NQ*, 8 (1853), 429–33; 2nd ser., 2 (1856), 161–63.

'On a Decimal Coinage', *Comp. Alm.* (1854), 5–15.

'Account of a Correspondence Between Mr George Barrett and Mr Francis Baily', *Assurance Magazine*, 4 (1854), 185–99.

'On the Demonstration of Formulae Connected with Interest and Annuities', *Assurance Magazine*, 4 (1854), 277–82.

'Geometrical Curiosity', *NQ*, 9 (1854), 14–15.

'*Book of Almanacs*', *NQ*, 9 (1854), 561.

'Mathematical Bibliography', *NQ*, 10 (1854), 47–48.

'Christopher Clavius', *NQ*, 10 (1854), 158–59.

'Southey and Voltaire', *NQ*, 10 (1854), 282.

'Boswell's Arithmetic', *NQ*, 10 (1854), 363–64.

'On Some Questions of Combination', *Assurance Magazine*, 5 (1855), 93–99.

'The Progress of the Doctrine of the Earth's Motion Between the Times of Copernicus and Galileo: Being Notes on the Ante-Galilean Copernicans', *Comp. Alm.* (1855), 5–25.

'*Memoirs of the Life, Writings and Discoveries of Sir Isaac Newton,* by Sir David Brewster [review]', *North British Review*, 23 (1855), 308–38.

'Arithmetical Notes, No. I', *NQ*, 11 (1855), 57–58.

'——, No. II', *NQ*, 12 (1855), 4–5.

'——, No. III', *NQ*, 12 (1855), 117–18.

'——, No. IV', *NQ*, 12 (1855), 237–38.

'New Moon', *NQ*, 11 (1855), 235.

'Books on Logic', *NQ*, 11 (1855), 332–33.

'Anticipated Inventions, Etc.', *NQ*, 11 (1855), 504–05.

'Original Correspondence', *NQ*, 12 (1855), 57–58.

'Ebrardus and Johannes de Garlandia', *NQ*, 12 (1855), 93.

'Fly-Leaves of Books: Reuben Burrow', *NQ*, 12 (1855), 142–43.

'Length of Miles', *NQ*, 12 (1855), 195.

'A Possible Test of Authorship', *NQ*, 12 (1855), 181–82.

'Reward for the Quadrature of the Circle', *NQ*, 12 (1855), 306–07.

'Selden's "Table-Talk"', *NQ*, 12 (1855), 426.

'Notes on the History of the English Coinage', *Comp. Alm.* (1856), 5–21.

'Bayle and His Continuers', *NQ*, 2nd ser., 1 (1856), 306.

'Musical Notation', *NQ*, 2nd ser., 2 (1856), 14–15.

'Means of Reading the Logic of Aristotle', *NQ*, 2nd ser., 2 (1856), 81–82.

'Pound and Mil Scheme', *NQ*, 2nd ser., 2 (1856), 112–13.

'Organ Tuning', *NQ*, 2nd ser., 2 (1856), 190.

'Corn Measures', *NQ*, 2nd ser., 2 (1856), 196–97.

'The Moon's Rotation', *NQ*, 2nd ser., 2 (1856), 208.

'Oxford Edition of Pappus', *NQ*, 2nd ser., 2 (1856), 227–28.

'The New Atlantis', *NQ*, 2nd ser., 2 (1856), 265.

'John Churchill and the Duchess of Cleveland', *NQ*, 2nd ser., 2 (1856), 463.

'On Some Points of the Integral Calculus', *TCPS*, 9 (1856), pt. 2, 107–38 (Read 24 Feb. 1851).

'On Some Points in the Theory of Differential Equations', *TCPS*, 9 (1856), pt. 4, 515–47 (read 27 Mar. 1854).

'On the Singular Points of Curves and on Newton's Method of Coordinated Exponents', *TCPS*, 9 (1856), pt. 4, 608–27 (read 21 May 1855).

'Notes on the State of the Decimal Coinage Question', *Comp. Alm.* (1857), 5–19.

'Newton's Nephew, the Rev. B. Smith', *NQ*, 2nd ser., 3 (1857), 41–42.

'———; The New Atlantis; Lord Halifax and Mrs. C. Barton', *NQ*, 2nd ser., 3 (1857), 250–52.

'Impossible Problems', *NQ*, 2nd ser., 3 (1857), 272–75.

'Abbreviation Wanted', *NQ*, 2nd ser., 4 (1857), 5.

'Musical Acoustics: Greek Geometers', *NQ*, 2nd ser., 4 (1857), 14.

'Quadrature of the Circle', *NQ*, 2nd ser., 4 (1857), 153; 7 (1859), 433.

'Divination', *NQ*, 2nd ser., 4 (1857), 186.

'Rue at the Old Bailey', *NQ*, 2nd ser., 4 (1857), 238.

'Butler's *Hudibras*', *NQ*, 2nd ser., 4 (1857), 229–30.

'Book-Dust', *NQ*, 2nd ser., 4 (1857), 241–43, 281–83, 301–02.

'Notes on Books', *NQ*, 2nd ser., 4 (1857), 305.

'Dr. Johnson and Dr. Maty', *NQ*, 2nd ser., 4 (1857), 341.

'Church Leases', *NQ*, 2nd ser., 4 (1857), 361.

'Hutchinsonianism', *NQ*, 2nd ser., 4 (1857), 386–87.

'On the Dimensions of the Roots of Equations', *QJPAM*, 1 (1857), 1-3, 80.

'On Fractions of Vanishing or Infinite Terms', *QJPAM*, 1 (1857), 204–09.

'Historical Note on the Theorem Respecting the Dimensions of Roots', *QJPAM*, 1 (1857), 232–35.

'Note on Euclid i, 47', *QJPAM*, 1 (1857), 236–37.

'"Rum"', *NQ*, 2nd ser., 5 (1858), 245.

'Desiderius Erasmus: The Ciceronianus', *NQ*, 2nd ser., 6 (1858), 8–9.

'*Epistolae obscurorum virorum*', *NQ*, 2nd ser., 6 (1858), 22–24, 41–42; 8 (1860), 375.

'Swift: *Gulliver's Travels*', *NQ*, 2nd ser., 6 (1858), 123–26, 251–53.

'Game of "One-and-Thirty"', *NQ*, 2nd ser., 6 (1858), 159.

'Newton's Apple', *NQ*, 2nd ser., 6 (1858), 169–71.

'Berners Street Hoax', *NQ*, 2nd ser., 6 (1858), 179.

'An Assailant of the Mathematical Sciences', *NQ*, 2nd ser., 6 (1858), 209.

'"P.M.A.C.F."', *NQ*, 2nd ser., 6 (1858), 279.

'The Midshipman's Three Dinners', *NQ*, 2nd ser., 6 (1858), 264–65.

'Greatness in Different Things', *NQ*, 2nd ser., 6 (1858), 292–94.

'Napier's Bones', *NQ*, 2nd ser., 6 (1858), 381.

'Chess Calculus', *NQ*, 2nd ser., 6 (1858), 435.

'Albini the Mathematician', *NQ*, 2nd ser., 6 (1858), 440.

'Something To Be Said on Both Sides', *NQ*, 2nd ser., 6 (1858), 480.

'On the Integrating Factor of $Pdx + Qdy + Rdz$', *QJPAM*, 2 (1858), 323–26.

'On the Classification of Polygons of a Given Number of Sides', *QJPAM*, 2 (1858), 340–41.

'Mr Hallam', *Athenæum*, 1632 (5 Feb. 1859), 188.

'Lexell's Comet', *NQ*, 2nd ser., 7 (1859), 13.

'Medicine', *NQ*, 2nd ser., 7 (1859), 23–24.

'Rising of the Lights', *NQ*, 2nd ser., 7 (1859), 138–39.

'Weapon Salve', *NQ*, 2nd ser., 7 (1859), 299–301, 402.

'Archbishop Neile', *NQ*, 2nd ser., 7 (1859), 346.

'Sundry Replies', *NQ*, 2nd ser., 8 (1859), 190.

'Synonymes', *NQ*, 2nd ser., 8 (1859), 224–25.

'The Great Exhibition of 1851', *NQ*, 2nd ser., 8 (1859), 299–301.

'Hypatia', *NQ*, 2nd ser., 8 (1859), 277.

'Bocardo', *NQ*, 2nd ser., 8 (1859), 270.

'Book-Markers', *NQ*, 2nd ser., 8 (1859), 301.

'Francis Burgersdicius', *NQ*, 2nd ser., 8 (1859), 327.

'Problem in Rhyme', *NQ*, 2nd ser., 8 (1859), 372.

'Arithmetical Notation', *NQ*, 2nd ser., 8 (1859), 460–61; 9 (1860), 52.

'On the Word Αριθμος', *Transactions of the Philological Society* (1859), 8–14.

'On the Determination of the Rate of Interest of an Annuity' *Assurance Magazine*, 8 (1860), 61–67.

'On A Statement Revived In Mr. Hodge's Paper On Interest, With Reference To The Authorship Of Graunt's *Observations*', *Assurance Magazine*, 8 (1860), 166–67.

'On a Property of Mr Gompertz's Law of Mortality', *Assurance Magazine*, 8 (1860), 181–84.

'On the Unfair Suppression of Due Acknowledgment to the Writings of Mr Benjamin Gompertz', *Assurance Magazine*, 9 (1860), 86–89.

'A Problem in Drawing', *Athenæum*, 1692 (31 Mar. 1860), 442.

'Hamilton's *Logic*', *Athenæum*, 1728 (8 Dec. 1860), 792–93.

'Rev. Thomas Bayes, Etc.', *NQ*, 2nd ser., 9 (1860), 9–10.

'A Question in Logic', *NQ*, 2nd ser., 9 (1860), 25, 184–85.

'John Gilpin', *NQ*, 2nd ser., 9 (1860), 33.

'Mariner's Compass', *NQ*, 2nd ser., 9 (1860), 62.

'Interest of Money', *NQ*, 2nd ser., 9 (1860), 216–17.

'Dedications to the Deity', *NQ*, 2nd ser., 9 (1860), 350–51.

'Drawing Society of Dublin', *NQ*, 2nd ser., 9 (1860), 444.

'Mathematical Bibliography', *NQ*, 2nd ser., 9 (1860), 449–50.

'Oliver Goldsmith', *NQ*, 2nd ser., 10 (1860), 206–07.

'Newton's *Treatise on Fluxions*', *NQ*, 2nd ser., 10 (1860), 232–33.

'Pencil Writing; Fire-Engine', *NQ*, 2nd ser., 10 (1860), 255, 457.

'Zinc', *NQ*, 2nd ser., 10 (1860), 248–49.

'Horrocks', *NQ*, 2nd ser., 10 (1860), 265.

'Sacheverell', *NQ*, 2nd ser., 10 (1860), 268.

'Leonard Euler' *NQ*, 2nd ser., 10 (1860), 272–74.

'The Tower Ghost', *NQ*, 2nd ser., 10 (1860), 277.

'Bullokar's *Bref Grammer*', *NQ*, 2nd ser., 10 (1860), 278.

'Dr Gowin Knight', *NQ*, 2nd ser., 10 (1860), 281–82.

'Versiera', *NQ*, 2nd ser., 10 (1860), 299–300.

'Value of Money', *NQ*, 2nd ser., 10 (1860), 311.

'Per Cent', *NQ*, 2nd ser., 10 (1860), 319.

'Character of the Germans', *NQ*, 2nd ser., 10 (1860), 330–31.

'Lagrange', *NQ*, 2nd ser., 10 (1860), 361–62.

'Ride v. Drive', *NQ*, 2nd ser., 10 (1860), 390–91.

'Changes of the Moon', *NQ*, 2nd ser., 10 (1860), 416.

'Newton's Table of Leases', *Assurance Magazine*, 9 (1861), 185–87.

'On Gompertz's Law of Mortality', *Assurance Magazine*, 9 (1861), 214–15.

'Authorship of the Treatise on Probability Published by the Society for the Diffusion of Useful Knowledge', *Assurance Magazine*, 9 (1861), 238.

'Mr Edmonds: College Life', *Assurance Magazine*, 10 (1861), 29–30.

'Hamiltonian Logic', *Athenæum*, 1759 (13 Jul. 1861), 51.

'——, no. 2', *Ibid.*, 1764 (17 Aug. 1861), 222.

'——, no. 3', *Ibid.*, 1775 (2 Nov. 1861), 582=–83.

'——, no. 4', *Ibid.*, 1783 (28 Dec. 1861), 883–84.

'——, [a further note]', *Ibid.*, 1825 (18 Oct. 1862), 496.

'——, [a response]', *Ibid.*, 1828 (8 Nov. 1862), 594.

'——, [another response]', *Ibid.*, 1831 (29 Nov. 1862), 698–99.

'——, [another response]', *Ibid.*, 1833 (13 Dec. 1862), 772.

'——, [a brief response]', *Ibid.*, 1835 (27 Dec. 1862), 845.

'Notes on the History of Perspective', *Athenæum*, 1771 (5 Oct. 1861), 446–47.

'——, no. 2', *Ibid.*, 1773 (19 Oct. 1861), 509–11.

'——, no. 3', *Ibid.*, 1774 (26 Oct. 1861), 544–45.

'——, no. 4', *Ibid.*, 1776 (9 Nov. 1861), 617–18.

'——, no. 5', *Ibid.*, 1777 (16 Nov. 1861), 652–53.

'——, no. 6', *Ibid.*, 1779 (30 Nov. 1861), 727–28.

'——, no. 7', *Ibid.*, 1860 (20 Jun. 1863), 812–13.

'——, no. 8', *Ibid.*, 1872 (12 Sep. 1863), 335–36.

'——, no. 9', *Ibid.*, 1873 (19 Sep. 1863), 368–69.

'The Alphabet', *NQ*, 2nd ser., 11 (1861), 209.

'The Word "America"', *NQ*, 2nd ser., 11 (1861), 264.

'Copernicus', *NQ*, 2nd ser., 11 (1861), 481–82.

'Slips of the Novelists', *NQ*, 2nd ser., 12 (1861), 7.

'Possible and Actual', *NQ*, 2nd ser., 12 (1861), 21.

'Fresnel', *NQ*, 2nd ser., 12 (1861), 169.

'[Isaac Newton's Descendants]', *NQ*, 2nd ser., 12 (1861), 315.

'India Rubber', *NQ*, 2nd ser., 12 (1861), 339.

'Bacon: Conference'; 'George Wharton'; 'Epigram on Sheepshanks', *NQ*, 2nd ser., 12 (1861), 358–59.

'Vossius *De Historicis Graecis*', *NQ*, 2nd ser., 12 (1861), 369; 3rd ser., 1 (1862), 74.

'Raining Cats and Dogs', *NQ*, 2nd ser., 12 (1861), 380–81.

'Sir I. Newton's Books', *NQ*, 2nd ser., 12 (1861), 440–41.

'Lost Passage of Aristotle', *NQ*, 2nd ser., 12 (1861), 443.

'Recovery of Things Lost', *NQ*, 2nd ser., 12 (1861), 506–07.

'Mr Woolhouse's recent paper', *Assurance Magazine*, 10 (1862), 237–38.

'On the Rejection of the Fractions of a Pound in Extensive Valuations', *Assurance Magazine*, 10 (1862), 247–51.

'A Query about Interest Accounts', *Assurance Magazine*, 10 (1862), 281–82.

'The Word "Any"', *NQ*, 3rd ser., 1 (1862), 23–24.

'Materials', *NQ*, 3rd ser., 1 (1862), 52.

'Michael Scott's Writings on Astronomy', *NQ*, 3rd ser., 1 (1862), 176.

'Colonel', *NQ*, 3rd ser., 1 (1862), 196.

'Not Too Good to be True', *NQ*, 3rd ser., 1 (1862), 245.

'Possession Nine Points of the Law', *NQ*, 3rd ser., 1 (1862), 388.

'Nullification', *NQ*, 3rd ser., 2 (1862), 85.

'Literature of Lunatics', *NQ*, 3rd ser., 2 (1862), 197.

'Fiddles, Flutes and Fancies', *NQ*, 3rd ser., 2 (1862), 206–07.

'Cut-throat Lane, Chalk Farm', *NQ*, 3rd ser., 2 (1862), 209.

'Essays on Assurance', *NQ*, 3rd ser., 2 (1862), 251–52.

'Galileo and the Telescope', *NQ*, 3rd ser., 2 (1862), 288–89.

'Andrew Horn(e)', *NQ*, 3rd ser., 2 (1862), 307.

'Algebra', *NQ*, 3rd ser., 2 (1862), 319.

'Pindar, Hallam, and Byron', *NQ*, 3rd ser., 2 (1862), 321–22.

'Gabriel Naudé', *NQ*, 3rd ser., 2 (1862), 332–33.

'Dog's Teeth: Pointing at Lightning', *NQ*, 3rd ser., 2 (1862), 342.

'Alchemy', *NQ*, 3rd ser., 2 (1862), 352–53.

'If Not', *NQ*, 3rd ser., 2 (1862), 384, 518.

'Butterfield of Paris', *NQ*, 3rd ser., 2 (1862), 398.

'On the Forms Under Which Barrett's Method is Presented, and on Changes of Words and Symbols', *Assurance Magazine*, 10 (1863), 301–12.

'Rules to be Observed in Converting the Parts of One Pound into Decimals', *Assurance Magazine*, 11 (1863), 53–54.

'A Budget of Paradoxes, No. I. Introduction', *The Athenæum*, 1876 (10 Oct. 1863), 466–68.

'——, No. II. 1503–1600', *Ibid.*, 1877 (17 Oct. 1863), 500–01.

'——, No. III. 1600–1658', *Ibid.*, 1878 (24 Oct. 1863), 534–35.

'——, No. IV. 1660–1668', *Ibid.*, 1879 (31 Oct. 1863), 573–74.

'——, No. V. 1676–1699', *Ibid.*, 1880 (7 Nov. 1863), 610–11.

'——, No. VI. 1705–1747', *Ibid.*, 1881 (14 Nov. 1863), 645–46.

'——, No. VII. 1747–1751', *Ibid.*, 1883 (28 Nov. 1863), 719–20.

'——, No. VIII. 1754–1792', *Ibid.*, 1885 (12 Dec. 1863), 800–01.

'——, No. IX. 1792–1802', *Ibid.*, 1887 (26 Dec. 1863), 878–79.

'——, No. X. 1803–1819', *Ibid.*, 1889 (9 Jan. 1864), 55–56.

'——, No. XI. 1819–1825', *Ibid.*, 1891 (23 Jan. 1864), 122–23.

'——, No. XII. 1825', *Ibid.*, 1893 (6 Feb. 1864), 195.

'——, No. XIII. 1825–1830', *Ibid.*, 1896 (27 Feb. 1864), 302–03.

'——, No. XIV. 1830–1833', *Ibid.*, 1899 (19 Mar. 1864), 408–09.

'——, No. XV. 1834–1835', *Ibid.*, 1901 (2 Apr. 1864), 474–76.

'——, ——. 1836–1839', *Ibid.*, 1906 (7 May. 1864), 647–48.

'——, ——. 1839–1840', *Ibid.*, 1914 (2 Jul. 1864), 20–21.

'——, No. XVI. 1842–1845', *Ibid.*, 1919 (6 Aug. 1864), 181–82.

'——, No. XVII. 1846–1847', *Ibid.*, 1921 (20 Aug. 1864), 246–47.

'——, No. XVIII. 1847–1849', *Ibid.*, 1922 (27 Aug. 1864), 276–77.

'——, No. XIX. 1849–1850', *Ibid.*, 1924 (10 Sept. 1864), 340–41.

'——, No. XX. 1851–1854', *Ibid.*, 1930 (22 Oct. 1864), 529.

'——, No. XXI. 1854–1855', *Ibid.*, 1942 (14 Jan. 1865), 54–55.

'——, No. XXII. 1855', *Ibid.*, 1947 (18 Feb. 1865), 236.

'——, No. XXIII. 1856', *Ibid.*, 1950 (11 Mar. 1865), 350.

'——, No. XXIV. 1856–1858', *Ibid.*, 1952 (25 Mar. 1865), 423–24.

'——, No. XXV. 1859', *Ibid.*, 1960 (20 May 1865), 685–86.

'——, No. XXVI. 1859', *Ibid.*, 1965 (24 Jun. 1865), 847–48.

'——, No. XXVII', *Ibid.*, 1967 (8 Jul. 1865), 52–53.

'——, No. XXVIII. 1859–1861', *Ibid.*, 1968 (15 Jul. 1865), 84–85.

'——, No. XXIX. 1862', *Ibid.*, 1970 (29 Jul. 1865), 149–50.

'——, No. XXX. 1862', *Ibid.*, 1973 (19 Aug. 1865), 248–49.

'——, No. XXXI. 1862–1863', *Ibid.*, 1975 (2 Sept. 1865), 312.

'——, No. XXXII. 1863', *Ibid.*, 1981 (14 Oct. 1865), 504–05.

'——, No. XXXIII', *Ibid.*, 1987 (25 Nov. 1865), 729–30.

'——, Supplement, no. I, *Ibid.*, 2008 (21 Apr. 1866), 531–32.

'——, ——, no. II, *Ibid.*, 2013 (26 May 1866), 706–07.

'——, ——, no. III, *Ibid.*, 2016 (16 Jun. 1866), 803–04.

'——, ——, no. IV, *Ibid.*, 2017 (23 Jun. 1866), 835–36.

'——, ——, no. V, *Ibid.*, 2018 (30 Jun. 1866), 868–69.

'——, ——, no. VI, *Ibid.*, 2019 (7 Jul. 1866), 19–20.

'——, ——, no. VII, *Ibid.*, 2020 (14 Jul. 1866), 52; 2029 (15 Sept. 1866), 336–37.

'——, ——, no. VIII, *Ibid.*, 2034 (20 Oct. 1866), 500–01.

'——, ——, no. IX, *Ibid.*, 2035 (27 Oct. 1866), 534–35.

'——, ——, no. X, *Ibid.*, 2040 (1 Dec. 1866), 718.

'——, ——, no. XI, *Ibid.*, 2046 (12 Jan. 1867), 51–52.

'——, ——, no. XII, *Ibid.*, 2047 (19 Jan. 1867), 89–90.

'——, ——, no. XIII, *Ibid.*, 2048 (26 Jan. 1867), 121.

'——, ——, no. XIV, *Ibid.*, 2054 (9 Mar. 1867), 324.

'——, ——, no. XV, *Ibid.*, 2057 (30 Mar. 1867), 422–23.

All Reprinted in *Assurance Magazine* (1863–1870), and as a monograph in 1872 (see above).

'Christmas Carol', *NQ*, 3rd ser., 3 (1863), 79.

'Reference to Preceding Authors', *NQ*, 3rd ser., 3 (1863), 223.

'The Rev. John Sampson', *NQ*, 3rd ser., 4 (1863), 24–25.

'On the Derivation of the Word Theodolite', *NQ*, 3rd ser., 4 (1863), 51–52.

'Apparitions', *NQ*, 3rd ser., 4 (1863), 68–69.

'Major-General John Lambert', *NQ*, 3rd ser., 4 (1863), 89.

'Jacob's Staff [and Theodolite]', *NQ*, 3rd ser., 4 (1863), 113–15.

'Spearman', *NQ*, 3rd ser., 4 (1863), 169.

'Maps [John Nicholson of Cambridge]', *NQ*, 3rd ser., 4 (1863), 170–71, 417.

'Epigram [on Newton Fellowes]', *NQ*, 3rd ser., 4 (1863), 174.

'Random', *NQ*, 3rd ser., 4 (1863), 183, 6 (1864), 183.

'Mistakes of the Novelists', *NQ*, 3rd ser., 4 (1863), 185.

'Riddle: Rhyme to Timbuctoo', *NQ*, 3rd ser., 4 (1863), 188, 338.

'I Know No More than the Pope', *NQ*, 3rd ser., 4 (1863), 217.

'Regiomontanus', *NQ*, 3rd ser., 4 (1863), 277.

'Counterfeit Ballads', *NQ*, 3rd ser., 4 (1863), 284–85.

'A Hint to Extractors', *NQ*, 3rd ser., 4 (1863), 286.

'Long Grass', *NQ*, 3rd ser., 4 (1863), 288.

'Sedechias', *NQ*, 3rd ser., 4 (1863), 309.

'Alexander the Great: Swift in the Nursery', *NQ*, 3rd ser., 4 (1863), 324.

'The Devil', *NQ*, 3rd ser., 4 (1863), 329–30.

'Notes on the Life of Robert Robinson (1735–1790)', *NQ*, 3rd ser., 4 (1863), 341–44, 481–82.

'Inkstand', *NQ*, 3rd ser., 4 (1863), 348, 462.

'Peter Walter', *NQ*, 3rd ser., 4 (1863), 348–49.

'The Kaleidoscope', *NQ*, 3rd ser., 4 (1863), 350.

'Quotation Wanted', *NQ*, 3rd ser., 4 (1863), 358.

'Heath Beer', *NQ*, 3rd ser., 4 (1863), 382–83.

'Cornelius Agrippa on the Morals of the Clergy', *NQ*, 3rd ser., 4 (1863), 387–88.

'Misuse of Words', *NQ*, 3rd ser., 4 (1863), 461.

'Text of Walter Scott's Novels', *NQ*, 3rd ser., 4 (1863), 470.

'Lord Bacon's Religious Faith', *Athenæum*, 1916 (16 Jul. 1864), 82.

'——, [a response]', *Athenæum*, 1917 (23 Jul. 1864), 115.

'History of the Signs + and –', *Athenæum*, 1925 (17 Sept. 1864), 82.

'Derivation of + and –', *Athenæum*, 1928 (8 Oct. 1864), 463.

'History of + and –', *Athenæum*, 1931 (29 Oct. 1864), 565.

'Scripture and Science', *Athenæum*, 1934 (19 Nov. 1864), 672.

'Publication of Diaries', *NQ*, 3rd ser., 5 (1864), 261–62, 361–63.

'Anonymous Contributions to N. & Q.', *NQ*, 3rd ser., 5 (1864), 307.

'Judicial Committee of Privy Council', *NQ*, 3rd ser., 5 (1864), 364.

'A Bull of Burke's', *NQ*, 3rd ser., 5 (1864), 366–67.

'Jeremiah Horrocks', *NQ*, 3rd ser., 5 (1864), 367.

'Miscellanea Curiosa', *NQ*, 3rd ser., 5 (1864), 387.

'John Bunyan', *NQ*, 3rd ser., 5 (1864), 455–56.

'Duchayla', *NQ*, 3rd ser., 5 (1864), 527–28.

'Jacob's Staff; Astrolabe; *Margarita Philosophica*', *NQ*, 3rd ser., 6 (1864), 51.

'Aristotle's *Politics*', *NQ*, 3rd ser., 6 (1864), 55.

'White Hats', *NQ*, 3rd ser., 6 (1864), 57.

'Unexpectedness of Phrase', *NQ*, 3rd ser., 6 (1864), 46.

'"Very Peacock": *Hamlet*', *NQ*, 3rd ser., 6 (1864), 66. Responding to a mistake in the index of vol. 5, attributing an earlier note on the subject (pp. 387–88), by 'Melites', to De Morgan.

'Two Suggestions on the Quadrature of the Circle', *NQ*, 3rd ser., 6 (1864), 67.

'Leland; Thomas Grynaeus', *NQ*, 3rd ser., 6 (1864), 83–84.

'Bale's *Scriptores*', *NQ*, 3rd ser., 6 (1864), 87–88, 154–55.

'Logical Bibliography', *NQ*, 3rd ser., 6 (1864), 101–03.

'Thomas Taylor's *Catalogue*', *NQ*, 3rd ser., 6 (1864), 117.

'Jacob's Staff; Atkinson's *Navigation*'; 'Sextant', *NQ*, 3rd ser., 6 (1864), 138.

'Penny Postage', *NQ*, 3rd ser., 6 (1864), 143.

'Is a Thing Itself, or Something Else?', *NQ*, 3rd ser., 6 (1864), 161.

'Papist', *NQ*, 3rd ser., 6 (1864), 175–76.

'Propositions', *NQ*, 3rd ser., 6 (1864), 181.

'"As Sure As Eggs Is Eggs"', *NQ*, 3rd ser., 6 (1864), 203.

'Pictorial Fiction', *NQ*, 3rd ser., 6 (1864), 207.

'"Miss Bailey", Latin Version', *NQ*, 3rd ser., 6 (1864), 218.

'Quarter-Sovereign', *NQ*, 3rd ser., 6 (1864), 226.

'Cheap Repository Tracts', *NQ*, 3rd ser., 6 (1864), 241–45, 353–55.

'Symbolization of Colours in Heraldry'; 'Negro New Testaments', *NQ*, 3rd ser., 6 (1864), 251.

'Detached Sheet', *NQ*, 3rd ser., 6 (1864), 266.

'Art Curiosity', *NQ*, 3rd ser., 6 (1864), 277.

'On the Question: What is the Solution of a Differential Equation? A Supplement to the ... Paper "On Some Points of the Integral Calculus"', *TCPS*, 10 (1864), 21–26 (read 28 Apr. 1856).

'On the Beats of Imperfect Consonances', *TCPS*, 10 (1864), 129–45 (read 9 Nov. 1857).

'A Proof of the Existence of a Root in Every Algebraic Equation: With an Examination and Extension of Cauchy's Theorem on Imaginary Roots ...', *TCPS*, 10 (1864), 261–70 (read 7 Dec. 1857).

'On the General Principles of Which the Composition or Aggregation of Forces is a Consequence', *TCPS*, 10 (1864), 290–304 (read 14 Mar. 1859).

'On the Theory of Errors of Observation', *TCPS*, 10 (1864), 409–27 (read 11 Nov. 1861).

'The Bases *Shall* be Equal', *Athenæum*, 1959 (13 May 1865), 653.

'Shall and Will', *Athenæum*, 1962 (3 June 1865), 758.

'A New Crotchet about 666', *Athenæum*, 1991 (23 Dec. 1865), 889.

'On a Problem in Annuities, and on Arbogast's Method of Development', *Assurance Magazine*, 12 (1865), 206–12.

'On the Summation of Divergent Series', *Assurance Magazine*, 12 (1865), 245–52.

'Speech of Professor De Morgan', *Proceedings of the London Mathematical Society*, 1 (1865), 1–9.

'A Proof That Every Function Has a Root', *Proceedings of the London Mathematical Society*, 1 (1865), 55–56.

'On the Calculation of Single Life Contingencies. Part I', *Assurance Magazine*, 12 (1866), 328–49.

'——, Part II', *Journal of the Institute of Actuaries*, 13 (1866), 129–49.

'The Portrait of Copernicus', *Gentleman's Magazine*, n.s., 1 (1866), 804–08.

'Letter to the President on the Foundation of the Society' *Mon. Notices Royal Astron. Soc.*, 26 (1866), suppl. Reprinted in Dreyer, J.L.E. and H.H. Turner, eds, *History of the Royal Astronomical Society* (London: Royal Astronomical Society, 1923), pp. 21–22.

'On the Conic Octagram', *Proceedings of the London Mathematical Society*, 2 (1866), 26–29.

'On Infinity and on the Sign of Equality', *TCPS*, 11 (1866), 145–89 (read 16 May 1864).

'A Theorem Relating to Neutral Series', *TCPS*, 11 (1866), 190–202 (read 16 May 1864).

'On the Root of Any Function and on Neutral Series, No. II', *TCPS*, 11 (1869), 239–66 (read 7 May 1866).

'Note on "A Theorem Relative to Neutral Series"', *TCPS*, 11 (1871), 447–60 (read 26 Oct. 1868).

'On the Early History of the Signs + and –', *TCPS*, 11 (1866), 203–12 (read 28 Nov. 1864).

'The Church Calendar', *Athenæum*, 2074 (27 Jul. 1867), 116.

'Pascal and Newton', *Athenæum*, 2079 (31 Aug. 1867), 273–74.

'Sir John Wilson', *Athenæum*, 2087 (26 Oct. 1867), 538.

'Value of a Policy—Formulæ—Milne', *JIA,* 14 (1867), 69–70.

'Note on the Annual Report', *Mon. Notices Royal Astron. Soc.* 27 (1867), 211.

'Pseudomath, Philomath, and Graphomath', *Athenæum*, 2097 (4 Jan. 1868), 21–22.

'Diogenes the Dog', *Athenæum*, 2101 (1 Feb. 1868), 173–74.

'An Old Song', *Athenæum*, 2103 (15 Feb. 1868), 254.

'Old Printing', *Athenæum*, 2120 (13 Jun. 1868), 832.

'Calculation', *Athenæum*, 2121 (20 June 1868), 864.

'[On Milton]', *Athenæum*, 2127 (1 Aug. 1868), 147.

'The Milton epitaph', *Athenæum*, 2128 (8 Aug. 1868), 179.

'Logic and Grammar', *Athenæum*, 2135 (26 Sept. 1868), 405–06.

'Fourier's Statistical Tables', *JIA*, 14 (1868), 89–90.

'On the Final Law of the Sums of Drawings', *JIA*, 14 (1868), 175–82.

'Some Account of James Dodson, F.R.S.', *JIA*, 14 (1868), 341–64.

'Remark on the Preceding Paper ["On General Numerical Solutions" by W.S.B. Woolhouse]', *Proceedings of the London Mathematical Society*, 2 (1868), 84–85.

   Reprinted in *JIA,* 15 (1870), 327–28.

'Private Life of Abraham de Moivre', *Athenæum*, 2149 (2 Jan. 1869), 21–22; no. 2150 (9 Jan. 1869), 57–58.

'The Warbling Lute', *Athenæum*, 2176 (10 July 1869), 51–52.

'Cocker', *Athenæum*, 2189 (9 Oct. 1869), 463–64.

'The Milton Difficulty', *Athenæum*, 2202 (8 Jan. 1870), 72.

List of Illustrations

Every effort has been made to trace all copyright holders, but if any have been inadvertently overlooked, the editors will be pleased to make the necessary arrangements at the earliest opportunity.

Frontispiece: Public domain, via MacTutor, https://mathshistory.st-andrews.ac.uk/Biographies/De_Morgan/pictdisplay/

Fig. 1 Public domain, via Wikimedia Commons, https://commons.wikimedia.org/wiki/File:Augustus_De_Morgan_1850s.jpg

Fig. 2 Public domain, via MacTutor, https://mathshistory.st-andrews.ac.uk/Biographies/De_Morgan/pictdisplay/

Fig. 3 Royal Astronomical Society Library and Archives, RAS MSS De Morgan 3

Fig. 4 UCL Library Services, Special Collections, MS ADD 7

Fig. 5 Royal Astronomical Society Library and Archives, RAS MSS De Morgan 3

Fig. 6 Senate House Library, University of London, MS 241

Fig. 7 Senate House Library, University of London, [DeM] L [Newton] SSR

Fig. 8 Public domain, via MacTutor, https://mathshistory.st-andrews.ac.uk/Biographies/De_Morgan/pictdisplay/

Fig. 9 UCL Library Services, Special Collections, MS ADD 7

Fig. 10 UCL Library Services, Special Collections, MS ADD 7

Fig. 11 By Nicholas Jackson - Own work, CC BY-SA 3.0, via Wikimedia Commons, https://commons.wikimedia.org/w/index.php?curid=26549887

Fig. 12 Public domain, from *Threescore Years and Ten. Reminiscences of the late Sophia Elizabeth De Morgan* (London: Richard Bentley, 1895).

Fig. 13 Public domain, via Wikimedia Commons, https://commons.wikimedia.org/wiki/File:Augustus_De_Morgan.jpg

Fig. 14A Senate House Library, University of London, [DeM] M [Hume] SSR

Fig. 14B Senate House Library, University of London, [DeM] M [Ursus] SSR

Fig. 15 Royal Society Archives, HS/6/183

Notes on Contributors

Rosemary Ashton OBE, FRSL, FBA is Emeritus Quain Professor of English Language and Literature and an Honorary Fellow of UCL. She has published critical biographies of Samuel Taylor Coleridge, George Eliot, G.H. Lewes, and Thomas and Jane Carlyle, studies of Anglo-German relations—*The German Idea* (1980) and *Little Germany* (1986)— and books on nineteenth-century cultural history, most recently *One Hot Summer* (2017), about the Great Stink of 1858. The author of *Victorian Bloomsbury* (2012), she was the leading investigator on the UCL Leverhulme-funded Bloomsbury Project, accessible at www.ucl.ac.uk/bloomsbury-project, which identifies and describes over 300 reforming institutions in the area.

Karen Attar is the Curator of Rare Books and University Art at Senate House Library, University of London, and was for many years a Research Fellow at the University's Institute of English Studies. Her publications cover various aspects of book collecting, library history and librarianship. They include several book chapters on Augustus De Morgan's library, which she also reconstituted within the University of London and catalogued. She is best known for the *Directory of Rare Book and Special Collections in the United Kingdom and Republic of Ireland* (3rd edn, 2016).

Daniel Belteki is a Research Fellow on the Congruence Engine Project at the Science Museum carrying out work on the applications of digital tools to the history of science and technology. His previous research focused on the history of astronomical instruments and the organisation of the astronomical community during the nineteenth century.

Olivier Bruneau is an associate professor of mathematics and its history (Université de Lorraine, Nancy, France). His research mainly focuses on mathematical circulation in France and Great Britain in the eighteenth and nineteenth centuries. He is particularly interested in the heritage of mathematics through encyclopedias in Great Britain during the same period. He recently published an article on mathematical knowledge presented in British encyclopedias between 1704 and 1850 (*Philosophia Scientiae*, 26:2 (2022), 67–90).

William Hale is Senior Rare Books Librarian at Christ Church, Oxford. This follows eighteen years as Assistant Under-Librarian in the Department of Rare Books and Early Manuscripts at Cambridge University Library; he has also worked in libraries in Manchester and London. He has previously contributed to *Treasures from Lord Fairhaven's Library in Anglesey Abbey* (Swindon, 2013) and *Emprynted in Thys Manere: Early Printed Treasures from Cambridge University Library* (Cambridge, 2014) and is working on a bibliography of the English carol.

Anna-Sophie Heinemann graduated in philosophy at the University of Jena, Germany, in 2007. In 2014, she received her Ph.D. from the University of Paderborn, Germany. From 2009 to 2020, she was affiliated with the University of Paderborn as a research and teaching assistant associated with the Chair of Philosophy of Science and Technology. Since 2020 she has been in charge of humanities at the head office of the Academic Advisory Commission of Lower Saxony (Wissenschaftliche Kommission Niedersachsen, www.wk.niedersachsen.de) located in Hanover, Germany.

Alexander Lock is Curator of Modern Archives and Manuscripts at the British Library, where he is responsible for collections dated 1600–1950. He is a specialist in early modern British history and co-curated the Library's best-selling exhibitions 'Magna Carta: Law, Liberty, Legacy' (2015) and 'Harry Potter: A History of Magic' (2017).

Katy Makin is an archivist at UCL Special Collections, where she has worked since 2011. She manages collections of donated and deposited archives covering all genres and dating from the medieval period to the present day.

Jane Maxwell has for several decades been a senior manuscripts curator in the Library of Trinity College Dublin, the University of Dublin.

Virginia Mills is the early collections archivist at the Royal Society, where she is responsible for curating pre-1900 material and the records of the past Fellowship. She has previously worked in scientific archives at the Royal Botanic Gardens, Kew, and the Natural History Museum, London.

Adrian Rice is the Dorothy and Muscoe Garnett Professor of Mathematics at Randolph-Macon College, Virginia, USA. He has also held visiting positions at the University of Virginia and the University of Oxford. His research focuses on the history of mathematics in the nineteenth and early twentieth centuries, with particular emphasis on the work of Augustus De Morgan. Previous books include *Mathematics in Victorian Britain*, co-edited with Raymond Flood and Robin Wilson (Oxford University Press, 2011) and *Ada Lovelace: The Making of a Computer Scientist*, co-authored with Christopher Hollings and Ursula Martin (Bodleian Library, 2018).

Joan L. Richards is Emerita Professor of History and Director of the Program of Science, Society, and Technology at Brown University, where she worked for over thirty-five years. Her work is infused with an abiding interest in the ways in which mathematics has served as a model of thinking that has developed in interaction with other approaches to the human mind, be they psychological, spiritual, physical, or even phrenological. Her book on the extended De Morgan family, *Generations of Reason: A Family's Search for Meaning in Post-Newtonian England*, was published by the Yale University Press in 2021.

Diana Smith is Assistant Archivist at Trinity College Library, Cambridge. She most recently contributed a chapter about the Whewell papers at Trinity College Library to *William Whewell: Victorian Polymath*, edited by Lukas Verburgt as part of the University of Pittsburgh Press series Science, Culture, and the Nineteenth Century (autumn 2024).

Christopher Stray is an Honorary Research Fellow in the Department of History, Heritage and Classics at Swansea University. His principal research interests are the history of classical scholarship and teaching, particularly at university level; his essay collection *Classics in Britain, 1800–2000* was published by Clarendon Press in 2018. He recently contributed to collaborative projects on William Whewell, Robert Leslie Ellis and Charles Babbage, and his edition of J.M.F. Wright's 1827 undergraduate memoir *Alma Mater; or, Seven Years at the University of Cambridge* appeared with University of Exeter Press in 2023.

Lukas M. Verburgt is currently an independent scholar based in the Netherlands. A fellow of the Royal Historical Society, he has held visiting research positions at Leiden University, Trinity College, Cambridge, the Department for the History and Philosophy of Science, Cambridge, and the Max Planck Institute for the History of Science, Berlin. Verburgt has published widely on the history of philosophy, science and mathematics in Victorian Britain and is (co-)editor of *A Prodigy of Universal Genius: Robert Leslie Ellis, 1817–1859* (Springer, 2022) as well as the forthcoming *Cambridge Companion to John Herschel* and *Cambridge Companion to Charles Babbage*.

Index

Abel, Niels Henrik 9
Academy of Arts 236
actuarial theory xii, xiii, xiv, xx, 3, 5–7, 85, 89, 98, 187, 260, 284
Adams, John Couch 63–65, 73, 129, 180
Affleck, Lady 297
Airy, George Biddell xvii, 59–60, 62–64, 73, 89, 186, 278, 287, 289, 293–296, 298
Aldrich, Henry 34, 36, 52
algebra xii, xv, xvi, xvii, xxi, xxv, xxxv, 2–3, 7, 9, 11–17, 20, 23–24, 61, 86, 98–100, 129, 154, 161, 191, 250, 263–264, 269–270, 284, 292
Amos, Andrew 212, 215
Analytical Society 59, 117
anti-Baconianism 112–114, 128–129, 132, 143
antiquarianism 58
Antrobus, Joan 229, 289
Appledore xv
Aquinas, Thomas xi
Arago, François 64, 72–73
Aristotle xi, 117–119, 121–122, 124, 135–137, 296
arithmetic xv, xvii, xxi, xxix, 4–5, 11–14, 19, 99, 101, 192, 215, 250, 257, 262–264, 269–270, 288, 291
Ashburner, John 234
Astronomers at War 61, 73
astronomy xii, xvii, xxviii, 7, 57–60, 65–67, 69–74, 77, 85, 89, 94, 97–102, 114, 122–123, 129, 156, 179, 182, 215, 222, 250, 253, 256, 263–264, 283–284, 294
Athenæum, The xxiv, xxviii, 5, 21, 63, 69, 72–75, 84, 87–88, 90, 102–103, 121, 132, 160, 171, 175–177, 184–186, 192, 295, 305
axiomatisation 23

Babbage, Charles xxx, 10, 17, 62, 64, 73, 89, 94, 112–113, 115, 123, 143, 159, 255, 258, 270, 273, 281, 287, 294, 298
Bacon, Francis xxvii, 107, 109–112, 114, 118–122, 124–130, 135–136, 138, 296
Baconianism xxviii, 107, 111–114, 117, 121, 128–130, 132, 143
Bagehot, Walter 17
Baily, Francis xxx, 60, 68–69, 254–256, 269, 283, 286, 289, 298
Barnes, Thomas 203
Barnstaple xv
Barton, Catherine 102
Bedford College xxi, 216, 229
Bedford, Duke of 198, 202, 211
Bedford Estate 198, 207, 211
Bellamy, John 222
Bell, E.T. 4, 191
Benson, James Bourne 17
Bernard, Claude 108
Bernoulli, Johann 98, 183, 250
bibliography xxix, xxxii, 52, 65, 85, 90, 99, 250, 258–259, 261, 264, 267, 281, 294, 302, 305
Bideford xv
Billingsley, Henry 253
Biot, Jean-Baptiste 254, 288
Birkbeck, George xxxi, 201–202, 204, 215
Bishop, George 61
Bland, Miles xvi
Bliss, Philip 281

Bloomsbury xxxi, 196–201, 204, 207–208, 210–212, 216–217, 265
Bloomsbury Group 198
Bolzano, Bernard 7
Boncompagni, Prince Baldassarre 299
Bonnycastle, John 250, 269–270
Book of Common Prayer 75–76
Boole, George xxvii, 4, 11, 16, 31, 51–52, 84, 140, 289, 292, 305
Boteler, William Fuller 255, 257
Bradley, James 67–68, 98, 286
Brahe, Tycho 98, 250, 256
Breakfast Club, Philosophical 117, 142
Brewster, David 14, 108, 112–113, 127, 288
Bridge, Bewick xvi
Bridgewater treatises 122, 124
British Almanac 67, 75–76, 85
British isolation from continental mathematics 7
British Museum xxix, 97, 197–198, 200–201, 209, 211–212, 217, 281
Broad, C.D. 142
Brougham, Henry xxxi, 95, 176, 201–206, 208, 210, 212–214, 289, 293
Brown, John 169
Buffon needle problem 179
Burke and Hare 213
Burton, James 199, 211–212
Byron, Lady xx, 17, 231

Cabinet Cyclopaedia 92, 94–95
calculus xxviii, 7–8, 11, 17, 22, 94, 100, 159, 250, 263–264, 285, 290
calendar xii, xxviii, 14, 74–77, 85, 97, 285
Cambridge xiv, xvi, xvii, xix, xxi, xxxi, 7, 17, 52, 58–60, 63, 86, 90, 95, 107, 114–115, 117, 132, 136, 153, 155, 157–170, 173, 202, 204, 207–210, 212, 216, 221–222, 234, 251–252, 274, 280, 283, 293–296, 301, 306

Cambridge Mathematical Journal 86, 90, 161
Cambridge Observatory 60
Cambridge Philosophical Society 52, 86, 90
Campbells are Coming, The 157–158
Campbell, Thomas 202, 238
Cape of Good Hope 71
Carnot, Lazare 94
Caroline of Brunswick-Wolfenbüttel, Queen 201–202
Carroll, Lewis xxii, 240, 271
Catholics 202, 206, 221
Cauchy, Augustin-Louis 7–8, 94, 99, 262
Cavalieri, Bonaventura 250, 265
Caxton, William 252
Cayley, Arthur 4, 11
Chadwick, Edwin 229
Chadwick, Rachel 229
Challis, James 63–64
Chambers, Robert 179–180
Christie, William 295
Christ's Hospital xiii
Church of England xix, 155, 202, 206–207, 210
Cincinnati Observatory 72
circle squaring 179, 183
Classics xvi, 154–155, 159, 166, 181
Clavius, Christoph 251
Clifford, William 177, 190
coach 17, 162, 165, 168
Cocker, Edward 250, 257, 263
Coleridge, Samuel Taylor 209
Comet, The 292
Cominale, Caelestino 182
Common Sense Realism 118
commutative law 12
Companion to the British Almanac 67, 75
Comte, Auguste 108
contraries 15, 37–41, 43–47, 51
controversy xxvi, 52, 61, 63, 73, 95, 179, 216, 288, 295
Copernicus 98, 188, 251, 256, 267, 269

copula, abstract 32, 34, 39, 51
Coram, Thomas 198
Coyne, Stirling xxiii
Cozens-Hardy, Herbert 17
Craig, John 181–182
cramming xxvi, 165
Crotchet Castle 205
Crown and Anchor 201–202
Cruikshank, Robert 206
Cubitt, Thomas 199
Cyclopaedia (Chambers) 91–92

Darwin, Charles and Emma 180, 212
Darwinism 176
Davies, Thomas Stephens xvi, 97, 99, 255
Da Vinci, Leonardo xi
Deal, Kent xv
decimal coinage 89, 179, 285
deduction 119–120, 122–124, 129, 131, 139, 142, 181, 184
de Fauré, J.P. 183
de Fermat, Pierre 250
Delambre, Jean-Baptiste 68, 75
De Morgan, Alice 223–227, 229–231, 233–234, 236, 238–239, 241, 289, 301
De Morgan, George xxiv, 215, 223, 236–238, 241
De Morgan, Helen Christiana (Chrissy) xxiv, 223, 237–238, 241
De Morgan, John xiii, xiv, xv, xviii, 291
De Morgan, Mary xxxi, 223, 237–241
De Morgan's Laws xii, xxvi, 15, 102–103, 175
De Morgan, Sophia 223
De Morgan, Sophia Elizabeth xvii, xx, xxxi, xxxii, 59–60, 69, 84, 97–98, 154–157, 165–166, 170, 177, 190, 197, 205, 210–211, 216, 220–238, 240–242, 283, 285, 289, 292, 296, 305
De Morgan, William xxxi, 215, 223–227, 230–231, 236–237, 239–243, 293

De Ricci, Seymour 252
Descartes, René xi, 68, 188
de Thou, Jacques-Auguste 271
Dewey, Melvil 263–264
De Worde, Wynkyn 252
Dibdin, Thomas Frognall 252
Dickens, Charles 200–201, 241
Diderot, Denis 91–92, 190–191
Dieudonné, Jean xi
dissenters 221
divergent series 8–10
Dixon, Edith Helen 239
Dixon, William Hepworth 184, 295
Dodgson, Charles Lutwidge 240, 270–271
Dodson, Elizabeth xiii
Dodson, James xiii, xiv, 5, 292
Dollond, John 68, 187
double algebra 11–12

Easter Day 74–77
Ecclesiastical History Society 75
École Polytechnique 85, 100
Edgeworth, Richard and Maria 224
Edinburgh University xxii, 213, 298
Educational Times, The 5, 171
Ellacombe, Rev. Henry Thomas 281–282
Ellis, Robert Leslie 116, 121, 140–141, 297
Ellis, Sir Henry 281
Elzevir 253, 269
Encyclopaedia Britannica 17, 91–92
Encyclopaedia Metropolitana xvii, 6, 34, 91–95
encyclopaedias xxiv, 23, 84, 91–94, 96
Encyclopédie (Diderot and D'Alembert) 91–92
English Encyclopaedia 94
enthymematic deduction 119
Equitable Life Assurance Society xiii
Euclid xvi, 14, 22, 154, 165, 184–185, 250, 253, 257, 263–264, 267, 269, 285

Euler, Leonhard 68, 123, 188, 190–191, 254
evangelicalism xviii, 155
Evening Hymn, the 157–158
examinations xiv, xviii, xxv, xxxiii, 155, 157, 160, 167–168, 262

Fermat's Last Theorem 193
Flamsteed, John 67–68
flute 60, 237
fluxions 93, 159, 286
Forman, Walter 182
Foundling Hospital 198, 211
Four-Colour Theorem 21, 228
Franklin, Benjamin xii
Freemasons' Tavern 207
Frend, William xx, 187, 210–211, 221–223, 231, 233, 243, 289, 292
friendship 60, 69, 73, 282, 294, 296
Fry, Elizabeth xx
functions 7, 264, 284

Galiano, Antonio Alcalà 209, 212
Galileo 117, 188, 282
Galloway, Thomas 98, 255–257, 269, 287
Gauss, Carl Friedrich 75
George IV 201, 206–207
Gilbert, W. S. xxii, 99
Girard, Albert 250
Gompertz, Benjamin 99, 187
goose vs duck 163
Grattan-Guinness, Ivor 31, 33
Graves, John Thomas xxx, 174, 237, 258, 289, 292, 302
Greatheed, Samuel 86
Gregory, Duncan 11, 86, 161
Gregory, Olinthus 92, 269, 289
Griffin, Charles 280
grinder 165
grinding 165

Halley, Edmond 67–68, 71, 292
Halliwell, James Orchard 252–253, 255, 257, 298–299
Hamilton, Mark 243

Hamilton, Sir William (philosopher) xxvii, 33, 38, 45, 117, 120, 295–297
Hamilton, Sir William Rowan (mathematician and astronomer) xxvii, 4, 13, 157, 185–186, 228, 287, 293, 300–302
Hardy, G. H. 8, 10, 19
Harriot, Thomas 250
Harrison, John 68
Hayden, Maria B. 233
heliocentricism 176
Helmholtz, Hermann von 108, 169
Hepburn, John Golch 290
Herbart, Johann Friedrich 181
Herschel, John Frederick William 10, 60, 65, 71, 89, 93–94, 108–109, 111–113, 120, 123, 128–129, 131, 141, 143, 155, 159, 262, 278, 283–285, 287, 294, 296, 298
Herschel, Sir William 67–68, 71, 98, 176
Higgins, Godfrey 222
Higgitt, Rebekah 69
Higman, John Philips xvii, 156, 298
Hill, Sir Rowland 281
Historical Society of Science 298
Hodder, James 250, 257
Hodgkin, Thomas 16
Hogarth, Catherine 200
Horrocks, Jeremiah 67
Hoskin, Michael 61
hospitals 211–213
Howitt, William and Mary 234
Huguenots xiv, xviii
Hume, Joseph 202
Hume's ghost, the paradox of 132
humour xxii, xxvi, xxvii, xxviii, 2, 56, 68, 116, 175, 189, 284, 300
Hurwitz, Hyman 209
Hustler, John 156
Hutton, Charles 254, 257, 262, 271

induction 101, 111–112, 119–125, 127–128, 132–137, 139, 141–142, 163
Infant School Society 201

infinite series 7–8, 100
Inns of Court 197, 200, 210, 212
Institute of Actuaries 88, 90
Insufficient Reason, Principle of 139
insurance xiii, xiv, 5–7, 59, 85, 264, 284
International Association for a Decimal System 89
International Short Title Catalogue 267

Jack, Richard 181
Jane, Duchess of Gordon xv
Jefferson, Thomas xii, 158
Jesus Christ xix
Jevons, William Stanley 17, 112, 132, 142, 169
Jews 197, 203, 221–222
John Bull 206, 213
Johnson, W.E. 142
Jones, Richard 121–122

Kant, Immanuel 109, 116, 124–125, 137
Karpinski, Louis 192
Kensal Green Cemetery xxiv, 238
Kepler, Johannes 126, 135, 250, 256
Keynes, J.M. 142
Key, Thomas 158, 289
King's College, London 168, 207–208
Kline, Morris 10
Knight, Charles 68, 96, 100, 280
König, Samuel 183
Kuhn, Thomas S. 117

Ladies' College, Bedford Square xxi
Lagrange, Joseph-Louis 68, 123, 159, 188
Lancet, The 212
Landseer, John 222
Lang, Andrew 252
language 21–23, 37, 40, 42, 52, 72, 88, 91, 162, 169, 208, 226–227, 268
Laplace, Pierre-Simon 6–7, 68, 94, 98–99, 123, 139–140, 142, 159, 188
Lardner, Dionysius 92, 94–95, 215

least squares, method of 6, 140, 287
Leibniz, Gottfried Wilhelm xii, xxviii, 7, 68, 100, 285
Le Verrier, Urbain 63–65, 180
Lewis, Clarence Irving 31, 33, 254
Lexicon Technicum 91–92
Leybourn, Thomas 87, 255–257
Libri, Guglielmo 73, 255, 257, 282
Liebig, Justus 108
limit, notion of a 7
Liston, Robert 212
Lloyd, Samuel, Lord Overstone xxx, 261
Loch, James 212
Locke, John 19, 22, 118, 125, 227
logic xii, xvii, xxvi, xxvii, 3, 15–16, 22–24, 31–38, 41, 44–48, 50–52, 84, 86, 88–89, 93, 98, 102, 112–114, 116–120, 122–124, 127–129, 132–139, 153, 163, 179, 281, 285, 294, 296, 302
 symbolic 16, 23, 31, 86
London Mathematical Society 18, 21, 23, 90, 189, 196, 215, 237, 287, 293
London Mechanics' Institution 201
London University xix, xxxi, 95–96, 115, 158, 164, 166–168, 214, 237, 260, 282
Long, George 86, 92, 96–97, 158, 164, 208
Lousada, Abigail Baruh 255, 270
Lovelace, Ada 17, 222, 270
Lucas, E.V. 197–198, 200, 203
Lushington, Stephen 202

Macaulay, Zachary 202
Macfarlane, Alexander 177
Madras xiv
Madura xiii, xiv
Magnus, Sir Philip 17
Malden, Henry 158
Mandelbrote, Giles 254
Manutius, Aldus 252–253
Marshalsea Prison 200
Martineau, Harriet 227

Martineau, James 216
Maseres, Francis 254
Maskelyne, Nevil 67–68
Masulipatam xiv
Mathematical Repository (Leybourn) 87
Mathematical Tripos 155–156, 159, 161–162
Mathematician, The 87
mathematics, history of 3, 14, 20, 24, 113–114, 183, 249, 259, 266, 269, 285
Maynard, Samuel 254, 299
Mechanics' Institution 201, 204
Merrington, Maxine 266, 286
mesmerism 231–233
meta-science xxvii, 108–109, 128, 133
methods, age of 110
Millais, John 212
Mill, James 202
Mill, John Stuart xx, xxi, 108, 112–113, 119–120, 124, 141, 292
Mitchel, Ormsby 72
Monk, J.H. 169
Montagu House 198
morbus cyclometricus 185
Morris, William 241–242
Murphy, Robert 11
Murphy's Law xxii
music 60, 157, 169, 175

Napier, John 187, 250
Naval Observatory, Washington DC 70
Neptune 58, 63–65, 73, 179–180, 295
Newton, Isaac xxviii, 7, 14, 68–69, 71, 92, 100, 102, 106, 109, 114, 116, 118, 123, 126, 129–130, 157–159, 179, 181–182, 188, 257, 269, 285–286, 288, 293, 296, 331
Noetics, Oriel 111, 122
nonconformism xviii, xx
Norfolk, Duke of 203
Northumberland, Duke of 71
Notes & Queries 84, 87–88, 90

numerically definite system 37–38, 47–48, 50, 52

optimes xviii
Oughtred, William 250
Overmier, Judith 255
Overstone, Lord. *See* Lloyd, Samuel, Lord Overstone

Pacioli, Luca 98, 250, 267
Panizzi, Antonio 208–209, 212, 281–282
Parsons, Rev. John xv, xvi, 154
Pascal, Blaise xii, 265
Pasley, Sir Charles 281
Pattison, Granville Sharp xx, 216, 288
Peacock, George xvii, 10–11, 20, 59, 89, 93–94, 99, 156, 159, 161, 163, 252–253, 286, 289, 297
Peacock, Thomas Love 205
Peirce, Charles Sanders xxvi, 110
Penny Cyclopaedia 67, 92, 94–102, 115, 294, 305
Perigal, Henry 99
Pestalozzi, Johann Heinrich 5
Phillipps, Sir Thomas 299
Philosophical Magazine 86, 251
Pickering, Evelyn 241
Pitt, William 211
Plantin 253, 269
Plato 234
Playfair, John 255–256
Poincaré, Henri xii, 108
poll men xviii
polymath xi, xxv, xxxiii, 59, 90, 107–108, 283
Pondicherry xiv
Pond, John 67
Portland Estate 199
positivism 19
Powell, Baden 108, 113, 166
Praed, Winthrop Mackworth 207
predicate, quantifying the xxvi, 16
principle of correspondence 236
priority disputes xxviii, 7, 100
private tutor 159, 168

probability 3, 5–7, 23, 50, 94–95, 99, 113–114, 121, 128, 131–133, 136–142, 178, 191, 250, 256, 263–264, 284
Pycior, Helena 20
Python, Monty xxii

quantification xxvii, 33, 38, 42, 48, 51–52
Quarterly Journal of Education 5, 85, 89–90, 163–164
quaternions 13, 23, 263, 300, 302
Queen Caroline. *See* Caroline of Brunswick-Wolfenbüttel, Queen

Ramchundra 270
Ranyard, Arthur Cowper 18, 237
Recorde, Robert 250, 257
Redland xv, 154
Reece, Robert xvi
Reeves, Charles 212
Reform Acts (1832, 1867) xii, xiii
Regent's Park 199
Reid, Elizabeth 216, 228, 230
Reid, Thomas 118
Ricardo, David 122
Richards, Joan xxxi, xxxii, 20
Robinson, James 212
Rosen, Friedrich 209, 216
Rouse Ball, Walter William xxiv
Routh, E.J. 17
Royal Astronomical Society xx, xxviii, 57, 59, 61–62, 64, 66–68, 77, 158, 182, 283–284, 292
Royal Mathematical School xiii
Royal Military Academy 97, 164
Royal Observatory, Greenwich 59, 67
Royal Society xxii, 68, 71, 90, 100, 115, 251, 282, 285–286, 294
Russell, Bertrand xii
Russell, Lord John 202
Rye, Reginald Arthur 261–264

Sabben, James 179
Sacrobosco, Joannes de 257, 262
scanners 154

Scharf, George 215
Second Scientific Revolution 107
Shaftesbury, Lord 229
Shakespeare, William 252, 267, 298–299
Sheepshanks, Richard 60, 62, 64, 73, 280, 294, 297–298
Shepheard, George 211
Simpson, Thomas 187
Smirke, Robert 198
Smith, Albert xxiii
Smith, Archibald 86
Smith, David Eugene xxx, 24, 175, 178, 191–192, 259, 261
Smith, James 183–186
Smyth, W.H. 65, 158, 289, 298
Society for the Diffusion of Useful Knowledge (SDUK) xx, 84–85, 90, 94–96, 201, 205–206, 290, 292
Society of Psychical Research 241
Somerville, Mary xii
sophister 155–156
South, James 62, 64, 73, 298
Spitalfields Mathematical Society 66, 187
steam engine 202, 205, 215
Stephen, Leslie 83, 198
Stephen, Vanessa 197
Stephen, Virginia 197
Stevin, Simon 250, 271
Stewart, Dugald 118
Stifel, Michael 250
Stinkomalee 207, 213
St John's College, Cambridge 155
Stowe, Harriet Beecher 229–230
string theory 9
Struik, Dirk 190, 192
Sturm's theorem 99–100
Sussex, Duke of 204
Sutherland, Duchess of 230
Swedenborg 234
Swift, Jonathan 179
syllogism 32, 36, 52, 117–120, 123, 136–138, 142, 296
Sylvester, James Joseph 3–4, 17

symbols 10–12, 22–23, 31, 162, 189, 227

Tait, Peter Guthrie 177, 189, 254
Tartaglia, Niccolò 250
Taunton xv
Taylor, Brook 100
Taylor, Sedley 169
Taylor's theorem 100
Taylor, Thomas 222
terminology 22, 39, 113, 183
Thackeray, William Makepeace 241
Thirty-Nine Articles 202
Thomson, William (Lord Kelvin) 295
Todhunter, Isaac 17, 250, 287, 297
Toplis, John 255, 257
Toronto Observatory 70
Transactions of the Cambridge Philosophical Society 86, 90
Trinity College, Cambridge xvi, xvii, 58–60, 114–115, 121, 143, 155–156, 158, 161, 208–209, 295–298, 301
triple algebra 12–13
trochoidal curves 99
Troughton and Simms 61–62, 73
Tunstall, Cuthbert 253, 257
Turing, Alan xii

Unitarianism 210
University College London xii, xix, xx, xxiii, xxvi, xxxi, xxxiii, 4, 16–18, 23, 95, 153, 165, 168, 204, 208, 223, 234, 236–238, 258, 260, 266, 288, 290–293
University College School 158, 169, 197, 215, 228, 236–237
University of London xii, xxx, xxxiii, 84, 86, 153, 165, 168, 202, 204, 208, 214–216, 258–262, 265–266, 286, 288

Uranus 67–68, 180

Venn, John 120, 142, 298
Viète, François 98, 250
Viga Ganita 99–101
Virginia, University of 158
Vlacq, Adriaan 250
von Mühlenfels, Ludwig 209
von Wowern, Johann xi
voting xxi, 64, 222

Wakley, Thomas 212
Walkingame, Francis 265
Wallace, Alfred Russel 281
Wallace, William 93, 255
Waller, Edmund 251
Weddle, Thomas 287
Weld, Charles 285
Wellington, Duke of 207
Wesley, Charles 158
Whately, Richard 34, 93, 112–113, 117–120, 122–124, 134, 136, 142
Wheatley, Henry 252
Whewell, William xvii, xxviii, 14, 20, 21, 59, 69, 106, 107, 108, 109, 110, 111, 112, 113, 114, 115, 116, 117, 120, 121, 122, 123, 124, 125, 126, 127, 128, 129, 130, 131, 132, 133, 134, 135, 136, 137, 138, 140, 141, 142, 143, 148, 157, 159, 163, 167, 227, 230, 231, 262, 287, 295, 296, 297
Wildig, George Burgess 255, 256
Wilkins, William 203, 206, 215
wranglers xviii, 162
Wright, J.M.F. 157, 169

Zitti Zitti 157, 158

About the Team

Alessandra Tosi was the managing editor for this book.

Jennifer Moriarty proofread this book. It was indexed by Anja Prichard and Adèle Kreager.

Jeevanjot Kaur Nagpal designed the cover. The cover was produced in InDesign using the Fontin font.

Cameron Craig typeset the book in InDesign and produced the paperback and hardback editions. The text font is Tex Gyre Pagella and the heading font is Californian FB.

Cameron also produced the PDF and HTML editions. The conversion was performed with open-source software and other tools freely available on our GitHub page at https://github.com/OpenBookPublishers.

Jeremy Bowman created the EPUB.

This book has been peer-reviewed by Prof. June Barrow-Green, the Open University and Dr Christopher Hollings, The Queen's College, Oxford. We thank them for their invaluable help.

This book need not end here...

Share

All our books — including the one you have just read — are free to access online so that students, researchers and members of the public who can't afford a printed edition will have access to the same ideas. This title will be accessed online by hundreds of readers each month across the globe: why not share the link so that someone you know is one of them?

This book and additional content is available at:
https://doi.org/10.11647/OBP.0408

Donate

Open Book Publishers is an award-winning, scholar-led, not-for-profit press making knowledge freely available one book at a time. We don't charge authors to publish with us: instead, our work is supported by our library members and by donations from people who believe that research shouldn't be locked behind paywalls.

Why not join them in freeing knowledge by supporting us:
https://www.openbookpublishers.com/support-us

Follow @OpenBookPublish

Read more at the Open Book Publishers BLOG

You may also be interested in:

Reign of the Beast
The Atheist World of W. D. Saull and his Museum of Evolution
Adrian Desmond

https://doi.org/10.11647/obp.0393

The Last Man Who Knew Everything
Thomas Young
Andrew Robinson

https://doi.org/10.11647/obp.0344

The Scientific Revolution Revisited
Mikuláš Teich

https://doi.org/10.11647/obp.0054

www.ingramcontent.com/pod-product-compliance
Lightning Source LLC
Chambersburg PA
CBHW040332300426
44113CB00021B/2737